Lecture Notes
in Control and Information Sciences 203

Editor: M. Thoma

Yury S. Popkov

Macrosystems Theory and its Applications

Equilibrium Models

Springer
London Berlin Heidelberg New York
Paris Tokyo Hong Kong
Barcelona Budapest

003
P82m

Series Advisory Board

A. Bensoussan · M.J. Grimble · P. Kokotovic · H. Kwakernaak
J.L. Massey · Y.Z. Tsypkin

Author

Yury S. Popkov
The Academy of Sciences of Russia, Institute for Systems Analysis,
9, Prospect 60-Let Octiabria, Moscow 117312, Russia

ISBN 3-540-19955-1 Springer-Verlag Berlin Heidelberg New York

British Library Cataloguing in Publication Data
A catalogue record for this book is available from the British Library

Apart from any fair dealing for the purposes of research or private study, or criticism or review, as permitted under the Copyright, Designs and Patents Act 1988, this publication may only be reproduced, stored or transmitted, in any form or by any means, with the prior permission in writing of the publishers, or in the case of reprographic reproduction in accordance with the terms of licences issued by the Copyright Licensing Agency. Enquiries concerning reproduction outside those terms should be sent to the publishers.

© Springer-Verlag London Limited 1995
Printed in Great Britain

The publisher makes no representation, express or implied, with regard to the accuracy of the information contained in this book and cannot accept any legal responsibility or liability for any errors or omissions that may be made.

Typesetting: Camera ready by author
Printed and bound at the Athenæum Press Ltd, Gateshead
69/3830-543210 Printed on acid-free paper

Contents

University Libraries
Carnegie Mellon University
Pittsburgh PA 15213-3890

Preface

The word "macrosystem" seems to be understandable for anybody. Really, "macro" is evidently something large, and the notion of "system" lies in the individual experience of each of us.

But don't hurry!

An increasing variety of objects studied by the modern science and, on the other hand, limited possibilities of language result in a polysemantic interpretation of terms used in scientific literature. "Macrosystem" is an example of such a polysemy.

The systems studied in this monograph do consist of a great number of elements. But this is not the number of elements that allows us to categorize macrosystems as a special class of systems. The feature that the behaviour of the elements differs from that of the system as a whole is more important. The behaviour of the elements is transformed into the system behaviour.

Such a transformation is often a consequence of "intermixing" or "averaging" effects. These are the case when a large number of elements exist in a system, so that the state and behaviour of each element infinitesimally influences the state and behaviour of the system as a whole.

There are many systems which contain a large number of elements, but cannot be classified as macrosystems.

Thus the national system of anti-aircraft defense consists of various radio-engineering devices, aircrafts, rockets etc., but independence in the behaviour of particular components is unlikely to be admitted in such a system. Another example is a modern FMS, where the behaviour of each component is predetermined by a program.

For the first time, physics came across the systems in which a large number of components is combined with the "freedom of the will" of each component. It was physics that found the material world around us to be constructed from elementary particles, namely atoms (and their aggregates, i.e. molecules). Each elementary particle moves according to the laws of classical mechanics. But if there are many particles, their interactions (for example, collisions) become non-deterministic because of reciprocal forces between the particles. Generally speaking, the trajectories of their future movement (after collision) form an everywhere dense continuum.

On the other hand, if we observe physical bodies that consist of moving particles, we can see that after a while the bodies reach a state (called equilibrium) which parameters (for example, temperature, volume and mass) remain constant.

Non-deterministic interactions between the particles and the equilibrium in their aggregate exist in the same physical body. The former seems to give rise to the latter ...but how?

For the first time, L. Boltzmann (1877) answered to this problem. He dropped the traditional approaches based on postulates of classical dynamics and built a probabilistic model of particle interactions. The main point of the model is that the equilibrium state of a physical body is generated by the distribution of points (representing the coordinates and velocities of each particle) in a six-dimensional phase space.

However, the notion of physical macrosystem appeared later. It have been shaped and broadened together with the appearance of new facts in studying the material world. In particular, in using the probabilistic model of Boltzmann, it was obtained that the model is applicable when the number of particles in the areas of phase space is smaller than the capacities of the areas. Therefore the Boltzmann model turned out to be too rough in investigating the substance at low temperature. Two new probabilistic models arose, viz the Fermi-Dirac model for particles having anti-symmetric wave functions and the Einstein-Boze model for particles having symmetric wave functions.

Further development of these models was associated with investigation of stochastic behaviour of particles distributed according to their quantum state and with determination of equilibrium state in a thermodynamic system (Landau, Lifshitz, 1964).

The theoretical results are confirmed by multiple experiments and are the basis of modern statistical thermodynamics, but they do not solve all problems in this field. To the contrary, they are the ground on which investigations of non-equilibrium thermodynamic processes are based (Prigogine, 1971; Haken, 1974).

Theory of macrosystems was considerably contributed by studying the distribution and exchange of economic resources. The behaviour of elements in thermodynamic systems is random, while it can be both random and regulated in exchanging resources between economic units. This depends on the degree of centralization in the given economic system (Rozonoer, 1973). Generally, the equilibrium conditions are investigated in such systems.

The laws inherent to the equilibrium states in economic exchange systems are similar to the laws which are the case in thermodynamic systems. This can be seen even in the names of some types of equilibrium, e.g. the Le-Chatelier-Samuelson equilibrium, the Carno-Hix equilibrium, etc.

Macrosystem theory was considerably developed by A. Wilson (for example, see 1967, 1970, 1981). He applied the macrosystem approach to investigation of urban, transport and regional systems, in which the spatial aspect is of vital importance.

The macrosystem approach is widely used in modeling equilibria and evolution of biological communities (Volterra, 1976; Volkenstein, 1981), in designing chemical technology equipment (Smith, 1982; Kafarov, 1988), and in reconstructing images (Gull, Skilling, 1984).

It was shown above that the fundamental peculiarity of macrosystems is in difference between "individual" properties of elements and "collective" properties of the system as a whole. This heterogeneity results in the effects which are not observed in other systems. In particular, it can produce quasi-stable structures from a large number of elementary chaotic movements.

Due to this reason, many problems embraced by macrosystem theory are close to problems of self-organization and deterministic chaos (Prigogine, Stengers 1984; Prigogine 1985).

Relations between the collective (cooperative) and the individual properties realized by macrosystem models of real-life processes is a part of general synergetic concept. This concept appeared in physics in early 1970's (Haken,

1974) and entered both the other natural sciences and the economical and social sciences (Weidlich, Haag, 1983). Many other links between macrosystem theory and other scientific fields may be demonstrated.

The special feature of the theory considered here is that it studies a class of systems (macrosystems) rather than a class of properties (self-structurization, chaos, periodical movements, order parameters and so on). Therefore many results of this theory can illustrate and sometimes extend the synergetic concept, theory of self-organization and theory of collective behaviour.

Now theory of macrosystems, as an independent branch of science, is worked out. Thus far it has been developed in three more or less independent directions. Those are associated with investigations of special classes of macrosystems, namely thermodynamic, economical (resource exchange) and territorial systems (urban, regional, global).

Many facts characterizing these macrosystems have been collected, and many methods of simulation, analysis and control have been developed for these systems.

Now a critical mass of rather diverse results is available. They should be theoretically generalized not only for the cognition purposes, but also as a source of scientific progress.

In this book we attempt to build one section of macrosystem theory. This section is devoted to investigation of equilibrium.

A great part of the monograph material is based on works of the author and the staff he guided in the Institute for System Analysis of Russian Academy of Sciences.

I am very thankful to my collaborators, viz to Prof. B. L. Shmulyan, Prof. Yu. A. Dubov, Prof. A. P. Afanasyev, Prof. Ye. S. Levitin, Dr. Sh. S. Imelbayev, Dr. L. M. Heifits, Dr. A. S. Aliev. Discussions of many parts of this monograph with them were extremely useful. Their benevolent criticism and constructive advice facilitated the improvement of the material presented.

Prof. A. G. Wilson played a special role in the appearance of this monograph. Our discussions during his visit to Moscow in 1978 as well as my ten-years participation in the editorial board of *Environment and Planning, A*, which he heads, have valuable contributed my understanding the problems of macrosystem theory and facilitated writing of this monograph.

I would like to thank my teacher, academician Ya. Z. Tsypkin, a great scientist and an eminent person.

Prof. Yu. Dubov, Dr. O. Agarkov and Dr. D. Vedenov mainly contributed the preparation of english version. I am very grateful to them.

Mrs E. Samburova and mrs L. Kuchkova prepared numerous versions of the text with enviable patience.

Moscow, 1993
Institute for System Analysis
9, Prospekt 60-let Oktyabrya,
117312 Moscow
Russia

CHAPTER

ONE

MACROSYSTEMS: DEFINITIONS AND EXAMPLES

1.1 Introduction

In the last 10–20 years the science tends to integrate knowledge and scientific disciplines. This reflects in investigating any object of nature or human society from the "wholestic" point of view. Such a tendency results from quite objective processes.

System analysis is one of the most effective examples of these integrative tendencies. This scientific direction is intensively developed now. Its characteristic feature is that it is oriented on the whole class of problems arising in connection with the phenomenon in hand. System analysis goes from a verbal description through cognition of the phenomenon essence to construction of mathematical models and creation of tools for research and control.

Particularly, system analysis can be applied to complex systems. In system analysis, the complexity is a qualitative category, rather than a quantitative one. It is not always associated with the system dimension, the number of factors influencing its behaviour or the number of its components. Nevertheless, all these characteristics are connected with the system complexity.

But another aspect is more important here. The main point is whether a system as a whole acquires new properties with respect to the properties of its parts. There exist many fundamental and applied problems where the behaviour of object as a whole significantly differs from that of its components.

For example, in investigating controllable dynamic systems, the situation is quite common where the system is stable as a whole, while some its parts are unstable, and vice versa. In a volume, gas molecules move chaotically, but the chaos results in a certain temperature of the gas. In the market economy, resource exchange is a stochastic phenomenon to a large extent, but a constant profit rate result from multiple individual exchanges.

Quite different objects are mentioned in the examples. However, a common thing is stressed, namely all objects as a whole acquire a behaviour different from that of their components.

Studied in this monograph are systems in which a stochastic behaviour of their components is transformed into a deterministic behaviour of the system as a whole. We shall call the systems with such properties *macrosystems.*

In the monograph, we investigate equilibria in macrosystems. Mathematical modeling which uses the principle of maximum entropy is involved. The latter is developed and generalized for systems with constrained resources. Models of stationary states based on this principle are proposed for macrosystems of various classes.

The parametric properties characterizing the model response to data variations are studied. Methods for qualitative analysis of parametric properties are developed. Quantitative estimates of sensitivity are given.

Computational methods oriented to the computer-aided realization of stationary state models are an important component of equilibrium theory for macrosystems. A new scheme is proposed for construction of iterative algorithms. The scheme uses a specific feature of stationary state model. The algorithms are qualitatively investigated, namely the convergency are studied. A computer realization is considered as well.

Methods of mathematical modeling the equilibrium states in macrosystems are applied to problems of managerial structures, interregional product exchanges and image reconstructions.

1.2 Macrosystem structure

In describing a macrosystem structure, we shall follow a concept of mathematical theory of systems. According to this concept, a system transforms input information, input substance or input energy into their output analogs.

Therefore various classes of systems have different properties of input and output processes and different rules of "input-output" transformation. Consider a system with the input $x(t, z)$ and the output $y(t, z)$, where $t \in \mathfrak{T}$ are time moments and $z \in \mathfrak{Z}$ are system locations.

$x(t, z)$ and $y(t, z)$ are elements of corresponding input and output spaces X and Y, respectively.

The "input-output" transformation is usually characterized by a mapping $\mathcal{H}$ (fig. 1.1);

$$X \overset{\mathcal{H}}{\Rightarrow} Y$$

In order to describe $\mathcal{H}$, this can be decomposed if an additional information about the system is available.

For example, there are many systems for which we can distinguish and study inner processes $u(t, z)$ (besides the output process $y(t, z)$). These can be considered as characteristics of system state if they have some special properties (Kalman, Falb, Arbib, 1969). The functions $u(t, z)$ are elements of state space U. Thus the mapping $\mathcal{H}$ is decomposed into two mappings. $\mathcal{H}_1$ is an "input-state" transformation and $\mathcal{H}_2$ is a "state-output" transformation (fig. 1.2). We assume that $\mathcal{H}_1$ and $\mathcal{H}_2$ characterize the transformations of deterministic processes, and thus they are homogeneous in this sense.

It was mentioned above that macrosystems are heterogeneous, i.e. the macrosystem elements have a non-deterministic behaviour, while the system as a whole has a deterministic one.

There exist many mathematical models of non-deterministic behaviour. One of them is a stochastic model. We shall assume that macrosystem elements have a stochastic type of behaviour. Thus a macrosystem have inner processes $g(t, z)$ characterizing its stochastic states. In order to describe this, the mapping $\mathcal{H}_1$ is subdivided into the mapping $\mathcal{H}_{11}$ ("input – stochastic state") and the mapping $\mathcal{H}_{12}$ ("stochastic state – state").

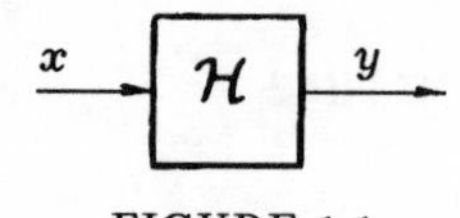

FIGURE 1.1.

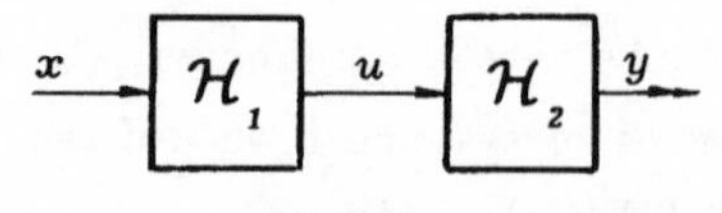

FIGURE 1.2.

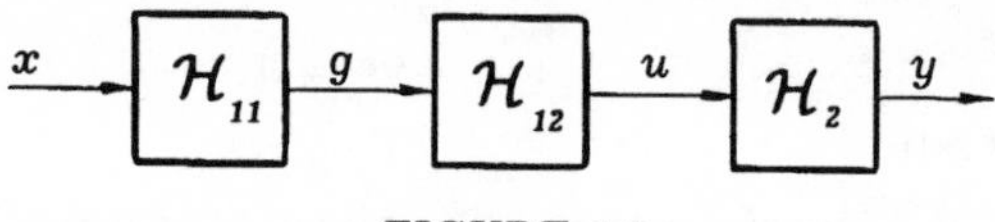

FIGURE 1.3.

The transformations performed by a macrosystem are illustrated in fig. 1.3 by the block-scheme. The mapping $\mathcal{H}_{11}$ characterizes the stochastic behaviour of macrosystem elements or a *microlevel* of the system.

The mapping $\mathcal{H}_{12}$ characterizes a transformation of stochastic state $g(t, z)$ to the deterministic one $u(t, z)$.

Finally, the relation between the state $u(t, z)$ and the output $y(t, z)$ is characterized by the deterministic mapping $\mathcal{H}_2$.

The composition of $\mathcal{H}_{12}$ and $\mathcal{H}_2$ characterizes a *macrolevel* of the system. Thus the stochastic state is generated on the system microlevel and is transformed to the deterministic state on the macrolevel.

1.3 Examples

Chemical technology. In designing various chemical reactors, processes of the molecular level are essential. Consider an absorber for separating multicomponent mixtures as an example. A gaseous mixture is fed to the absorber and come into contact with a liquid absorbent. The absorber scheme is shown in fig. 1.4. $(x_{\text{inp}}, z_{\text{inp}})$ and $(x_{\text{out}}, z_{\text{out}})$ are mole portions of gas and absorbent in the input and output flows, respectively;

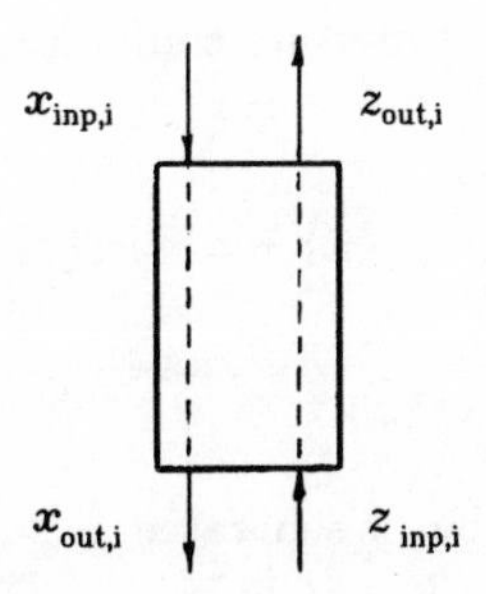

FIGURE 1.4.

$(x_{inp,i}, z_{inp,i})$ and $(x_{out,i}, z_{out,i})$ are mole portions of the i-th component of gas and absorbent in the input and output flows, respectively.

Let us introduce the following notations. $n_{x_{inp,i}}, n_{z_{inp,i}}$ and $(n_{x_{out,i}}, n_{z_{out,i}})$ are the numbers of the i-th component molecules which enter (leave) the absorber together with the gas and the liquid absorbent per unit of time; $N_{x_{inp}}$, $N_{z_{inp}}$ and $(N_{x_{out}}, N_{z_{out}})$ are the numbers of molecules in the gas flow and liquid absorbent which enter (leave) the absorber per unit of time. It is obvious that

$$N_{x_{inp}} = \sum_{i=1}^{m} n_{x_{inp,i}}; \quad N_{z_{inp}} = \sum_{i=1}^{m} n_{z_{inp,i}};$$

$$N_{x_{out}} = \sum_{i=1}^{m} n_{x_{out,i}}; \quad N_{z_{out}} = \sum_{i=1}^{m} n_{z_{out,i}}.$$

If N is the total number of molecules entering and leaving the absorber per a unit of time then

$$N = N_{x_{inp}} + N_{z_{inp}} = N_{x_{out}} + N_{z_{out}}.$$

Weighted average mole ratio at the absorber output characterizes the process of separating the mixture, viz

$$a = \frac{I}{N} \sum_{i=1}^{m} (n_{x_{out,i}} b_i + n_{z_{out,i}} c_i),$$

where b_i and c_i are phenomenological coefficients characterizing the molecules of the i-th component.

By using the mole parts of mixture components, we can present the previous equations in the form

$$\sum_{i=1}^{m}(x_{\text{out,i}}b_i + z_{\text{out,i}}c_i) = a,$$
$$\sum_{i=1}^{m}(z_{\text{out,i}} + x_{\text{out,i}}) = 1. \qquad (*)$$

The mixture in the absorber is separated according to the following physical scheme. Consider a chamber divided into two parts by a semipermeable partition. Molecules of different kinds are put into the chamber, with the number of molecules of every kind being known in each part of the chamber.

After a while the molecules are distributed according to their properties in both parts of the chamber. In order to define this equilibrium distribution, we will carry out a "mental" experiment. It consists in successive extracting the molecules from both parts of the chamber. In doing so, a value b_1 is attributed to the experimental result if a molecule of the first kind is extracted from one part of the chamber; a value b_2 is attributed if the molecule is of the second kind, and so on. Values $c_1, c_2, \dots, c_m$ are obtained in similar situations for the other chamber part. The experiment data are summed and then divided on the total number of tests. Thus the average estimate a of the whole experiment is obtained.

If the tests are independent, the most probable state in the experiment corresponds to the maximum of entropy

$$H = -\sum_{i=1}^{m} x_{\text{out,i}}\, ln\, x_{\text{out,i}} - \sum_{i=1}^{m} z_{\text{out,i}}\, ln\, z_{\text{out,i}}$$

subject to constraints (*) (Moikov 1978).

This model is used to calculate the absorber parameters. Here a random choice of elements (molecules) is transformed into a quite deterministic distribution of mole parts.

Economical exchange. Resource exchange and distribution are major in any economic system. Let us consider a system of n consumer groups and m resource kinds. Denote the number of consumers in the i-th group by N_i and denote the number of the j-th resource by M_j.

After the exchanges, the resources are distributed among the consumers. The distribution at the time moment t is given by the matrix $T(t)$. The element T_{ij} of the matrix is the amount of the j-th resource possessed by the i-th group consumers.

In general, not all resources should be distributed among the consumers, and not all consumers can acquire a resource. Denote the number of consumers having no resources by R_i and denote the number of j-th resource which is not distributed among the consumers by V_j. Then the following balance relations should hold at any moment of time:

$$M_j = V_j + \sum_{i=1}^{m} T_{ij}.$$

Not all the consumers redistribute the resources and not all the resources are redistributed.

Denote an "active part" of the group T_{ij} by S_{ij} and denote an "active part" of the group R_i by Q_i. S_{ij} can be interpreted as the amount of the j-th resource which belongs to the consumers of the i-th group taking part in the redistribution. Q_i is the number of consumers who have no resource but want to acquire it.

Denote "passive parts" of resources and consumers by $\widetilde{T}_{ij}$ and $\widetilde{R}_i$, respectively.

Thus there are four types of group with the capacities S_{ij} and Q_i, $\widetilde{T}_{ij}$ and $\widetilde{R}_i$. If a consumer passes from the group $\widetilde{T}_{ij}$ to S_{ij}, he becomes "active", while his resource remain unchangeable. If a consumer passes from the group $\widetilde{R}_i$ to Q_i then he becomes "active", but has no resource. If a consumer passes from $\widetilde{T}_{ij}$ to $\widetilde{R}_i$ then he rejects buying the resource. Active consumers redistribute their unused resources. After acquiring a suitable resource, they becomes passive and come to one of the groups $\widetilde{T}_{ij}$.

All transitions between the groups are caused by the respective motives. The motivation is the basis of any decision and is essentially non-deterministic. A large number of producers, consumers and units of exchangeable resources are involved in economic exchanges. Therefore non-deterministic character of transition motivations can be described in stochastic terms.

A scheme illustrating possible transitions between the groups is shown in fig. 1.5. Each of these transitions is characterized by the corresponding probability w.

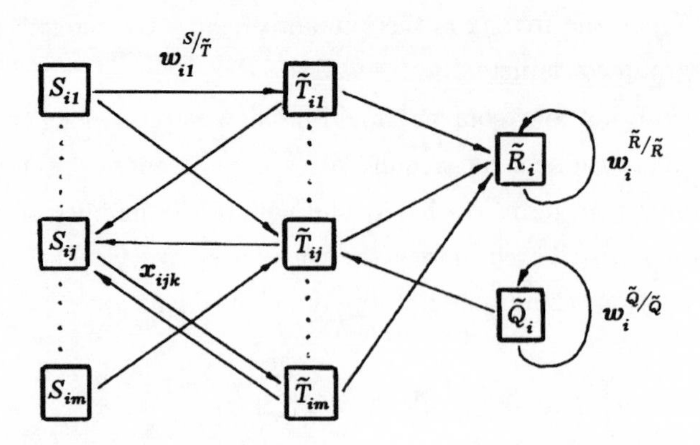

FIGURE 1.5.

The variables N, M, V, R, T, S, Q, $\widetilde{T}$ and $\widetilde{R}$ give a macrodescription of given system of resource exchanges. We can see from the scheme that they result from random pair exchanges. Many exchanges produce a certain observable system state. This state is corresponded by the distribution of consumers in the groups. The distribution is characterized by the number of transitions from one group to another. Thus we can speak about the most probable distribution of transition numbers x_{ij} (the indices show the groups between which a transition is done). It follows from the transition scheme that

$$x_{ij}^{\widetilde{T}/\widetilde{R}} + x_{ij}^{\widetilde{T}/\widetilde{S}} + x_{ij}^{\widetilde{T}/\widetilde{T}} = T_{ij},$$

$$\sum_{k=1}^{m} x_{ijk}^{S/\widetilde{T}} + x_{ij}^{S/S} = S_{ij},$$

$$x_i^{\widetilde{R}/\widetilde{Q}} + x_i^{\widetilde{R}/\widetilde{R}} = \widetilde{R}_i.$$

We assume that a consumer passes from one group to another because of two reasons. The first one is associated with the consumer's own choice of one or another resource. The second one is associated with the possibility to acquire the resource. If we denote the choice probability by p_{ij} and the acquisition probability by α_j, the transition probability can be presented as

$$w_{ij} = p_{ij}\alpha_j.$$

The number of transitions x_{ij} can be expressed (Ivandikov, 1981) in terms of the probabilities p_{ij} and α_j, while the evolution model for the macroparameters T, S, $\widetilde{R}$ and Q can be presented as

$$\dot{T}_{ij} = \alpha_j p_{ij} \left(\sum_{j=1}^{m} S_{ij} + Q_i \right) - S_{ij} \sum_{k=1}^{m} \alpha_k p_{ik} - \beta_{ij}^{R}(T_{ij} - S_{ij}),$$

$$\dot{S}_{ij} = \beta_{ij}^{S}(T_{ij} - S_{ij}) - S_{ij} \sum_{k=1}^{m} \alpha_k p_{ik},$$

$$\dot{R} = \sum_{k=1}^{m} (T_{ik} - S_{ik})\beta_{ik}^{R} - Q_i \sum_{k=1}^{m} \alpha_k p_{ik},$$

$$\dot{Q}_i = \beta_i^{Q}(R_i - Q_i) - Q_i \sum_{k=1}^{m} \alpha_k p_{ik},$$

where β_{ij}^{R} and β_{ij}^{S} are probabilities of coming from the groups T to $\widetilde{R}$ and S; β_i^{Q} are the probabilities of coming from $\widetilde{R}$ to Q.

This model describes the dynamics of economic exchange in terms of capacities of corresponding groups. It can be seen from the equations that the transition probabilities are parameters of the model. The probabilities characterize the stochastic behaviour of elements (producers, consumers and resources) in the system of economic exchange.

Transport flows. Transportation systems are important communication means. Its modern conception consists of three components, namely a network, modes and resources. The last two components produce the flows in the network.

Let a resource be produced and consumed in a set of regions. A transportation network links the regions and transfers the resource. We assume that the resource is discrete, and its quantity is measured in portions.

Let P_i be an amount of the resource in the producing region, let Q_i be a capacity of the consuming region, and let x_{ij} be a flow of the resource transferred from i to j in a time unit. Then the following balance relations hold:

$$\sum_i x_{ij} \leq Q_j, \quad j \in \overline{1,n}; \quad \sum_j x_{ij} \leq P_i, \quad i \in \overline{1,m};$$

where n is the number of consuming regions and m is the number of producing regions.

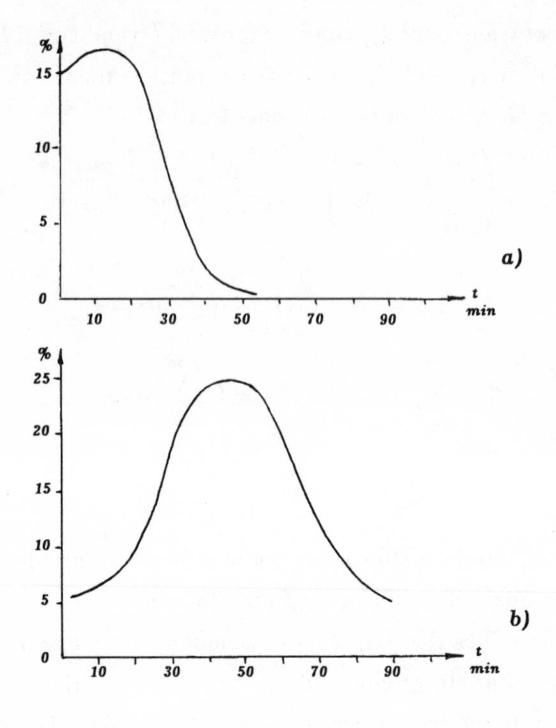

FIGURE 1.6.

Various mechanisms form transport flows for various kinds of the resources transported.

In the passenger transportation system, the elements have a considerable "freedom of the will" in choosing the routes. This is more evident in an urban transportation system, since the latter has a denser network. The urban traffic service allows a passenger to choose various routes depending on his own preferences and external influences.

Panel studies of urban population show that any individual decision about the way to follow is caused by a number of factors and thus is deterministic. But different people obtain different decisions by using the same set of factors. Therefore when we consider a passenger as a universal element of transported resource, his choice turns out to be non-deterministic. Multiple realizations of these individual choices generate a distribution of flows in a transportation network.

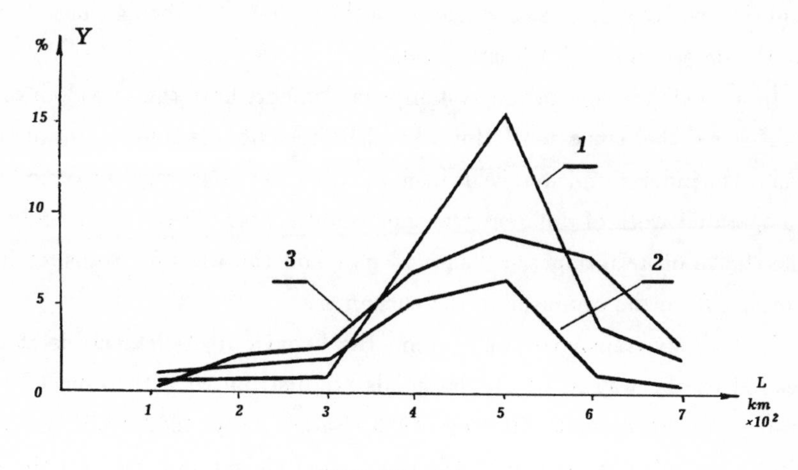

FIGURE 1.7.

Since the flow is a sum characteristic of the system, it is more deterministic than the behaviour of an element of the transported resource.

Real observations of daily flow distribution confirm this property. In fact, the flow distribution in a fixed time interval is practically the same from day to day. Logically, in this case we suppose that the transportation network and the population allocation are not changed.

So-called population-allocation curve is an aggregated characteristics of the flow distribution in a transportation network. The curve shows how the relative quantities of passengers are distributed with respect to trip durations. Population-allocation curves are shown in fig. 1.6 for small towns (100,000 – 150,000 inhabitants, fig. 1.6(a)) and large cities (more than 5 million inhabitants, fig. 1.6(b)). We can see from these diagrams that the inhabitants of small towns and those of large cities behave in the different ways. In the towns, the factor of "home-to-work" distance appears to be more important than in the cities.

Statistical analysis of population-allocation functions shows (Goltz, 1981) that qualitative morphological features and some characteristics of these functions (e.g. mean value and dispersion) are quite stable not only for a particular city, but also for cities with similar size.

Thus there are two levels in a passenger transportation system, namely the level of elements (passengers) having a non-deterministic type of behaviour, and the level of system as a whole, with the system state being characterized by the deterministic flow distribution.

In a goods transportation system, the elements have the "freedom of the will" to a rather less extent. However, their behaviour is non-deterministic as well. The indeterminism reveals itself in "producer-consumer" links, in badly co-operated work of different transport modes, etc. These factors influence the choice of a "producer-consumer" pair and the ways for transporting a product from the producer to the consumer.

In a goods transportation system, the behaviour of elements is characterized by the way in which the goods are distributed with respect to the transportation distance (Batten, 1983; Batten, Roy, 1982; Wilson, 1970). This distribution is similar to the population-allocation curves for the passenger transportation system. Kotel'nikov (Kotel'nikov, 1985) analyzed the transportation of construction materials by the USSR railway network during 1972–1983 years.

It is shown in fig. 1.7 how relative amounts of construction material are distributed among the distances (1 means cement; 2 means keramsite; 3 means reinforced concrete). These characteristics are quite stable for the most intensive work periods of the transportation system. During twelve years the maximum mean-square deviations from the average values of transportation distances were 5% for the cement, 11% for the keramsite and 6% for the reinforced concrete.

The quite stable distribution of the goods confirms implicitly the stability of goods flows in the transportation network.

Thus the goods transportation system has dual characteristics, namely indeterminism of the element level and determinism of the system as a whole.

CHAPTER

TWO

MODELS OF STATIONARY STATES FOR HOMOGENEOUS MACROSYSTEMS

In order to determine stationary states of homogeneous macrosystems, we develop mathematical models based on the principle of entropy maximization. A macrosystem stationary state is characterized by the way in which the elements are distributed among the groups of close states and depends on the occupation mechanism. Three classes of models corresponding to the Boltzmann, Fermi-Dirac and Boze-Einstein statistics are considered.

Models of stationary states (MSS) are characterized by constrained extremum problems having entropy objective functions. This feature is substantially used in the optimality conditions of corresponding models. Useful interpretations of the results can be obtained from these conditions.

2.1 Phenomenological scheme

Consider an abstract macrosystem consisting of Y undistinguishable elements having a stochastic behavior. The states of each element fall in one of ρ classes, viz $\mathcal{K}_1, \ldots, \mathcal{K}_\rho$. The classes do not intersect.

Denote by $\mathfrak{S}^1, \ldots, \mathfrak{S}^\rho$ the sets of states, where $\mathfrak{S}^i \in \mathcal{K}^i$ $(i \in \overline{1, \rho})$. Further we assume that the sets $\mathfrak{S}^i$ are discrete and finite.

The states are considered as elements of $\mathfrak{S}^1, \ldots, \mathfrak{S}^\rho$ and can be of two types, namely (a) Fermi-type (each state can contain only one element) and (b) Einstein-type (each state can contain any number of elements).

Let the set $\mathfrak{S}^i$ ($i \in \overline{1,\rho}$) contain only the states of one type. We can distinguish subsets $\mathfrak{S}^i_n$ ($n \in \overline{1,m_i}$) of "close" states which satisfy the following conditions

$$\mathfrak{S}^i = \bigcup_{n=1}^{m_i} \mathfrak{S}^i_n, \qquad \mathfrak{S}^i_{n_1} \bigcap \mathfrak{S}^i_{n_2} = \varnothing,$$

for any $n_1, n_2 \in \overline{1,m_i}$. If the states can be grouped in such a way then each state from $\mathfrak{S}^i$ should have a set of numerical characteristics which can be used to disaggregate $\mathfrak{S}^i$ and to form the subsets $\mathfrak{S}^i_1, \ldots, \mathfrak{S}^i_{m_i}$.

The subsets $\mathfrak{S}^i_n$ are characterized by their capacity G^i_n which is the number of states in $\mathfrak{S}^i_n$.

In this chapter, we consider ***homogeneous macrosystems*** or systems with elements occupying only the states of one class.

Such a system has one set of states $\mathfrak{S}$. The subsets $\mathfrak{S}_1, \ldots, \mathfrak{S}_m$ do not intersect and their sum gives $\mathfrak{S}$.

To illustrate these definitions, we consider an urban system, with residents being the system elements. State classes for such an element can be places of residence, work, and service (cultural, shopping, medical). Thus, the states are divided into five classes. Obviously, they do not intersect, since they are generated by different functions.

For this system, the states can be classified in another way. For example, we can interpret the classes as the pairs, namely the place of residence — the place of work, the place of residence — the place of service, the place of work — the place of service, and the place of residence — the place of rest. Here we have only four classes. Note that for one and the same system the state classes and their number can be defined in various way depending on the purposes of the model.

Now let us return to the first example and consider two of the state classes, viz the place of residence and the place of work. Elements of these classes are somehow distributed over an urban area, i.e. there are the set $\mathfrak{S}^1$ of the first class states (the places of residence) and the set $\mathfrak{S}^2$ of the second class states (the places of work). If the elements of the sets are not differentiated, the sets are characterized by their capacities G^1 and G^2. But the states from $\mathfrak{S}^2$, for example, can be characterized by the professional abilities of workers. For the simplicity sake, we assume that the professional abilities are measured in a discrete and finite scale. Then the numerical characteristic breaks the states from $\mathfrak{S}^2$ into groups of states (job positions) with the same value of professional ability value. These groups are the subsets $\mathfrak{S}_1^2, \ldots, \mathfrak{S}_{m_2}^2$. By construction, the sum of these subsets is $\mathfrak{S}^2$, while any two subsets do not intersect. Here the capacities $G_1^2, \ldots, G_{m_2}^2$ of subsets are the numbers of working places with professional ability levels $1, 2, \ldots, m_2$. All these definitions concern the concept of macrosystem element state.

Now we proceed to introducing the concept of macrosystem state, with only homogeneous macrosystem being considered. The elements of a system can occupy any state from the subsets $\mathfrak{S}_n$ randomly and independently. For any fixed subset $\mathfrak{S}_n$, an element has just two possibilities, namely to occupy any state from $\mathfrak{S}_n$ with the apriori probability a_n or to keep out this subset with the probability $1 - a_n$.

Note that the apriori probabilities $a_1, \ldots, a_m$ characterize unoccupied subsets $\mathfrak{S}_1, \ldots, \mathfrak{S}_m$ (in this situation the states of the subsets contain no elements).

The probabilities $a_1, \ldots, a_n$ should satisfy the obvious condition $\sum_{n=1}^{m} a_n \leq 1$. If $\sum_{n=1}^{m} a_n = 1$ then any element occupies one of the subsets $\mathfrak{S}_1, \ldots, \mathfrak{S}_m$ with the probability equal to 1. If $\sum_{n=1}^{m} a_n < 1$ then an element can keep out all subsets $\mathfrak{S}_1, \ldots, \mathfrak{S}_m$ with the probability $p_{m+1} = 1 - \sum_{n=1}^{m} a_n > 0$. In this case, it is useful to introduce an additional subset $\mathfrak{S}_{m+1}$ of fictitious states which can be occupied with the apriori probability p_{m+1}.

We consider the distribution of all Y elements among the states from subsets $\mathfrak{S}_1, \ldots, \mathfrak{S}_m$ as system *microstate*. Hence the microstate is a list containing the element number and the subset occupied by this element, i.e. element 1 is in the subset $\mathfrak{S}_2$, element 2 is in $\mathfrak{S}_1; \ldots;$ element Y is in $\mathfrak{S}_{m-1}$.

Since the macrosystem elements are not distinguishable, the microstates having N_1 elements in $\mathfrak{S}_1$, N_2 elements in $\mathfrak{S}_2, \ldots,$ and N_m elements in $\mathfrak{S}_n$ are also non-distinguishable.

Therefore we say that the system *macrostate* is the way in which the macrosystem elements are distributed among the subsets $\mathfrak{S}_1, \ldots, \mathfrak{S}_m$.

Macrostate is characterized by the collection $N_1, \ldots, N_m$ of the numbers of elements occupying the corresponding subsets $\mathfrak{S}_1, \ldots, \mathfrak{S}_m$. If these subsets contain Einstein-states, their capacities $G_1, \ldots, G_m$ do not restrict the numbers of elements which can occupy them. Therefore if the overall number of elements Y is fixed then $0 \leq N_n \leq Y$ $(n \in \overline{1,m})$.

If the subsets $\mathfrak{S}_1, \ldots, \mathfrak{S}_m$ contain Fermi-states, the capacities $G_1, \ldots,$ G_m bound the numbers of elements which can occupy the subsets, i.e. $0 \leq$ $N_n \leq G_m$ $(n \in \overline{1,m})$. Hence if the number Y of elements is fixed, the Fermi-states can exist only if certain relations between the subsets capacities $G_1, \ldots, G_m$ and the apriori probabilities $a_1, \ldots, a_m$ hold.

According to the definitions, both macrostates and microstates are not unique. But by observing the macrostates for various macrosystems, we can see that the equilibrium state is corresponded by the unique macrostate $N^0 = (N_1^0, \ldots, N_m^0)$ which is called *realizable macrostate*.

The distribution of macrosystem elements among the subsets $\mathfrak{S}_1, \ldots,$ $\mathfrak{S}_m$ is subjected to the constraints caused by balance relations between $N_1^0, \ldots, N_m^0$ and consumption of various resources.

The simplest example of the balance constraint is the requirement for the number of elements to be constant $(\sum_{i=1}^{m} N_i^0 = Y)$. A typical resource constraint in thermodynamic system is conservation of sum energy of the elements; usual constraint in a transportation system is restriction of the average trip cost.

In order to illustrate the phenomenology described above, let us consider an ideal system consisting of Y particles (Reif 1977), each having the spin 1/2 and the magnetic moment μ_0. Electrons or atoms with odd number of electrons can be examples of such a system. Once a particle has a spin, the component of movement moment (measured in given direction) can be equal only to two values, namely $+\frac{1}{2}(\frac{\mathfrak{h}}{2\pi})$ or $-\frac{1}{2}(\frac{\mathfrak{h}}{2\eta})$. (Here $\mathfrak{h}$ is the Planck constant).

The system of Y particles with spins is like Y magnetic bars which have the magnetic moment μ_0 directed upwards or downwards (fig. 2.1). Assume that the particles are quite far from each other, their interaction is negligible, and the directions of their magnetic moments change randomly.

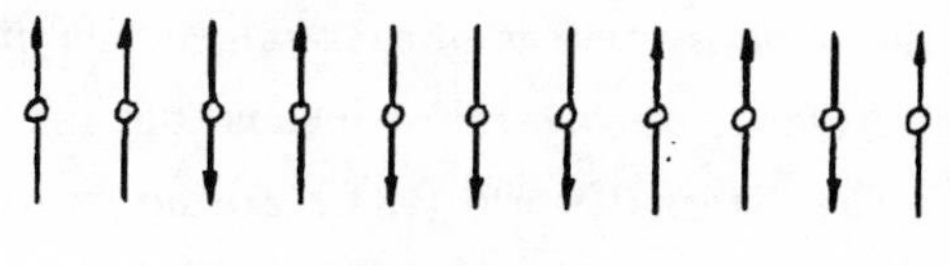

FIGURE 2.1.

Consider the spin orientation as a particle state. Then for the system elements (particles) we have two classes of states, namely the states with the spin $+\frac{1}{2}$ and those with the spin $-\frac{1}{2}$. Each class can include all particles.

A set of states corresponds to each class. In our case, there are two such sets, viz $\mathfrak{S}_1$ (the states with the spin $+\frac{1}{2}$) and $\mathfrak{S}_2$ (the states with the spin $-\frac{1}{2}$). The states in each of the sets are non-distinguishable. If the states in $\mathfrak{S}_1$ and $\mathfrak{S}_2$ are of Einstein-type, the capacities G_1 and G_2 can be arbitrary. If the states in $\mathfrak{S}_1$ and $\mathfrak{S}_2$ are of Fermi-type then $G_1 = G_2 = Y$.

Because the magnetic moments are oriented randomly, the set $\mathfrak{S}_1$ can contain N_1 particles (with the spin $+\frac{1}{2}$), and the set $\mathfrak{S}_2$ can contain $N_2 = N - N_1$ particles (with the spin $-\frac{1}{2}$). The macrostate of this system is described by the vector $N = (N_1, N_2)$.

To illustrate the microstate concept, let us mark the particles. Obviously, a macrostate N can be produced by various distributions of marked particles among the sets $\mathfrak{S}_1$ and $\mathfrak{S}_2$. In particular, for $N = 4$ each of the macrostates (4,0) and (0,4) is produced by one microstate, the macrostates (3,1) and (1,3) are produced by 4 microstates, respectively, and the macrostate (2,2) is produced by 6 macrostates.

2.2 Stochastic characteristics of macrostates for unconstrained system

Random distribution of macrosystem elements generates the set $\mathfrak{N}$ of feasible macrostates. Any macrostate is characterized by the vector $N = \{N_1, \ldots, N_m\}$, with the coordinates being the numbers of elements occupying the subsets $\mathfrak{S}_1, \ldots, \mathfrak{S}_m$. These numbers can be equal to any value from 0 till Y for Einstein-states and to any value from 0 till G_n for Fermi-states, with a non-zero probability.

Assume that the macrosystem is unconstrained, i.e. it is not influenced from the outside. Absence of constraints means that the macrosystem does not interact with the external world (there are no exchanges of elements, energy, resources, etc.). In this case, the set $\mathfrak{N}$ includes all the possible macrostates. Such macrosystems are said to be *isolated.*

To study the properties of $\mathfrak{N}$, it is useful to determine some stochastic characteristics of this set.

We begin with the probability distribution function $P(N)$, where $N \in \mathfrak{N}$. Since the sets $\mathfrak{S}_1, \ldots, \mathfrak{S}_m$ do not intersect and the stochastic mechanism is such that the element behavior is independent, the probability distribution function is

$$P(N) = \prod_{n=1}^{m} P_n(N_n), \tag{1}$$

and

$$\sum_{N_1,\ldots,N_m=1}^{Y} P(N_1, \ldots, N_m) = 1, \tag{1'}$$

where $N = (N_1, \ldots, N_m)$,

$P_n(N_n)$ is the probability of N_n elements occupying subset $\mathfrak{S}_n$. To determine the probability distribution function $P_n(N_n)$, we consider the mechanism of occupying the states in $\mathfrak{S}_n$.

(a) *A macrosystem having the Fermi-states.* Here each state in $\mathfrak{S}_n$ can be occupied by only one element with the probability a_n (or an element can keep out $\mathfrak{S}_n$ with the probability $1 - a_n$). Elements choose states in $\mathfrak{S}_n$ independently.

Note that the apriori probability a_n is the same for all states in the subset $\mathfrak{S}_n$. Therefore the events "an element occupies the subset $\mathfrak{S}_n$" and "an element occupies a state in the subset $\mathfrak{S}_n$" are equivalent.

Since the capacity G_n of the subset $\mathfrak{S}_n$ is assumed to be prescribed, N_n can vary in the interval $0 \le N_n \le G_n$. Thus if $\sum_{n=1}^{m} a_n = 1$, i.e. if all the elements are distributed among the subsets $\mathfrak{S}_1, \ldots, \mathfrak{S}_m$, then $\sum_{n=1}^{m} G_n \ge Y$. Otherwise (if $\sum a_n < 1$) $\sum_{n=1}^{m} G_n$ can be less than Y.

Let N_n be fixed. Then the probability that N_n elements occupy $\mathfrak{S}_n$ and $G_n - N_n$ elements keep out of $\mathfrak{S}_n$ is equal to $a_n^{N_n}(1 - a_n)^{G_n - N_n}$.

The numbers N_n can be produced by various configurations or distributions of marked elements among N_n states in $\mathfrak{S}_n$ (an analog of system microstate). For example, for $G_n = 4$ and $N_n = 2$ we have 6 configurations, namely (1,2), (1,3), (1,4), (2,3), (2,4) and (3,4). In general, the number of configurations with the same N_n which contain different elements is equal to $C_{G_n}(N_n)$. Hence the probability $P_n(N_n)$ in (1) (N_n elements in $\mathfrak{S}_n$ and $G_n - N_n$ elements outside of $\mathfrak{S}_n$) equals the probability of the first configuration, or that of the second configuration,..., or, at last, that of the configuration $C_{G_n}(N_n)$. Since all these configurations have the same probability, we obtain

$$P_n(N_n) = P_n^F(N_n) = C_{G_n}(N_n) a_n^{N_n}(1 - a_n)^{G_n - N_n}, \tag{2}$$

$$P_n^F(N_n) = \frac{G_n!}{N_n!(G_n - N_n)!} a_n^{N_n}(1 - a_n)^{G_n - N_n}. \tag{3}$$

The distribution function coincides with the classical binomial distribution (the Bernoulli distribution) (Prochorov, 1967). It follows from (3) that $P_n(N_n)$ is determined by two parameters G_n and a_n, and

$$\mathcal{M}(N_n) = G_n a_n; \quad \mathcal{M}[N_n - \mathcal{M}(N_n)]^2 = G_n a_n(1 - a_n), \tag{4}$$

where $\mathcal{M}$ is the operator of mathematical expectation. In this case,

$$\mathcal{M}(\cdot) = \sum_{N_n=1}^{G_n} (\cdot) P_n(N_n). \tag{5}$$

Since the macrosystem contains a large number of elements and the capacity G_n of the subset $\mathfrak{S}_n$ are also large, asymptotic forms of distribution (3) are of interest.

If $G_n \to \infty$ and $G_n a_n \to \infty$ then the normal approximation holds for the binomial distribution, namely

$$P(N_n) = \frac{1}{\sqrt{2\pi}\sigma_n} e^{-\frac{(N_n - \overline{N}_n)^2}{2\sigma_n^2}}, \tag{6}$$

where

$$\overline{N}_n = G_n a_n; \quad \sigma_n = \sqrt{a_n(1 - a_n)G_n}. \tag{7}$$

Note that the asymptotically normal approximation for (3) can be applied when the average number of elements occupying the subset $\mathfrak{S}_n$ increases along with the capacity G_n so that $\overline{N}_n/G_n = a_n = \text{const}$. If the latter does not hold, for example as $G_n \to \infty$, then the Poisson approximation holds for (3)

$$P_n(N_n) = \frac{\lambda_n^{N_n}}{N_n!} e^{-\lambda_n}. \tag{8}$$

In this case, $\frac{\overline{N}_n}{G_n} \to 0$ i.e. the Poisson approximation characterizes the macrosystem with average values $\overline{N}_n$ smaller than the capacity G_n.

Now, by substituting (4) in (1), we obtain the probability distribution function for possible macrostates in an isolated macrosystem with the Fermi-states

$$P_F(N) = \prod_{n=1}^{m} \frac{G_n!}{N_n!(G_n - N_n)!} \tilde{a}_n^{N_n} (1 - a_n)^{G_n}, \tag{9}$$

where

$$\tilde{a}_n = \frac{a_n}{1 - a_n}. \tag{10}$$

Note that the probability distribution function is derived for the unconstrained set $\mathfrak{N}$ of possible macrostates, in particular, the sum number of elements $Y = \sum N_n$ is not fixed. For a macrosystem with the Fermi-states, Y is the number of elements needed to produce a macrostate with the given collection $N_1, \ldots, N_m$.

A ***physical entropy*** is important in studying macrosystems. It is defined by using the statistical weight $\Delta\Gamma(N) = AP(N)$ (Landau 1964), where A is a norming constant. The physical entropy is

$$E(N) = k \ln \Delta\Gamma(N) = E_0 + k \ln P(N), \tag{11}$$

where

$$E_0 = k \ln A, \tag{12}$$

k is the parameter defining the unit in which the entropy is measured (for thermodynamic systems, k is Boltzmann's constant).

For a macrosystem with the Fermi-states, the physical entropy is

$$E_F(N) = k \sum_{n=1}^{m} (\ln G_n! + N_n \ln \tilde{a}_n + G_n \ln(1 - a_n) - \ln N_n! - \ln(G_n - N_n)!) + E_0.$$

If G_n and N_n are sufficiently large, the expression for $E_F(N)$ can be simplified by using the Stirling approximation

$$\ln x! \cong x(\ln x - 1).$$

We obtain

$$E_F(N) = C - k\sum_{n=1}^{m} N_n \ln\frac{N_n}{\widetilde{a}_n} + (G_n - N_n)\ln(G_n - N_n), \tag{13}$$

where

$$C = k\sum_{n=1}^{m} G_n(\ln G_n + \ln(1 - a_n)) + E_0. \tag{13'}$$

Let us call the function

$$H_F(N) = -\sum_{n=1}^{m} N_n \ln\frac{N_n}{\widetilde{a}_n} + (G_n - N_n)\ln(G_n - N_n), \tag{14}$$

the generalized information entropy of Fermi-Dirac. Denote by

$$\mathfrak{Z}_F = \{N:\ 0 \le N_n \le G_n, \quad n \in \overline{1,m}\} \tag{14'}$$

the set on which $H_F(N)$ is defined. Introduce relative occupation numbers

$$y_n = N_n/G_n; \quad 0 \le y_n \le 1. \tag{15}$$

Then

$$\widetilde{H}_F(y) = -\sum_{n=1}^{m} G_n \ln G_n + \widehat{H}_F(y), \tag{16}$$

where

$$\widehat{H}_F(y) = -\sum_{n=1}^{m} G_n \left(y_n \ln\frac{y_n}{\widetilde{a}_n} + (1 - y_n)\ln(1 - y_n)\right) \tag{17}$$

We see from expressions (9), (14) and (17) that in order to determine the probability distribution function, or the entropy, we should know the capacities G_n and the apriori probabilities a_n for an element to occupy the set $\mathfrak{S}_n$.

(b) *A macrosystem having the Einstein-states.* Here each state can be occupied by an arbitrary number of elements. It means that the subset $\mathfrak{S}_n$ with the capacity G_n can contain N_n elements, and N_n can be produced by different configurations. For example, all N_n elements can occupy state 1, while the others $G_n - 1$ states in $\mathfrak{S}_n$ can be empty. Or $N_n/2$ elements can occupy state 1, $N_n/2$ can occupy state 3, and the others $G_n - 2$ states can be empty. From this we can see that the number of configurations producing N_n is the number of different ways to distribute N_n balls among G_n cells.

In (Landau, Lifshitz. 1964) this problem is interpreted quite obviously. Let us identify the balls with the sequence of N_n points and let us identify the cell boundaries with vertical dashes separating the points. There are $G_n - 1$ dashes. For example, the sequence ...|.|....|.|.. shows how 11 balls are distributed among 5 cells. The sells contain 3, 1, 4, 1 and 2 balls, respectively. The overall number of positions in this sequence equals $N_n + G_n - 1 = 16$, while the number of distributions of the balls among the cells is the number of ways to choose $G_n - 1$ positions for dashes, i.e. $\binom{G_n-1}{N_n+G_n-1}$.

The latter means that distribution of N_n elements in the subset $\mathfrak{S}_n$ having Einstein-states and capacity G_n is equivalent to distribution of N_n elements in an auxiliary subset $\overline{\mathfrak{S}}_n$ having the Fermi-states and the capacity $R_n = N_n + G_n - 1$.

A macrosystem element occupies the subset $\overline{\mathfrak{S}}_n$ with the apriori probability a_n. It occupies $\overline{\mathfrak{S}}_n$, which has the Fermi-states, with the same probability. $\overline{\mathfrak{S}}_n$ always has free states, since its capacity R_n depends on the number of elements occupying this state. Thus, the probability for N_n elements to occupy $\overline{\mathfrak{S}}_n$ and for $R_n - N_n$ elements to keep out $\overline{\mathfrak{S}}_n$ is equal to

$$a_n^{N_n}(1-a)^{G_n-1}. \tag{18}$$

Following the approach applied to the macrosystems with the Fermi-states, it is easy to show that the probability of N_n elements occupying $\overline{\mathfrak{S}}_n$ equals the sum of probabilities (18), i.e.

$$P_n^E(N_n) = \frac{(G_n + N_n - 1)!}{N_n!(G_n - 1)!} a_n^{N_n}(1-a_n)^{G_n-1}. \tag{19}$$

The probability distribution function arrived at coincides with the classical Pascal distribution, which is related to the Bernoulli trials as well as the binomial distribution (Prochorov 1967).

It is seen from (19) that the function $P_n^E(N_n)$ is determined by two parameters G_n and a_n, and

$$\mathcal{M}(N_n) = G_n \frac{a_n}{(1-a_n)^2}; \quad \mathcal{M}(N_n - \mathcal{M}(N_n))^2 = G_n \frac{a_n^2}{(1-a_n)^2}. \tag{20}$$

By substituting (19) into (1), we obtain the probability distribution function for possible macrostates in an isolated macrosystem with the Einstein-states

$$P_E(N) = \prod_{n=1}^{m} \frac{(G_n + N_n - 1)}{N_n!(G_n - 1)!} a_n^{N_n}(1 - a_n)^{G_n - 1}. \tag{21}$$

According to (11), the physical entropy for this macrosystem is

$$E_E(N) = C - k\sum_{n=1}^{m} N_n \ln\frac{N_n}{a_n} - (G_n + N_n)\ln(G_n + N_n), \tag{22}$$

where

$$C = k\sum_{n=1}^{m} G_n(\ln(1 - a_n) - \ln G_n) + E_0. \tag{22'}$$

We call the function

$$H_E(N) = -\sum_{n=1}^{m} N_n \ln\frac{N_n}{a_n} - (G_n + N_n)\ln(G_n + N_n), \tag{23}$$

the *generalized information entropy of Einstein-Boze.* Denote by

$$\mathfrak{Z}_E = \{N : \; N_n \geq 0, \quad n \in \overline{1,m}\} \tag{23'}$$

the set on which the function $H_E(N)$ is defined.

Introduce the relative variables y_n

$$y_n = \frac{N_n}{G_n} \geq 0. \tag{24}$$

We obtain from (23)

$$\widetilde{H}_E(y) = \sum_{n=1}^{m} G_n \ln G_n + \widehat{H}_E(y), \tag{25}$$

where

$$\widehat{H}_E(y) = -\sum_{n=1}^{m} G_n \left(y_n \ln\frac{y_n}{a_n} - (1 + y_n)\ln(1 + y_n) \right) \tag{26}$$

Now let us consider some limit properties of the probability distribution functions $P_F(N)$ and $P_E(N)$ from (9) and (21) and the entropy functions $H_F(N)$ and $H_E(N)$ from (14) and (23).

(c) *Macrosystems with the Boltzmann-states.* Let us assume that $N_n \ll G_n$ and $a_n G_n = \text{const}$. Then

$$\frac{G_n!}{(G_n - N_n)!} \approx G_n^{N_n}, \quad (1 - a_n)^{G_n - N_n} \approx (1 - a_n)^{G_n}$$

and

$$\frac{(G_n + N_n)!}{(G_n - 1)!} \approx G_n^{N_n}, \quad (1 - a_n)^{G_n - 1} \approx (1 - a_n)^{G_n}.$$

In this case, the functions $P_n^F(N)$ and $P_n^E(N)$ coincide and are

$$P_n^B(N_n) = A_n \frac{(G_n a_n)^{N_n}}{N_n!} (1 - a_n)^{G_n},$$

where A_n is norming constant determined from

$$\sum_{N_n=0}^{Y} P_n^B(N_n) = 1.$$

Here we obtain

$$A_n (1 - a_n)^{G_n} \sum_{N_n=0}^{Y} \frac{q_n^{N_n}}{N_n!} = 1, \quad q_n = a_n G_n.$$

It is easy to see that the series is the expansion of the exponent e^{qx} at the point $x = 1$. Therefore

$$A_n = \frac{e^{-a_n G_n}}{(1 - a_n)^{G_n}}.$$

Then the probability distribution function for the macrostates is

$$P_B(N) = \prod_{n=1}^{m} e^{-a_n G_n} \frac{(a_n G_n)^{N_n}}{N_n!}. \tag{27}$$

The entropy of distribution (27) (after applying the Stirling approximation) is

$$E_B(N) = C - k \sum_{n=1}^{m} N_n \ln \frac{N_n}{a_n G_n e}, \tag{28}$$

where

$$C = -k \sum_{n=1}^{m} G_n a_n + E_0. \tag{28'}$$

Let us call the function

$$H_B(N) = -\sum_{n=1}^{m} N_n \ln \frac{N_n}{a_n G_n e} \tag{29}$$

the *generalized information entropy of Boltzmann.*

Denote by

$$\mathfrak{Z}_B = \{N : \ N_n \geq 0, \quad n \in \overline{1, m}\} \tag{29'}$$

the set on which $H_B(N)$ is defined. For relative variables (15) we obtain

$$\widehat{H}_B(y) = -\sum_{n=1}^{m} y_n G_n \ln \frac{y_n}{a_n e}. \tag{30}$$

Note that here $0 \le y_n \ll 1$.

If the occupation numbers N_n are significantly less than the capacity G_n of the subset $\mathfrak{S}_n$, such macrosystems are called the macrosystems with Boltzmann-states.

Let us consider another limit case $N_n \gg G_n$. Then

$$\frac{(G_n + N_n - 1)!}{N_n!} \approx N_n^{G_n}$$

and (19) is like

$$P_E^{\infty}(N) = \prod_{n=1}^{m} \frac{N_n^{G_n}}{(G_n - 1)!} a_n^{N_n} (1 - a_n)^{G_n - 1}. \tag{31}$$

The entropy of distribution (31) is

$$E_E^{\infty}(N) = C + \sum_{n=1}^{m} G_n \ln N_n + N_n \ln a_n, \tag{32}$$

where

$$C = -\sum_{n=1}^{m} (\ln(G_n - 1)! - (G_n - 1)\ln(1 - a_n)) + E_0.$$

The function

$$H_E^{\infty}(N) = \sum_{n=1}^{m} G_n \ln N_n + N_n \ln a_n \tag{33}$$

is *the limit generalized information entropy of Einstein-Boze.*
In terms of relative variables (15), the above expression is

$$\widetilde{H}_E^{\infty}(y) = \sum_{n=1}^{m} G_n \ln G_n + \widehat{H}_E^{\infty}(y),$$

where

$$\widehat{H}_E^{\infty}(y) = \sum_{n=1}^{m} G_n (\ln y_n + y_n \ln a_n). \tag{34}$$

Table 1 contains stochastic characteristics of unconstrained homogeneous macrosystems.

A. *Boltzmann – statistics*, $N_n \ll G_n$

Probability distribution function
$$P_B(N) = \prod_{n=1}^{m} \frac{G_n^{N_n}}{N_n!} a^{N_n}(1-a_n)^{G_n}$$
Information entropy
$$E_B(N) = -\sum_{n=1}^{m} a_n G_n + H_B(N) + E_0$$ $$H_B(N) = -\sum_{n=1}^{m} N_n \ln(N_n/ea_n G_n)$$ $$\widehat{H}_B(y) = -\sum_{n=1}^{m} G_n y_n \ln(y_n/ea_n)$$

B. *Fermi-Dirac statistics*, $N_n \leq G_n$

Probability distribution function
$$P_F(N) = \prod_{n=1}^{m} \frac{G_n!}{N_n!(G_n-N_n)!} \widetilde{a}_n^{N_n}(1-a_n)^{G_n}$$ $$\widetilde{a}_n = \frac{a_n}{1-a_n}$$
Information entropy
$$E_F(N) = \sum_{n=1}^{m} G_n(\ln G_n + \ln(1-a_n)) + H_F(N) + E_0$$ $$H_F(N) = -\sum_{n=1}^{m} N_n \ln \frac{N_n}{\widetilde{a}_n} + (G_n - N_n)\ln(G_n - N_n)$$ $$\widetilde{H}_F(y) = -\sum_{n=1}^{m} G_n \ln G_n + \widehat{H}_F(y)$$ $$\widehat{H}_F(y) = -\sum_{n=1}^{m} G_n(y_n \ln \frac{y_n}{\widetilde{a}_n} + (1-y_n)\ln(1-y_n))$$

C. *Einstein -Boze statistics*, $N_n \geq 0$

Probability distribution function
$P_E(N) = \prod_{n=1}^{m} \frac{(G_n + N_n - 1)!}{N_n!(G_n - 1)!} a_n^{N_n}(1 - a_n)^{G_n - 1}$
Information entropy
$E_E(N) = \sum_{n=1}^{m} G_n(\ln(1 - a_n) - \ln G_n) + H_E(N) + E_0$ $H_E(N) = - \sum_{n=1}^{m} N_n \ln \frac{N_n}{a_n} - (G_n + N_n)\ln(G_n + N_n)$ $\widetilde{H}_E(y) = \sum_{n=1}^{m} G_n \ln G_n + \widehat{H}_E(y)$ $\widehat{H}_E(y) = - \sum_{n=1}^{m} G_n(y_n \ln \frac{y_n}{a_n} - (1 + y_n)\ln(1 + y_n))$

TABLE 1. The stochastic characteristics of a homogeneous macrosystem.

These characteristics (the probability distribution function and the generalized information entropy) take into account the properties of phenomenological scheme of homogeneous macrosystem appearing because the system elements choose the corresponding states in the subsets $\mathfrak{S}_n$ $(n \in \overline{1,m})$ with different probabilities.

Schemes used by statistical thermodynamics (Landau, 1964) usually assume that the particles are distributed among the states with equal probabilities. The expressions for the probability distribution function and the information entropy are transformed into "thermodynamic" expressions if the apriori probabilities are equal.

2.3 Morphological properties of stochastic characteristics

The stochastic characteristics of a homogeneous macrosystem (viz the probability distribution function and the entropy function) are functions of macrostates, i.e. of the occupation numbers $N_1, \ldots, N_m$ for the corresponding subsets $\mathfrak{S}_1, \ldots, \mathfrak{S}_m$ in the set of element states.

Now some properties of the characteristics are interesting, namely the existence of the maximum and its "sharpness". Because both are related to the form of the probability distribution function or that of the entropy function, they are called morphological.

While the existence of maximum can be determined unambiguously, the "sharpness" is estimated by various numerical characteristics. This appears to be because the "sharpness" is a qualitative property of the maximum. We tend to consider the maximum as "sharp" if small deviations of the argument give rise to significant decreases of the function values.

In order to estimate the maximum "sharpness" for the probability distribution function, various quantitative characteristics are used, e.g. the ratio of dispersion and maximum value (Kittel, 1969), the width of rectangle having the length equal to the maximum value and the unit area (Landau, 1964), the value of the second derivative at the maximum point (in this case we assume that N is large so that it can be replaced by a real variable), etc.

All these characteristics of maximum "sharpness" allow us to estimate the fluctuations from the most probable states.

This estimation can be even simpler. The probability distribution function $p(x)$ of a scalar random variable x is shown in fig. 2.2(a). Let $p(x_0) = \max_x p(x)$. The maximum "sharpness" for this function means that $p(x_0, \varepsilon) = \max(p(x_0 + \varepsilon),\ p(x_0 - \varepsilon))$ is significantly less than $p(x_0)$ for small deviations $\pm\varepsilon$ from x_0. Note that if the function $p(x)$ is symmetrical in the neighborhood $x_0 \pm \varepsilon$ then $p(x_0, \varepsilon) = p(x_0 + \varepsilon) = p(x_0 - \varepsilon)$. Below we assume that there exists a neighborhood of the maximum point in which the probability distribution of macrostates is symmetrical.

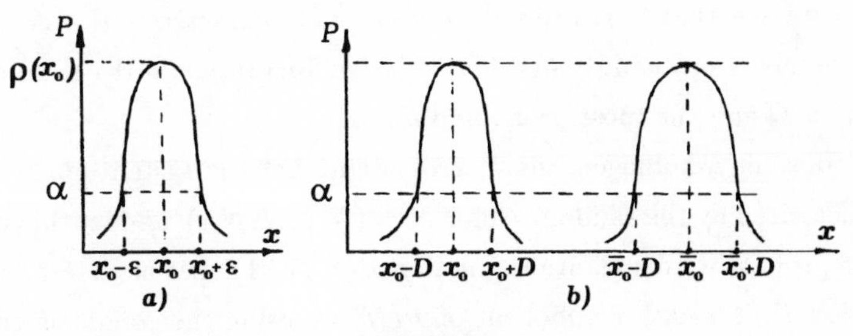

FIGURE 2.2.

The obvious way for characterizing the "sharpness" is based on the quantitative measurement of how $p(x_0 \pm \varepsilon)$ is "negligible" with respect to $p(x_0)$ and on the estimate of corresponding ε; i.e. ε is obtained from

$$\frac{p(x_0)}{p(x_0 \pm \varepsilon)} = a,$$

where a is a given constant $(a > 1)$.

Usually we consider the fluctuation ε decreasing the probability at e times. Denote this value by D; i.e. D is obtained from the equation

$$\frac{p(x_0)}{p(x_0 \pm D)} = e. \tag{1}$$

Note that if $p(x)$ is the normal distribution then $D = \sigma$, where σ is the mean-squared deviation of the random variable x.

But D characterizes only an absolute value of the fluctuation, and therefore it cannot characterize the maximum "sharpness". Two probability distribution functions $p_1(x)$ and $p_2(x)$ are shown in fig. 2.2(b). They have the same maxima $p_{\max}$, the same fluctuations D (1) and the same form, but their most probable values are different, so that $\overline{x}_0 \gg x_0$. For the sake of clarity we assume that $x_0 = 5D$ and $\overline{x}_0 = 10^3 \cdot 5D$. Then for the random variable having the probability distribution function $p_1(x)$, the values $\mu = (5 \pm 1)D$ are realized with the probability $p_{\max}/e$. For the random variable having the probability distribution function $p_2(x)$, the values $\nu = (5000 \pm 1)D$ are realized with the same probability. Hence, although the probabilities μ and ν are equal, the deviation on $\pm D$ in the first case is more significant than in the second.

Hence we see that a ***relative fluctuation*** can characterize the deviation having the given probability (1). The relative fluctuation is the ratio of the fluctuation D and the most probable value x_0.

Now consider a homogeneous macrosystem. Let the system macrostates be characterized by the random vector $N = (N_1, \ldots, N_m)$, and let this vector have the probability distribution function $P(N)$ (2.1). Let us determine the fluctuation D_n for each component of $P(N)$ by using the condition similar to (1), namely

$$\frac{P_n(N_n^0)}{P_n(N_n^0 + D_n)} = e^{\frac{1}{m}}, \quad n \in \overline{1,m}. \tag{2}$$

It follows from (2.1) that

$$\frac{P(N_1^0, \ldots, N_m^0)}{P(N_1^0 + D_1, \ldots, N_m^0 + D_m)} = e. \tag{3}$$

The equations (2) determine the absolute values $D_1, \ldots, D_m$ of fluctuations. Denote

$$D = \max D_n. \tag{4}$$

Now if $D_1, \ldots, D_m$ in (2), and (3) are replaced by D from (4), then these expressions become the inequalities

$$\frac{P_n(N_n^0)}{P_n(N_n^0 + D)} \geq e^{\frac{1}{m}}, \quad n \in \overline{1,m}; \qquad \frac{P(N_1^0, \ldots, N_m^0)}{P(N_1^0 + D, \ldots, N_m^0 + D)} \geq e. \tag{5}$$

The value of D (4) is the guaranteed absolute fluctuation. For this fluctuation, the probability of macrostate with the vector N deviating from the most probable vector in exactly D units is less than the maximal value in more than e times (an equality is the case only if $D_1 = D_2 = \ldots = D_m = D$).

We define the ***guaranteed relative fluctuation*** as

$$\rho = \max \frac{D_n}{N_n^0}, \tag{6}$$

and say that it characterize the maximum "sharpness".

Hence to estimate the guaranteed relative fluctuation ρ, we need the absolute fluctuation D and the most probable macrostate N^0.

Let us determine D in terms of the entropy. Turn to the equality (2) and express it in terms of corresponding entropy functions. We obtain

$$\ln P_n(N_n^0) - \ln P_n(N_n^0 + D_n) = \frac{1}{m},$$

$$E_n(N_n) = k\ln A_n + k\ln P_n(N_n),$$

where A and k are norming constants.

By substituting this expression into the previous equality, we obtain

$$E_n(N_n^0) - E_n(N_n^0 + D_n) = \frac{k}{m}, \quad n \in \overline{1,m}. \tag{7}$$

The sum of these equalities is

$$E(N^0) - E(N_1^0 + D_1, \ldots, N_m^0 + D_m) = k.$$

Thus the fluctuations $D_1, \ldots, D_m$ give rise to the entropy decrease at k units.

If the guaranteed absolute fluctuation is used, the relations similar to (5) can be obtained in terms of entropy

$$E_n(N_n^0) - E_n(N_n^0 + D) \geq \frac{k}{m};$$

$$E(N^0) - E(N_1^0 + D, \ldots, N_m^0 + D) \geq k.$$

Consider a *macrosystem with the Fermi-states.* According to the formula for the physical entropy (Table 1), the equations (7), which determine the absolute fluctuations D_n, are

$$H_n^F(N_n^0) - H_n^F(N_n^0 + D_n) = \frac{k}{m}, \quad n \in \overline{1,m}, \tag{8}$$

where

$$H_n^F(N_n^0) = -N_n \ln\frac{N_n}{\tilde{a}_n} - (G_n - N_n)\ln(G_n - N_n).$$

The stationary point of the generalized information entropy of Fermi-Dirac (2.14) is obtained from the equations

$$\frac{\partial H_F}{\partial N_n} = -\left(\ln\frac{N_n}{\tilde{a}_n} - \ln(G_n - N_n)\right) = 0, \quad n \in \overline{1,m}.$$

By solving them for N_n, we obtain

$$N_n^0 = a_n G_n. \tag{9}$$

It is easy to see that the point $N^0 = (N_1^0, \ldots, N_m^0)$ is the maximum point, because

$$\frac{\partial^2 H_F}{\partial N_i \partial N_n} = \begin{cases} 0 & (i \neq n); \\ -\left(\frac{1}{N_n^0} + \frac{1}{G_n - N_n^0}\right) = \frac{-1}{a_n G_n(1-a_n)} < 0 & (i = n). \end{cases} \tag{10}$$

Hence it follows that $H_F(N)$ is strictly concave on the set $0 \leq N_n \leq G_n$.

Let us return to the equation (8) and, assuming that D_n are not large, replace the left-hand sides of these equations by the 2-nd order approximations. By virtue of (10), we obtain

$$H_n^F(N_n^0) - H_n^F(N_n^0 + D_n) = \frac{D_n^2}{2a_n G_n(1-a_n)} = \frac{k}{m}.$$

Hence the absolute fluctuation is

$$D_n = \kappa_n \sqrt{G_n},$$

where

$$\kappa_n = \sqrt{2\frac{k}{m} a_n(1-a_n)}.$$

According to (6) and (9), the guaranteed relative fluctuation is

$$\rho = \max \frac{\beta_n}{\sqrt{G_n}}, \tag{11}$$

where

$$\beta_n = \kappa_n / a_n.$$

If $G_n \rightarrow \infty$ then $\rho \rightarrow 0$, i.e. the relative fluctuations of the macrostate parameters from their most probable values tend to zero.

It is seen from (9) that $N_n^0 \rightarrow \infty$ and $Y = \sum N_n^0 \rightarrow \infty$ as $G_n \rightarrow \infty$. Therefore it follows from $\lim\limits_{G_n \rightarrow \infty} \rho = 0$ that the "*sharpness*" *of the unique maximum of generalized information entropy* (2.14) *increases when the number of system elements increases.*

Consider a *macrosystem with the Einstein-states.* The equations (7) for D_n are

$$H_n^E(N_n^0) - H_n^E(N_n^0 + D_n) = \frac{k}{m}, \quad n \in \overline{1,m}; \tag{12}$$

where

$$H_n^E(N_n) = -N_n \ln \frac{N_n}{a_n} + (G_n + N_n)\ln(G_n + N_n). \tag{12'}$$

The stationary point of generalized information entropy of Boze-Einstein (2.23) is obtained from the equations

$$\frac{\partial H_E}{\partial N_n} = -\ln \frac{N_n}{a_n} + \ln(G_n + N_n) = 0, \quad n \in \overline{1,m}.$$

Hence

$$N_n^0 = \frac{a_n}{1-a_n} G_n, \quad n \in \overline{1,m}. \tag{13}$$

The Hessian elements are

$$\frac{\partial^2 H_E}{\partial N_i \partial N_n} = \begin{cases} 0 & (i \neq n); \\ -\dfrac{1}{N_n^0} + \dfrac{1}{G_n + N_n^0} = \dfrac{(1-a_n)^2}{a_n G_n} < 0 & (i = n). \end{cases} \tag{14}$$

Note that the point N^0 (13) is the maximum point for $N_n^0 \geq 0$, and the function $H_E(N)$ is strictly concave on the set $N_n \geq 0 \ (n \in \overline{1,m})$.

Assuming that D_n are not large in (12), we use the 2-nd order approximations for the function in the left-hand side of the equation, namely

$$H_n^E(N_n^0) - H_n^E(N_n^0 + D_n) = \frac{1}{2} \frac{D_n(1-a_n)^2}{2a_n G_n} = \frac{k}{m}.$$

Thus the absolute fluctuations are

$$D_n = \omega_n \sqrt{G_n},$$

where

$$\omega_n = \frac{1}{1-a_n} \sqrt{2 \frac{k}{m} a_n};$$

and the guaranteed relative fluctuation (6) is

$$\rho = \max \frac{\gamma_n}{\sqrt{G_n}}, \tag{15}$$

where

$$\gamma_n = \frac{(1-a_n)\omega_n}{a_n}.$$

From (15) we see that $\rho \to 0$ as $G_n \to \infty$.

Therefore *both the physical entropy* (2.22) *and the generalized information entropy* (2.23) *of a macrosystem with the Einstein-states have the unique maximum and its "sharpness" increases when the number of system elements increases (see* (13)).

Now consider a *macrosystem with the Boltzmann-states.* It is characterized by the generalized information entropy of Boltzmann (2.29).

The equation (7) is

$$H_n^B(N_n^0) - H_n^B(N_n + D_n) = \frac{k}{m}, \quad n \in \overline{1,m}; \tag{16}$$

where

$$H_n^B(N_n) = -N_n \ln \frac{N_n}{e a_n G_n}. \tag{16'}$$

The stationary point of (2.29) is determined from the equations

$$\frac{\partial H_B}{\partial N_n} = -\ln \frac{N_n}{a_n G_n} = 0; \quad n \in \overline{1,m}.$$

from this

$$N_n^0 = a_n G_n. \tag{17}$$

Note that the arguments of the maximum are formally the same for both the generalized information entropy of Boltzmann (2.29) and that of Fermi-Dirac (2.14). But $N_n^0 \ll G_n$ in (17), and therefore $a_n \ll 1$.

The Hessian, as well as above, is a diagonal matrix with the elements

$$\frac{\partial^2 H_B}{\partial N_n^2} = -\frac{1}{N_n^0} = -\frac{1}{a_n G_n}, \quad n \in \overline{1,m}. \tag{18}$$

Hence it follows that the function $H_B(N)$ is strictly concave on the set $N_n \geq 0$ $(n \in \overline{1,m})$.

If the fluctuations D_n are small, (16) may be rewritten as

$$H_n^B(N_n^0) - H_n^B(N_n^0 + D_n) = \frac{D_n^2}{2a_n G_n} = \frac{k}{m}.$$

From this the absolute fluctuations are

$$D_n = \nu_n \sqrt{G_n},$$

where

$$\nu_n = \sqrt{2\frac{k}{m}a_n}$$

and the guaranteed relative fluctuation (6) becomes

$$\rho = \max \frac{\alpha_n}{\sqrt{G_n}}, \tag{19}$$

where

$$\alpha_n = \nu_n / a_n.$$

It is obvious that $\rho \to \infty$ as $G_n \to 0$.

Thus the *generalized information entropy* (2.29) *for a macrosystem with the Boltzmann-states has the unique maximum. According to* (17), *the maximum "sharpness" increases when the number of system elements increases.*

2.4 Set of feasible macrostates

According to the phenomenological scheme, various resources are consumed in distributing the elements of a homogeneous macrosystem among the subsets of close states. If the resources are unconstrained, the set $\mathfrak{N}$ contains all possible macrostates generated by the stochastic mechanism for distributing the elements among the states.

The set of possible macrostates $\mathfrak{N}$ coincides with the sets on which the generalized information entropy functions are defined. Thus for macrosystems with the Fermi-states

$$\mathfrak{N} = \mathfrak{Z}_F, \tag{1}$$

where $\mathfrak{Z}_F$ is given by equality (2.14');
for macrosystems with the Einstein- and Boltzmann-states

$$\mathfrak{N} = \mathfrak{Z}_B = \mathfrak{Z}_E, \tag{2}$$

where $\mathfrak{Z}_B$ is given by equality (2.29').

If the resources are constrained, not all the macrostates from $\mathfrak{N}$ are feasible. The resource constraints give a subset $\mathfrak{D}$ of feasible macrostates in $\mathfrak{N}$.

For example, in distributing the elements of a thermodynamic system, the number of elements is constant and the energy is expended (Uhlenbeck, Ford 1963). Given the energy e_n needed for moving one element, the energy consumed by all elements in occupying all the states is equal to $\sum e_n N_n$. Usually it is assumed that $\sum e_n N_n = R$, where $R = \text{const}$.

This equality corresponds to a hyperplane in $\mathfrak{N}$, with all feasible macrostates containing in it.

Consider another example, namely the transportation macrosystem (Imelbayev, Shmulyan, 1978). Recall that for this the subset $\mathfrak{S}_n$ of close states is the set of communications (i, j) (or links between the regions i and j). For such a system, a macrostate is characterized by the matrix $N = [N_{ij} | i \in \overline{1,m};\ j \in \overline{1,m}]$. In distributing the elements among the communications, the number of elements is constant and units of costs (travel cost) are expended. Usually an individual travel cost c_{ij} is introduced. It estimate many factors, namely the travel time, the trip comfort, the number of changes and so on. The individual travel cost can depend on the number of passengers moving in the communication $c_{ij}(N_{ij})$, since it is this number that influences the above-mentioned factors. The sum travel cost equals $\sum c_{ij}(N_{ij})N_{ij}$. In this case the subset of feasible macrostates $\mathfrak{D} = \{N : \sum c_{ij}(N_{ij})N_{ij} = T\}$ is selected in $\mathfrak{N}$.

Only one type of the resources (the energy or the overall cost) was consumed in all these examples. In general, r types of resources can be used in distributing. Let $q_1, \ldots, q_k$ be stocks of the resources.

It can be seen from the examples that the consumption φ_k of the resource k depends on the occupation numbers $N_1, \ldots, N_m$ for the subsets $\mathfrak{S}_1, \ldots, \mathfrak{S}_m$.

We call a macrosystem the *macrosystem with a non-linear consumption of resources* if the way in which the consumption depends on the occupation number is described by non-linear functions $\varphi_1(N), \ldots, \varphi_r(N)$.

We call a macrosystem the *macrosystem with a linear consumption of resources if* the way in which the consumption depends on the occupation number is described by linear functions $\varphi_1(N), \ldots, \varphi_r(N)$.

The stock q_k of the k-th resource can be consumed both completely and incompletely.

For *a macrosystem with the complete consumption of resources* we have

$$\varphi_k(N) = q_k, \quad k \in \overline{1,r}. \tag{3}$$

If the consumption is linear, (3) are linear constraints of equality type, viz

$$\sum_{n=1}^{m} t_{kn} N_n = q_k, \quad k \in \overline{1,r}. \tag{3'}$$

For *a macrosystem with the incomplete consumption of resources* we have

$$\varphi_k(N) \le q_k, \quad k \in \overline{1,r} \tag{4}$$

or

$$\sum_{n=1}^{m} t_{kn} N_n \le q_k, \quad k \in \overline{1,r}, \tag{4'}$$

for macrosystems with linear consumption.

Finally, some resources (for example, of the first p types) can be consumed completely, while the other resources are consumed incompletely.

For *a macrosystem with the mixed consumption* we have

$$\begin{aligned} \varphi_k(N) &= q_k, \quad k \in \overline{1,p}, \\ \varphi_k(N) &\leq q_k, \quad k \in \overline{p+1,r}. \end{aligned} \tag{5}$$

If the consumption in the macrosystem is linear, (5) are transformed into a linear system of equalities and inequalities, namely

$$\begin{aligned} \sum_{n=1}^{m} t_{kn} N_n &= q_k, \quad k \in \overline{1,p}; \\ \sum_{n=1}^{m} t_{kn} N_n &\leq q_k, \quad k \in \overline{p+1,r}. \end{aligned} \tag{5'}$$

Relative variables $y_n = N_n/G_n$ are often useful.

Denote

$$\varphi_k(y_1, \ldots, y_m) = \varphi_k(G_1 y_1, \ldots, G_m y_m) \tag{6}$$

$$\widetilde{t}_{kn} = t_{kn} G_n.$$

Table 2.2 contains the possible variants of resource constraints arising from the introduced classification.

The nonempty subset $\mathfrak{D}$ of $\mathfrak{N}$ given by the resource constraints, namely

$$\mathfrak{D} = \left\{ N : \varphi_k(N) \overset{=}{\underset{\leq}{}} q_k, \quad k \in \overline{1,r} \right\} \subset \mathfrak{N},$$

is called the *feasible set*.

The schemes of resource consumption represent the constraints which determine the sets of feasible macrostates. While any macrostate can be produced for an unconstrained macrosystem, only the feasible macrostates can be produced for a constrained macrosystem. Because of this reason, the criteria for the feasible set to be not empty are very important.

Consider a macrosystem with mixed linear consumption of the resources (2B, Table 2.2). The feasible set for this macrosystem in terms of relative variables is

$$\mathfrak{D} = \left\{ y : \sum_{n=1}^{m} \widetilde{t}_{kn} y_n = q_k, \ k \in \overline{1,p}; \ \sum_{n=1}^{m} \widetilde{t}_{kn} y_n \leq q_k, \ k \in \overline{p+1,r} \right\} \tag{7}$$

	Nonlinear consumption	*Linear consumption*
A	$\varphi_k(N_1,\ldots,N_m) = q_k; \quad k \in \overline{1,r}$	$\sum_{n=1}^{m} t_{kn} N_n = q_k; \quad k \in \overline{1,r}$
B	$\varphi_k(N_1,\ldots,N_m) \leq q_k; \quad k \in \overline{1,r}$	$\sum_{n=1}^{m} t_{kn} N_n \leq q_k; \quad k \in \overline{1,r}$
C	$\varphi_k(N_1,\ldots,N_m) = q_k; \quad k \in \overline{1,p}$ $\varphi_k(N_1,\ldots,N_m) \leq q_k; \quad k \in \overline{p+1,r}$	$\sum_{n=1}^{m} t_{kn} N_n = q_k; \quad k \in \overline{1,p}$ $\sum_{n=1}^{m} t_{kn} N_n \leq q_k; \quad k \in \overline{p+1,r}$

A. Complete resource consumption;
B. Incomplete resource consumption;
C. Mixed resource consumption;

TABLE 2.2. Types of the resource constraints.

Denote

$$l_j(y) = \sum_{n=1}^{m} a_{jn} y_n; \quad h_j(y) = \sum_{n=1}^{m} b_{jn} y_n; \qquad j \in \overline{1,r-p}; \quad j \in \overline{1,p}; \tag{8}$$

$$a_{jn} = \tilde{t}_{j+r-p,n}; \quad b_{jn} = \tilde{t}_{jn};$$

$$a_j = q_{j+r-p}; \quad b_j = q_j;$$

ρ is the rank of the matrix $T = [\tilde{t}_{kn}], \quad k \in \overline{1,r}; \quad n \in \overline{1,m};$

π is the rank of the matrix $T^1 = [\tilde{t}_{kn}], \quad k \in \overline{1,p}; \quad n \in \overline{1,m}.$

In terms of new notation the set $\mathfrak{D}$ is determined by the following system of equalities and inequalities

$$\begin{aligned} l_j(y) - a_j \leq 0; \quad j \in \overline{1,r-p}; \\ h_j(y) - b_j = 0; \quad j \in \overline{1,p}. \end{aligned} \tag{9}$$

Let $\pi \leq \rho \leq r$. Denote by

$$\Delta = \begin{vmatrix} a_{j_1 n_1} & \dots & a_{j_1 n_\rho} \\ \dots & \dots & \dots \\ a_{j_s n_1} & \dots & a_{j_s n_\rho} \\ \dots & \dots & \dots \\ b_{j_{s+1} n_1} & \dots & b_{j_{s+1} n_\rho} \\ \dots & \dots & \dots \\ b_{j_{s+\pi} n_1} & \dots & b_{j_{s+\pi} n_\rho} \end{vmatrix} \tag{10}$$

the minor of order ρ. It contains π rows and they are constructed from the coefficients of functions $h_{j_{s+1}}(y), \dots, h_{j_{s+\pi}}(y)$ from (9). Here $(n_1, \dots, n_\rho) \in \overline{1,m}$; $(j_1, \dots, j_s) \in \overline{1, r-p}$; $(j_{s+1}, \dots, j_{s+\pi}) \in \overline{1,p}$; $s = \rho - \pi$;

$$\Delta_j^{(1)} = \begin{vmatrix} \boxed{\quad \Delta \quad} & \begin{matrix} a_{j_1} \\ \dots \\ a_{j_s} \\ b_{j_{s+1}} \\ \dots \\ b_{j_{s+\pi}} \end{matrix} \\ a_{j_{n_1}} \quad \dots \quad a_{j_{n_\rho}} & a_j \end{vmatrix} \tag{11}$$

is the accompanying $(\rho + 1)$-order minor corresponding to the inequalities in (9) $(j \in \overline{1, r-p})$ and

$$\Delta_j^{(2)} = \begin{vmatrix} \boxed{\quad \Delta \quad} & \begin{matrix} a_{j_1} \\ \dots \\ a_{j_s} \\ b_{j_{s+1}} \\ \dots \\ b_{j_{s+\pi}} \end{matrix} \\ b_{j_{n_1}} \quad \dots \quad b_{j_{n_\rho}} & b_j \end{vmatrix} \tag{12}$$

is the accompanying $(\rho + 1)$-order minor corresponding to the equations in (9) $(j \in \overline{1,p})$.

Theorem 1. *For the system* (9) *to be consistent it is sufficient that there exist a minor* $\Delta \neq 0$ (10) *and minors* $\Delta_j^{(1)}$ (11) *and* $\Delta_j^{(2)}$ (12) *satisfying the following conditions*

$$\Delta_j^{(1)}/\Delta \geq 0, \quad j \in \overline{1, r-p}; \tag{13}$$

$$\Delta_j^{(2)} = 0, \quad j \in \overline{1, p}. \tag{14}$$

This problem is considered in (Chernikov, 1968). Theorem 1.6 on consictence is proved there, but it should be specified. Such a specification is given by Theorem 1. This results in the consistence conditions different from those in the book cited.

Proof. Reduce the mixed system (9) to the system of inequalities

$$\begin{gathered} l_j(y) - a_j \leq 0, \quad j \in \overline{1, r-p}; \\ h_j(y) - b_j \leq 0, \\ -h_j(y) + b_j \leq 0, \quad j \in \overline{1, p}. \end{gathered} \tag{15}$$

(a) *Necessity.* Let the system in (15) be consistent. Then it has at least one subsystem of active inequalities (equalities) which has the rank ρ and is such that any of its solutions satisfies the systems (15) or (9). Hence from the subsystem we can choose $s = \rho - \pi$ equalities

$$l_{j_v}(y) - a_{j_v} = 0, \quad v \in \overline{1, s}; \tag{16}$$

corresponding to the inequalities in (9) and π equalities

$$h_{j_s+\mu}(y) - b_{j_s+\mu} = 0, \quad \mu \in \overline{1, \pi}; \tag{17}$$

corresponding to the equalities in (9).

Let the functions $l_{j_1}(y), \ldots, l_{j_s}(y)$, $h_{j_s+1}(y), \ldots, h_{j_s+\pi}(y)$ be linearly independent.

The determinant of system (16)–(17) is as in (10). Consider the determinants which are identically zero

$$\begin{vmatrix} \boxed{\quad \Delta \quad} & \begin{matrix} l_{j_1} \\ \cdots \\ l_{j_s} \\ h_{j_s+1} \\ \cdots \\ h_{j_s+\pi} \end{matrix} \\ a_{j_{\pi_1}} \quad \cdots \quad a_{j_{\pi_\rho}} & l_j \end{vmatrix} \equiv 0; \tag{18}$$

$$j \in \overline{1, r-p};$$

$$\begin{vmatrix} & & & l_{j_1} \\ & & & \cdots \\ & \Delta & & l_{j_s} \\ & & & h_{j_s+1} \\ & & & \cdots \\ & & & h_{j_s+\pi} \\ b_{j n_1} & \cdots & b_{j n_\rho} & h \end{vmatrix} \equiv 0;$$

and $j \in \overline{1,p};$ (18′)

$$\begin{vmatrix} & & & l_{j_1} \\ & & & \cdots \\ & \Delta & & l_{j_s} \\ & & & h_{j_s+1} \\ & & & \cdots \\ & & & h_{j_s+\pi} \\ -b_{j n_1} & \cdots & -b_{j n_\rho} & -h_j \end{vmatrix} \equiv 0.$$

$j \in \overline{1,p}.$

Let us use them to transform the inequalities (15). Here we illustrate the transformation idea for one group of determinants in (18) (for example, for the first one). By decomposing the determinant in terms of the $(\rho+1)$-th column elements, we obtain

$$t_j = -\frac{1}{\Delta}\begin{vmatrix} & & & l_{j_1} \\ & & & \cdots \\ & \Delta & & l_{j_s} \\ & & & h_{j_s+1} \\ & & & \cdots \\ & & & h_{j_s+\pi} \\ a_{j n_1} & \cdots & a_{j n_\rho} & 0 \end{vmatrix}, \qquad j \in \overline{1,r-p}$$

Next, we substitute these expressions in the first group of inequalities in (15) and represent them as

$$-\frac{1}{\Delta}\begin{vmatrix} & & & l_{j_1} \\ & & & \cdots \\ & \Delta & & l_{j_s} \\ & & & h_{j_s+1} \\ & & & \cdots \\ & & & h_{j_s+\pi} \\ a_{j n_1} & \cdots & a_{j n_\rho} & a_j \end{vmatrix} \leq 0, \tag{19}$$

$j \in \overline{1,r-p}.$

By substituting $a_{j_1}, \dots, a_{j_s}$ and $b_{j_{s+1}}, \dots, b_{j_{s+\pi}}$ for the elements of the $(\rho+1)$-th column (except the last element), we obtain the condition (13) of theorem by virtue of (16)–(17).

If the second and third groups of determinants (18′) are transformed in such a way, we obtain

$$-\frac{1}{\Delta}\begin{vmatrix} & & & l_{j_1} \\ & & & \dots \\ & \Delta & & l_{j_s} \\ & & & h_{j_{s+1}} \\ & & & \dots \\ & & & h_{j_{s+\pi}} \\ b_{j_{n_1}} & \dots & b_{j_{n_\rho}} & b_j \end{vmatrix} \leq 0, \qquad j \in \overline{1,p}; \tag{20}$$

$$-\frac{1}{\Delta}\begin{vmatrix} & & & l_{j_1} \\ & & & \dots \\ & \Delta & & l_{j_s} \\ & & & h_{j_{s+1}} \\ & & & \dots \\ & & & h_{j_{s+\pi}} \\ b_{j_{n_1}} & \dots & b_{j_{n_\rho}} & -b_j \end{vmatrix} \leq 0, \qquad j \in \overline{1,p}; \tag{21}$$

The condition (14) follows immediately from this.

(b) *Sufficiency.* Let the conditions (13) and (14) be satisfied. Then the system (19), (20), (21) has the solution

$$(l_{j_1}, \dots, l_{j_s}, h_{j_{s+1}}, \dots, h_{j_{s+\pi}}) = (a_{j_1}, \dots, a_{j_s}, b_{j_{s+1}}, \dots, b_{j_{s+\pi}}).$$

This equality is equivalent to the equations (16) and (17). The determinant Δ (10) is the ρ-order minor of the system matrix. Since the rank of the matrix equals the number of equations, the system (16), (17) is consistent and its solution y^* solves the system (15), (9).

2.5 Variational principle. Models of stationary states (MSS)

The phenomenological scheme of macrosystems assumes that the macrosystem elements are distributed randomly and independently among the states, with the apriori probabilities a_n. The set of states consists of close subsets of the states. Some amount of resources enter the system from outside and are consumed during the distribution process. As a result, a stationary macrostate is reached after a while. The macrostate is characterized by the constant occupation numbers $N_1^*, \ldots, N_m^*$ for the subsets $\mathfrak{S}_1, \ldots, \mathfrak{S}_m$.

General premises to develop a constructive method for determining the stationary state are given by the second law of thermodynamics discovered by R. Klausius. The law characterizes the way in which the system macrostates evolve in time. If a system is in a nonstationary state at a moment of time then the monotonic increase of entropy is the most probable during the subsequent movement of the system.

Therefore if the state with the maximum entropy is reached, no deviations from this state will occur, i.e. the macrostate with the maximum entropy is the stationary state N^* realized in the macrosystem.

Although these considerations are quite logical, the physical nature of processes and mathematical problems arisen here are still far from being completed (Landau, Lifshitz, 1964, Prigogine, 1985).

The entropy maximization principle (or the principle of the most probable macrostate) generates an efficient procedure for computing this macrostate. The first stage consists in considering an abstract macrosystem. Let the elements be distributed among the states of the macrosystem as randomly as in a real macrosystem, but let the resources be not constrained. Any macrostate is feasible for such a macrosystem. The probability distribution function, or entropy (Section 2), is defined on the set of feasible macrostates.

At the second stage, the resource constraints are involved. They determine the subset of macrostates for which the resource consumption is consistent with the resource limit.

Some comments are needed here. At the first glance, it seems that by introducing the resource constraints, we significantly change the distribution mechanism. Hence we change the entropy function and the probability distribution function obtained at the first stage.

But in a macrosystem, an element slightly contributes a macrostate, and any macrostate is produced by a great number of elements. Therefore the resource constraints are, in fact, the constraints for "average" recourse characteristics.

Since there are many elements in each subset of the set of states, and the contribution of any element is small, the constraints for the "average" resource insignificantly influence the element behaviour; the latter is still random and independent.

These considerations allow us to assume that the second law of thermodynamics holds not only for an opened set of feasible macrostates, but also for any its subset.

This assumption implies that if a system is in a macrostate from the feasible set then the entropy increase is most probable during the future system dynamics. The stationary state is reached only when the entropy attains its maximum on given set of feasible macrostates.

Actually, this hypothesis formulate the ***variational principle*** on which models of stationary states for homogeneous macrosystems are based, namely

$$N^* = \arg\max_{N \in \mathfrak{D}} P(N) = \arg\max_{N \in \mathfrak{D}} E(N), \tag{1}$$

where

$$\mathfrak{D} \subset \mathfrak{N}, \tag{2}$$

$\mathfrak{N}$ is the set of possible macrostates and

$\mathfrak{D}$ is the set of feasible macrostates with constrained resources.

Besides the "physical" considerations confirming the principle (1), there exists pure mathematical principles. The latter are connected with the morphological properties of $P(N)$ and $E(N)$ (namely with the maximum "sharpness" of these functions) (see Section 3). But in case (1), the maximums of $P(N)$ or $E(N)$ should be "sharp" on the constrained set $\mathfrak{D}$.

To determine the stationary macrostate corresponding to (1) and to study its properties, it is more convenient to use the generalized information entropy, instead of the physical one. These differ only by a constant.

The set $\mathfrak{D}$ in R^m is given by one of the constraint groups (4.1)–(4.6).

Taking into account the notes above, we can represent the *model of stationary states* (MSS) on the basis of variational principle (1) in the following way

$$H(N) \Rightarrow \max, \tag{3}$$

$$N \in \mathfrak{D} \subset \mathfrak{N}. \tag{4}$$

Note that the feasible set $\mathfrak{D}$ is a (nonempty) subset of the possible macrostate set $\mathfrak{N}$ which coincides with the definition domain of corresponding entropy functions (2.14′, 2.23′, 2.29′).

Model (3, 4) is described either by a mathematical programming problem (for macrosystems having incomplete or mixed resource consumption, Tab. 2.2 (B, C)) or by a constrained extremum problem (for the macrosystems having complete resource consumption, Tab. 2.2(A)).

If a macrosystem with the Fermi-states is considered then $H(N) = H_F(N)$, where $H_F(N)$ is the generalized information entropy of Fermi-Dirac (2.14). The corresponding MSS is called the *F-model.*

For a macrosystem with the Fermi-states and mixed resource consumption (4.5) the F-model is

$$H_F(N) = -\sum_{n=1}^{m} N_n \ln \frac{N_n}{\tilde{a}_n} + (G_n - N_n) \ln (G_n - N_n) \Rightarrow \max, \tag{5}$$

where

$$N \in \mathfrak{D} \subset \mathfrak{N}; \tag{6}$$

$$\mathfrak{D} = \{N : \varphi_k(N_1, \dots, N_m) = q_k, \quad k \in \overline{1,p}; \quad \varphi_k(N_1, \dots, N_m) \leq q_k, \quad k \in \overline{p+1,r}\} \text{ for a nonlinear consumption} \tag{7}$$

and

$$\mathfrak{D} = \left\{N : \sum_{n=1}^{m} t_{kn} N_n = q_k, \quad k \in \overline{1,p}; \quad \sum_{n=1}^{m} t_{kn} N_n \leq q_k, \quad k \in \overline{p+1,r}\right\} \text{ for a linear consumption.} \tag{8}$$

The macrostate N^* given by F-model belongs to the interior of the feasible macrostate set $\mathfrak{N} = \mathfrak{Z}_F$ (2.14′) under quite general assumptions on the properties of $\mathfrak{D}$.

Lemma 1. *Let the feasible set* $\mathfrak{D}$ *from* (7) *have the following properties:*
(a) $\mathfrak{D} \neq \varnothing$,
(b) $\mathfrak{D}$ *is not a set of isolated points.*
Then the solution of problem (5)–(7) *satisfies the condition*

$$0 < N_n^* < G_n, \quad n \in \overline{1,m}. \tag{9}$$

In order to prove the Lemma, we consider the point

$$\widetilde{N} = \{\widetilde{N}_1, \ldots, \widetilde{N}_{i-1}, 0, \widetilde{N}_{i+1}, \ldots, \widetilde{N}_m\},$$

where $0 < \widetilde{N}_n < G_n$, $n \in \overline{1,m}$ and $n \neq i$. Such a point always exists by condition (a) of the Lemma.

Let us consider the vectors p and $(\widetilde{N} + \alpha p) \in \mathfrak{N}$, where $\alpha \geq 0$. The value of $H_F(N)$ at the point $\widetilde{N} + \alpha p$ is equal to

$$\begin{aligned} H_F(\widetilde{H} + \alpha p) = & -\sum_{n \neq i} \Big[\Big(\widetilde{N}_n + \alpha p_n\Big) \ln \frac{\widetilde{N}_n + \alpha p_n}{\widetilde{a}_n} + \\ & + (G_n - \widetilde{N}_n - \alpha p_n) \ln (G_n - \widetilde{N}_n - \alpha_n) \Big] - \\ & - \alpha p_i \ln \frac{\alpha p_i}{\widetilde{a}_i} - (G_i - \alpha p_i) \ln (G_i - \alpha p_i), \quad \alpha \geq 0. \end{aligned}$$

Hence the derivative in the direction p at the point $\widetilde{N}$ is

$$\left. \frac{\partial H_F}{\partial \alpha} \right|_{\alpha=0} = +\infty.$$

By (b), the point $\widetilde{N}$ is not isolated, therefore there exists a neighborhood of $\widetilde{N}$ in $\mathfrak{N}$ such that $H_F(N) > H_F(\widetilde{N})$ at each point $N \neq \widetilde{N}$ of $\mathfrak{N}$. Thus the left-hand inequality in (9) is proved.

To prove the right-hand inequality in (9), let us consider the point $\widehat{N} = \{\widehat{N}_1, \ldots, \widehat{N}_{i-1}, G_i, \widehat{N}_{i+1}, \ldots, \widehat{N}_m\}$, where $0 < \widehat{N}_n < G_n$, $n \in \overline{1,m}$, and $n \neq i$, and the vectors p and $(\widehat{N} + \alpha p) \in \mathfrak{N}$, where $\alpha \geq 0$.

We obtain

$$\begin{aligned} H_F(\widehat{N} + \alpha p) = & -\sum_{n \neq i} \Big[(\widehat{N}_n + \alpha p_n) \ln \frac{\widehat{N}_n + \alpha p_n}{\widetilde{a}_n} + (G_n - \widehat{N}_n - \alpha p_n) \cdot \\ & \cdot \ln (G_n - \widehat{N}_n - \alpha p_n) \Big] - (G_i + \alpha p_i) \cdot \\ & \cdot \ln \frac{G_i + \alpha p_i}{\widetilde{a}_i} + \alpha p_i \ln (-\alpha p_i). \end{aligned}$$

The derivative in the direction p at $\widehat{N}$ is

$$\left.\frac{\partial H_F}{\partial \alpha}\right|_{\alpha=0} = -\infty.$$

Hence, by (b), there exists a neighborhood of $\widehat{N}$ in $\mathfrak{N}$ such that $H_F(N) > H_F(\widehat{N})$ at any point $N \neq \widehat{N}$. This finishes the proof.

Consider a macrosystem with the Einstein-states. In (3, 4), $H(N) = H_E(N)$, where $H_E(N)$ is the generalized information entropy of Einstein-Boze. The corresponding MSS-model is called the *E-model*. For the mixed resource consumption the model is

$$H_E(N) = -\sum_{n=1}^{m} N_n \ln\frac{N_n}{a_n} - (G_n + N_n)\ln(G_n + N_n) \Rightarrow \max, \tag{10}$$

where

$$N \in \mathfrak{D} \subset \mathfrak{N}, \tag{11}$$

$$\mathfrak{D} = \{N : \varphi_k(N_1, \ldots, N_m) = q_k, \quad k \in \overline{1,p}; \quad \varphi_k(N_1, \ldots, N_m) \leq q_k, \\ k \in \overline{1+p,r}\} \quad \text{for a nonlinear consumption} \tag{12}$$

and

$$\mathfrak{D} = \left\{N\colon \sum_{n=1}^{m} t_{kn}N_n = q_k;\ k \in \overline{1,p};\ \sum_{n=1}^{m} t_{kn}N_n \leq q_k,\ k \in \overline{1+p,r}\right\} \\ \text{for a linear resource consumption.} \tag{13}$$

Lemma 2. *Let the conditions of Lemma 1 be satisfied. Then the solution of problem* (10)–(12) *is such that*

$$N_n^* > 0, \quad n \in \overline{1,m}. \tag{14}$$

The proof is similar to that of Lemma 1.

The third type of MSS describes macrosystems with the Fermi-states but with small occupation numbers. Such macrosystems are characterized by the generalized Boltzmann entropy. MSS with $H(N) = H_B(N)$ (2.29) is called the *B-model*. For the mixed consumption of resources, B-model is

$$H_B(N) = -\sum_{n=1}^{m} N_n \ln(N_n/ea_nG_n) \Rightarrow \max, \tag{15}$$

where

$$N \in \mathfrak{D} \subset \mathfrak{N} \tag{16}$$

$$\mathfrak{D} = \{N : \varphi_k(N_1, \ldots, N_m) = q_k, \quad k \in \overline{1,p}; \quad \varphi_k(N_1, \ldots, N_m) \le q_k,$$
$$k \in \overline{1+p,r}\} \quad \text{for a nonlinear resource consumption} \tag{17}$$

and

$$\mathfrak{D} = \left\{N : \sum_{n=1}^{m} t_{kn} N_n = q_k; \quad k \in \overline{1,p}; \quad \sum_{n=1}^{m} t_{kn} N_n \le q_k,\right.$$
$$\left. k \in \overline{1+p,r}\right\} \quad \text{for a linear resource consumption.} \tag{18}$$

Lemma 3. *Let the conditions of Lemma 1 be satisfied. Then the solution of problem* (15)–(17) *is such that*

$$N_n^* > 0, \quad n \in \overline{1,m}. \tag{19}$$

The proof is similar to that of Lemma 1.

2.6 Optimality conditions for MSS

The models of stationary states given by (5.5)–(5.8), (5.10)–(5.13) and (5.15)–(5.18) are mathematical programming problems, with the objective functions $H_F(N)$ (5.5), $H_E(N)$ (5.10) and $H_B(N)$ (5.15) being strictly concave on their definition sets (2.14′), (2.23′) and (2.29′), respectively.

Consider the optimality conditions for these mathematical programming problems (Polyak,1983).

Recall these conditions in terms considered in MSS. The general formulation of MSS is

$$H(N) \Rightarrow \max; \tag{1}$$

$$\varphi_k(N) = q_k, \quad k \in \overline{1,p}; \tag{2}$$

$$\varphi_k(N) \le q_k, \quad k \in \overline{p+1,r}. \tag{3}$$

Here we assume that the objective function $H(N)$ either maps $R_+^m \Rightarrow R^1$ for B- and E-models (the Boltzmann and Einstein-Boze entropies, respectively), or maps $\prod_+^m \Rightarrow R^1$ ($\prod_+^m = \mathfrak{Z}_F = \{N: 0 \le N_n \le G_n, n \in \overline{1,m}\}$ is a rectangle in R_+^m) for the F-model (the Fermi-Dirac entropy, (5.5)).

Recall also that the feasible set

$$\mathfrak{D}=\{N:\ \varphi_k(N)=q_k,\quad k\in\overline{1,p};\quad \varphi_k(N)\le q_k,\quad k\in\overline{p+1,r};\}\tag{4}$$

is a nonempty subset of the set $\mathfrak{N}$ of possible macrostates (2.14′), (2.23′) and (2.29′).

Let us consider the Lagrange function

$$L(N,\lambda)=H(N)+\sum_{k=1}^{p}\lambda_k(q_k-\varphi_k(N))+\sum_{k=p+1}^{r}\lambda_k(q_k-\varphi_k(N));\tag{5}$$

then the necessary maximum conditions in (1)–(3) are given by the following theorem.

Theorem 1. *Let* N^* *be a local maximum point in* (1)–(3); *let the functions* $\varphi_k(N)$ $(k\in\overline{1,r})$ *be continuously differentiable in a neighborhood of* N^*; *and let the gradients* $\nabla\varphi_k(N^*)$ *be linearly independent. Then there exist numbers* $\lambda_k^*\ge 0$ $(k\in\overline{p+1,r})$ *such that*

$$\nabla_{N_n}L(N^*,\lambda^*)=0,\quad n\in\overline{1,m},\tag{6}$$

$$\nabla_{\lambda_k}L(N^*,\lambda^*)=0,\quad k\in\overline{1,p},\tag{7}$$

$$\begin{aligned}&\nabla_{\lambda_k}L(N^*,\lambda^*)\ge 0,\quad k\in\overline{p+1,r},\\ &\lambda_k^*\nabla_{\lambda_k}L(N^*,\lambda^*)=0,\quad k\in\overline{p+1,r};\end{aligned}\tag{8}$$

where

$$\nabla_{N_n}L=\frac{\partial H}{\partial N_n}+\nabla_{N_n}\Phi;\quad \nabla_{N_n}\Phi=-\sum_{k=1}^{r}\lambda_k\frac{\partial\varphi_k}{\partial N_n},\tag{9}$$

$$\nabla_{\lambda_k}L=q_k-\varphi_k(N).\tag{10}$$

If we take into account the properties of the objective functions of MSS under these general conditions then we arrive at the system of equalities and inequalities which should be satisfied in a realizable macrostate N^*.

The optimality conditions for F-model are

$$-\left(\ln\frac{N_n^*}{\tilde{a}_n}-\ln(G_n-N_n^*)\right)+\nabla_{N_n}\Phi(N^*,\lambda^*)=0,\quad n\in\overline{1,m};\tag{11}$$

$$q_k-\varphi_k(N^*)=0,\quad k\in\overline{1,p};\tag{12}$$

$$q_k-\varphi_k(N^*)\ge 0,\quad \lambda_k^*\ge 0,\quad \lambda_k^*(q_k-\varphi_k(N^*))=0,\quad k\in\overline{p+1,r}.\tag{13}$$

The optimality conditions for E-model are

$$-\left(\ln\frac{N_n^*}{a_n} - \ln(G_n + N_n^*)\right) + \nabla_{N_n}\Phi(N^*, \lambda^*) = 0, \quad n \in \overline{1,m}; \tag{14}$$

$$q_k - \varphi_k(N^*) = 0, \quad k \in \overline{1,p}; \tag{15}$$

$$q_k - \varphi_k(N^*) \geq 0, \quad \lambda_k^* \geq 0, \quad \lambda_k^*(q_k - \varphi_k(N^*)) = 0, \quad k \in \overline{p+1,r}. \tag{16}$$

If we take into account the specific features of the generalized information entropy of Boltzmann, we obtain *the optimality condition for B-model*, namely

$$-\ln\frac{N_n^*}{a_n G_n} + \nabla_{N_n}\Phi(N^*, \lambda^*) = 0, \quad n \in \overline{1,m}; \tag{17}$$

$$q_k - \varphi_k(N^*) = 0, \quad k \in \overline{1,p}; \tag{18}$$

$$q_k - \varphi_k(N^*) \geq 0, \quad \lambda_k^* \geq 0, \quad \lambda_k^*(q_k - \varphi_k(N^*)) = 0, \quad k \in \overline{p+1,r}. \tag{19}$$

In (11)–(19), $\nabla_{N_n}\Phi$ is determined by (9).

The optimality conditions (10)–(19) correspond to the general situations in F-, E-, and B-model, i.e. they correspond to arbitrary feasible sets of macrostates. In this case, these conditions are known to be necessary and local. If the set $\mathfrak{D}$ is convex then (11)–(19) are necessary and sufficient. This is the case both for a linear consumption (table 2.2, (2A)–(2C)) and for a non-linear resource consumption for macrosystems with incomplete consumption and convex consumption functions (table 2.2, (1B)).

Models of macrosystems with *linear consumption functions and complete consumption of resources* (table 2.2, (2A)) are the most interesting. Consider the optimality conditions for this class of macrosystems.

Recall that F-model with such a consumption of resources is

$$H_F(N) \Rightarrow \max, \quad \sum_{n=1}^{m} t_{kn}N_n = q_k, \quad k \in \overline{1,r}, \tag{20}$$

where $H_F(N)$ is given by (5.5).

Contrary to the mathematical programming problem (1)–(3), (20) is a constrained extremum problem.

The Lagrange function for (20) is

$$L_F(N, \lambda) = H_F(N) + \sum_{k=1}^{r}\lambda_k\left(q_k - \sum_{n=1}^{m} t_{kn}N_n\right). \tag{21}$$

Since $H_F(N)$ is strictly concave and the feasible set

$$\mathfrak{D} = \left\{ N : \sum_{n=1}^{m} t_{kn} N_n = q_k, \quad k \in \overline{1,r} \right\}$$

is convex, the necessary and sufficient optimality conditions for (20) are

$$\frac{\partial L_F}{\partial N_n} = \ln \frac{G_n - N_n}{N_n} \tilde{a}_n - \sum_{k=1}^{r} \lambda_k t_{nk} = 0, \quad n \in \overline{1,m}; \tag{22}$$

$$\frac{\partial L_F}{\partial \lambda_k} = q_k - \sum_{n=1}^{m} t_{kn} N_n = 0, \quad k \in \overline{1,r}. \tag{23}$$

A special feature of (20) is that the first group of equations (22) can be solved with respect to N_n, namely

$$N_n^* = \frac{G_n}{1 + b_n \exp\left(\sum\limits_{j=1}^{r} \lambda_j t_{jn}\right)}, \quad n \in \overline{1,m}; \tag{24}$$

where $b_n = 1/\tilde{a}_n$. It follows from this that $N_n^* \leq G_n$.

By substituting this expression in (23), we obtain the equations for the Lagrange multipliers $\lambda_1^*, \ldots, \lambda_r^*$

$$\sum_{n=1}^{m} \frac{t_{kn} G_n}{1 + b_n \exp\left(\sum\limits_{j=1}^{r} \lambda_j t_{jn}\right)} = q_k, \quad k \in \overline{1,r}. \tag{25}$$

Consider E-model for a macrosystem with linear and complete resource consumption

$$H_E(N) \Rightarrow \max, \quad \sum_{n=1}^{m} t_{kn} N_n = q_k, \quad k \in \overline{1,r}, \tag{26}$$

where $H_E(N)$ is given by (5.10).

The Lagrange function for (26) is

$$L_E(N) = H_E(N) + \sum_{k=1}^{r} \lambda_k \left(q_k - \sum_{n=1}^{m} t_{kn} N_n \right). \tag{27}$$

The maximum in (26) exists because the system of equations

$$\frac{\partial L_E}{\partial N_n} = \ln \frac{G_n + N_n}{N_n} a_n - \sum_{k=1}^{r} \lambda_k t_{kn} = 0, \quad n \in \overline{1,m}; \tag{28}$$

$$\frac{\partial L_E}{\partial \lambda_k} = q_k - \sum_{n=1}^{m} t_{kn} N_n = 0, \quad k \in \overline{1,r} \tag{29}$$

can be solved. We obtain from (28)

$$N_n^* = \frac{G_n}{c_n \exp\left(\sum_{j=1}^{r} \lambda_j t_{jn}\right) - 1}, \quad n \in \overline{1,m}, \tag{30}$$

where $c_n = 1/a_n$.

By substituting this expression in (29), we obtain the system for determining the Lagrange multipliers $\lambda_1^*, \ldots, \lambda_r^*$:

$$\sum_{n=1}^{r} \frac{t_{kn} G_n}{c_n \exp\left(\sum_{j=1}^{r} \lambda_j t_{jn}\right) - 1} = q_k, \quad k \in \overline{1,r}. \tag{31}$$

Now consider B-model for a macrosystem with linear and complete resource consumption:

$$H_B(N) \Rightarrow \max, \quad \sum_{n=1}^{m} t_{kn} N_n = q_k, \quad k \in \overline{1,r}, \tag{32}$$

where $H_B(N)$ is given by (5.15).

The Lagrange function for (32) is

$$L_B(N, \lambda) = H_B(N) + \sum_{k=1}^{r} \lambda_k \left(q_k - \sum_{n=1}^{m} t_{kn} N_n \right). \tag{33}$$

As in the previous two classes of MSS, the optimality conditions for (32) are reduced to the solvability of the system

$$\frac{\partial L_B}{\partial N_n} = -\ln \frac{N_n}{a_n G_n} - \sum_{k=1}^{r} \lambda_k t_{kn} = 0, \quad n \in \overline{1,m}; \tag{34}$$

$$\frac{\partial L_B}{\partial \lambda_k} = q_k - \sum_{n=1}^{m} t_{kn} N_n = 0, \quad k \in \overline{1,r}. \tag{35}$$

The maximum point is determined from (34):

$$N_n^* = a_n G_n \exp\left(-\sum_{j=1}^{r} \lambda_j t_{jn} \right), \quad n \in \overline{1,m}. \tag{36}$$

By substituting these expressions in (35), we obtain the equations for determining the Lagrange multipliers, namely

$$\sum_{n=1}^{m} a_n t_{kn} G_n \exp\left(-\sum_{j=1}^{r} \lambda_j t_{jn}\right) - q_k = 0, \quad k \in \overline{1,r}. \tag{37}$$

These optimality conditions allow us to reduce the problem of determining the realizable macrostate to solution of a corresponding system of equations.

2.7 MSS of homogeneous macrosystems with linear resource consumption ("sharpness" of maximum)

The variational principle used in constructing MSS is based on the maximum "sharpness" of the macrostate distribution function (or the generalized information entropy). The validity of this property for the Fermi-Dirac, Boze-Einstein and Boltzmann entropy functions describing random distribution of elements among the states of unconstrained macrosystems was established in Section 3.

Let us consider how the constrained resource consumption influences the distribution mechanism and the maximum "sharpness".

We investigate this problem for MSS with linear and complete resource consumption:

$$H(N) \Rightarrow \max, \tag{1}$$

$$\mathfrak{D} = \left\{ n : \sum_{n=1}^{m} t_{kn} N_n = q_k, \quad k \in \overline{1,r} \right\}, \tag{2}$$

where $H(N)$ is the entropy objective function and is strictly concave on the set $\mathfrak{Z}$ (2.14′, 2.23′, 2.29′).

In the sequel, we assume that the matrix $T = [t_{kn}]$ has the rank $r < m$.

The set $\mathfrak{D}$ is convex; therefore if $\mathfrak{D} \cap \mathfrak{Z} \neq \varnothing$ then the problem (1,2) has the unique maximum point N^* determined from the equations

$$\frac{\partial H}{\partial N_n} = \sum_{k=1}^{r} \lambda_k t_{kn}, \quad n \in \overline{1,m}, \tag{3}$$

$$\sum_{n=1}^{m} t_{kn} N_n = q_k, \quad k \in \overline{1,r}, \tag{4}$$

where $\lambda_1, \ldots, \lambda_r$ are the Lagrange multipliers.

Denote

$$\Delta N_n = N_n - N_n^*, \tag{5}$$

where $N \in \mathfrak{D}$ and $N \neq N^*$. By taking (5) into account, we obtain from (2)

$$\sum_{n=1}^{m} t_{kn} \Delta N_n + \sum_{n=1}^{m} t_{kn} N_n^* = q_k, \quad k \in \overline{1,r}.$$

It follows from this that the deviations ΔN_n from the maximum point should satisfy the system of equations

$$\sum_{n=1}^{m} t_{kn} \Delta N_n = 0, \quad k \in \overline{1,r}. \tag{6}$$

Let us determine the 2-nd order approximation $\widehat{H}(N)$ of $H(N)$ for the deviation ΔN from the maximum, and let this deviation satisfy (6). It follows from the concavity of $H(N)$ that

$$H(N) \leq \widehat{H}(N) = H(N^*) + \sum_{n=1}^{m} \left.\frac{\partial H}{\partial N_n}\right|_* \Delta N_n + \frac{1}{2} \sum_{n=1}^{m} \left.\frac{\partial^2 H}{\partial N_n^2}\right|_* \Delta N_n^2. \tag{7}$$

In what follows, we investigate the maximum "sharpness" for $\widehat{H}(N)$, and (7) will help us to draw conclusions on the asymptotic behaviour of $H(N)$ maximum.

In (7), we take into account that the Hessian of $H(N)$ is a diagonal matrix. By substituting $\frac{\partial H}{\partial N_n}$ from (3) in (7), we obtain

$$\widehat{H}(N) = H(N^*) + \sum_{k=1}^{r} \lambda_k \sum_{n=1}^{m} t_{kn} \Delta N_n + \frac{1}{2} \sum_{n=1}^{m} \left.\frac{\partial^2 H}{\partial N_n^2}\right|_* \Delta N_n^2.$$

Since the deviations ΔN_n from the maximum point satisfy (6), the 2-nd order approximation of the function $\widehat{H}(N)$ is

$$\widehat{H}(N) = H(N^*) + \frac{1}{2} \sum_{n=1}^{m} \left.\frac{\partial^2 H}{\partial N_n^2}\right|_* \Delta N_n^2. \tag{8}$$

Represent (6) as

$$T_1 \Delta N_{(1)} + T_2 \Delta N_{(2)} = 0, \tag{9}$$

where

$$T_1 = \left[t_{kn}; \quad k \in \overline{1,r}; \quad n \in \overline{1,r}\right],$$

$$T_2 = \left[t_{kn}: \quad k \in \overline{1,r}; \quad n \in \overline{r+1,m}\right], \tag{10}$$

$$\Delta N_{(1)} = \{\Delta N_1, \ldots, \Delta N_r\}; \quad \Delta N_{(2)} = \{\Delta N_{r+1}, \ldots, \Delta N_m\}.$$

We obtain from (9)

$$\Delta N_{(1)} = C \Delta N_{(2)}, \tag{11}$$

where

$$C = -T_1^{-1} T_2 \quad \text{is } (r \times (m-r)) \text{ matrix.} \tag{12}$$

The inverse matrix exists because the matrix T is of full rank. Let us substitute (11) in (8)

$$\widehat{H}(N) = H(N^*) + \frac{1}{2} \Delta N'_{(2)} G \Delta N_{(2)}, \tag{13}$$

where

$$G = \begin{bmatrix} \ddots & & \\ & \ddots & \begin{array}{c} \sum_{k=1}^{r} \beta_k c_{ki} c_{kj} \\ j > i \end{array} \\ & \beta_{j+r} + \sum_{k=1}^{r} \beta_k c_{kj}^2 & \\ \begin{array}{c} \sum_{k=1}^{r} \beta_k c_{kj} c_{ki} \\ j > i \end{array} & & \ddots \end{bmatrix} \tag{13'}$$

c_{kj} are elements of the matrix C (12);

$$\beta_n = \left. \frac{\partial^2 H}{\partial N_n^2} \right|_*, \quad n \in \overline{1,m}. \tag{13''}$$

In Section 3, the guaranteed relative fluctuation was introduced as a characteristic of the maximum "sharpness" for $H(N)$. The guaranteed relative fluctuation is the maximum ratio of the absolute fluctuation and the argument of the entropy function maximum.

Recall that the absolute fluctuation D characterizes the decrease of $H(N)$ on k units with respect to the maximum value (for thermodynamic systems k is the Boltzmann constant (see 2.11)), i.e.

$$\widehat{H}(N^*) - \widehat{H}(N^* + D) \geq k.$$

By using (13), D can be determined from the condition

$$D'_{(2)}\widetilde{G}D_{(2)} \geq 2k, \tag{14}$$

where

$$D_{(2)} = \{D_{r+1}, \ldots, D_m\}, \quad \widetilde{G} = -G, \tag{15}$$

$$D_n = \Delta N_n, \quad n \in \overline{r+1, m};$$

Consider the set

$$\mathfrak{Q} = \left\{D_{(2)} : D'_{(2)}\widetilde{G}D_{(2)} > 2k\right\} \tag{16}$$

in R^{m-r} and its complement

$$\overline{\mathfrak{Q}} = \left\{D_{(2)} : D'_{(2)}\widetilde{G}D_{(2)} \leq 2k\right\}. \tag{17}$$

The latter is closed and bounded if the matrix $\widetilde{G}$ is positive-definite. The boundary $\widehat{\mathfrak{Q}}$ of this set is determined from the equality

$$D'_{(2)}\widetilde{G}D_{(2)} = 2k. \tag{18}$$

The equality (18) determines a 2-nd order surface in R^{m-r}. If $\widetilde{G}$ is positive-definite, this surface is a real ellipsoid.

Let a and b be minimum and maximum half-axis of this ellipsoid, respectively. Consider two balls

$$\mathfrak{S}_{\max} = \left\{D_{(2)}; D'_{(2)}ED_{(2)} \leq 2kb^2\right\} \tag{19}$$

and

$$\mathfrak{S}_{\min} = \left\{D_{(2)} : D'_{(2)}ED_{(2)} \leq 2ka^2\right\}. \tag{19'}$$

where E is the unit matrix. Then the following inclusions hold

$$\mathfrak{S}_{\min} \subseteq \overline{\mathfrak{Q}} \subseteq \mathfrak{S}_{\max}. \tag{20}$$

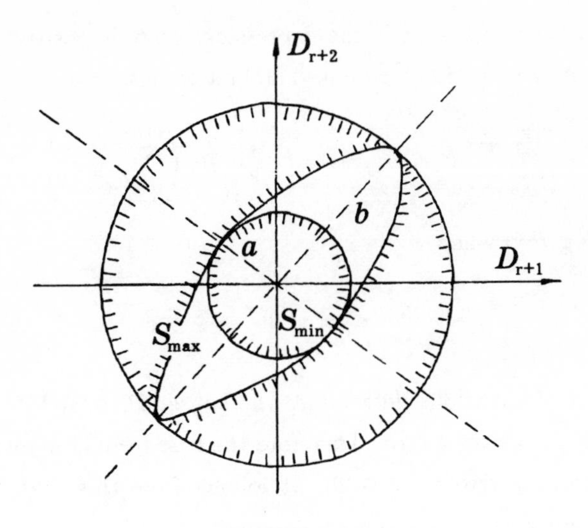

FIGURE 2.3.

The situation considered here is illustrated by fig 2.3. From (20) and fig. 2.3 it can be seen that a and b are the lower and upper estimate of the absolute fluctuation, namely

$$D_{\text{min}} = a, \quad D_{\text{max}} = b. \tag{21}$$

Thus to determine these estimates we should establish that $\widetilde{G}$ (18) is positive-definite.

Lemma 1. *Let the matrix* $T = [t_{kn} | k \in \overline{1,r};\ n \in \overline{1,m}]$ *in* (2) *have a full rank and let the function* $H(N)$ (1) *be strictly concave. Then the matrix* $\widetilde{G}$ *from* (13′) *is positive-definite.*

To prove the lemma, let us consider the quadratic form K defined for vectors $z \in R^{m-r}$

$$\begin{aligned} K &= \langle \widetilde{G} z, z \rangle = \\ &= -2 \sum_{i,j=1;i>j}^{m-r} \sum_{n=1}^{r} \beta_n c_{nj} c_{ni} z_i z_j - \sum_{j=1}^{m-r} \left(\beta_{j+r} + \sum_{n=1}^{r} \beta_n c_{nj}^2 \right) z_j^2 \\ &= -\sum_{i,j=1}^{m-r} \sum_{n=1}^{r} \beta_n c_{ni} c_{nj} z_i z_j - \sum_{j=1}^{m-r} \beta_{j+r} z_j^2. \end{aligned}$$

It follows from the strict concavity of $H(N)$ that

$$\beta_n = \left. \frac{\partial^2 H}{\partial N^2} \right| < 0, \quad n \in \overline{1,m}.$$

Therefore the second term in the expression for K is positive and becomes zero only for $z_j \equiv 0$; $j \in \overline{1, m-r}$. The first term is

$$-\sum_{i,j=1}^{m-r} \sum_{n=1}^{r} \beta_n c_{ni} c_{nj} z_i z_j = -\sum_{n=1}^{r} \beta_n \left(\sum_{i=1}^{m-r} c_{ni} z_i \right)^2 \geq 0,$$

with it being zero when

$$\sum_{i=1}^{m-r} c_{ni} z_i = 0, \quad n \in \overline{1, r}.$$

This system of equations determines the subset $\mathfrak{W} \subset R^{m-r}$ on which the first term of K becomes zero. Therefore the first term is more than zero for all $z \notin \mathfrak{W}$ and is zero for $z \in \mathfrak{W}$. It follows from this that $K > 0$ for all $z \in R^{m-r}$ and $K = 0$ for $z \equiv 0$.

Let us proceed to determining a and b. The maximum and minimum axes of the ellipsoid are known to depend on the minimum and maximum eigenvalues of the matrix $\widetilde{G}$ (13', 15). Consider the quadratic form $\widetilde{K} = D'_{(2)} \widetilde{G} D_{(2)}$. The matrix $\widetilde{G}$ has $(m-r)$ eigenvalues $\lambda_1, \ldots, \lambda_{m-r}$. There exists a real orthogonal matrix O satisfying the matrix equality

$$\Lambda = O^{-1} \widetilde{G} O, \quad \Lambda = [\text{diag} \lambda_i, \; i \in \overline{1, m-r}]$$

(Gantmacher, 1966). The matrix O determines the transformation of variables, namely

$$D_{(2)} = Ox.$$

In terms of these variables, the form $\widetilde{K}$ becomes

$$\widetilde{K} = x^T \Lambda x.$$

It follows from this expression and Lemma 1 that all the eigenvalues of $\widetilde{G}$ are real and positive.

Therefore the ellipsoid (18) may be described by the equation

$$x' \Lambda x = 2k,$$

or

$$\sum_{i=1}^{m-r} \lambda_i x_i^2 = 2k. \tag{22}$$

By reducing the latter equation to the canonic form, we obtain

$$\sum_{i=1}^{m-r} \frac{1}{\mu_i^2} x_i^2 = 1,$$

where

$$\mu_i = \sqrt{\frac{2k}{\lambda_i}}, \quad (i \in \overline{1, m-r}) \quad \text{are the half-axes of ellipsoid.}$$

Then

$$\begin{aligned} a &= \min_i \mu_i = \sqrt{\frac{2k}{\lambda_{\max}}}, \\ b &= \max_i \mu_i = \sqrt{\frac{2k}{\lambda_{\min}}}, \end{aligned} \tag{23}$$

where $\lambda_{\max}$ and $\lambda_{\min}$ are the minimum and maximum eigenvalues of $\widetilde{G}$, respectively.

In practice, it is rather difficult to determine the eigenvalues of $\widetilde{G}$ (13′, 15). Therefore the eigenvalues should be estimated.

The upper and lower estimates of the minimum absolute fluctuation are determined by the minimum sizes of the balls in (19, 20). To obtain them, the upper estimates for $\lambda_{\max}$ and $\lambda_{\min}$ (23) are necessary.

To determine the latter, we use the Gershgorin theorem (Gantmacher, 1966). The matrix $\widetilde{G}$ (13′, 15) has the eigenvalues $\lambda_i > 0$, $i \in \overline{1, m-r}$, and for these the following estimates hold

$$|\widetilde{g}_{ii} - \lambda| \leq W_i, \quad i \in \overline{1, m-r}; \tag{24}$$

where

$$\begin{aligned} \widetilde{g}_{ii} &= -\beta_{i+r} - \sum_{n=1}^{r} \beta_n c_{ni}^2 > 0, \\ W_i &= \sum_{\substack{j=1 \\ j \neq i}}^{m-r} \left| \sum_{n=1}^{r} \beta_n c_{nj} c_{ni} \right|. \end{aligned} \tag{25}$$

We assume that

$$\widetilde{g}_{ii} - W_i \geq \varepsilon_i > 0, \quad i \in \overline{1, m-r}.$$

Then for the positive eigenvalues, the system (24) is equivalent to

$$\widetilde{g}_{ii} - W_i \leq \lambda \leq \widetilde{g}_{ii} + W_i, \quad i \in \overline{1, m-r}. \tag{26}$$

Thus the estimates for the maximum and minimum eigenvalues are

$$\min_i (\tilde{g}_{ii} - W_i) \leq \lambda_{\max} \leq \max_i (\tilde{g}_{ii} + W_i), \tag{27}$$

$$\min_i (\tilde{g}_{ii} - W_i) \leq \lambda_{\min} \leq \min_i (\tilde{g}_{ii} + W_i). \tag{27'}$$

The minimums of the upper and lower estimates of the absolute fluc- tuation D (21) are

$$D_{\min} = a \geq D^{-}_{\min} = \sqrt{\frac{2k}{A^{+}}}, \tag{28}$$

$$D_{\max} = b \leq D^{+}_{\max} = \sqrt{\frac{2k}{A^{-}}}, \tag{28'}$$

where

$$A^{+} = \max_i (\tilde{g}_{ii} + W_i); \quad A^{-} = \min_i (\tilde{g}_{ii} - W_i).$$

From these expressions and from (25) it can be seen that $D^{-}_{\min}$ and $D^{+}_{\max}$ depend on the maximum and minimum values of $\beta_n = \left.\frac{\partial^2 H}{\partial N_n^2}\right|_*$ at the point N^* which corresponds to the maximum of entropy H on the set $\mathfrak{D}$ from (1,2). Thus $D^{-}_{\min} = D^{-}_{\min}(N^*)$ and $D^{+}_{\max} = D^{+}_{\max}(N^*)$.

Usually, the values of $D^{-}_{\min}$ and $D^{+}_{\max}$ themselves are not as interesting as their behaviour when the number Y of macrosystem elements increases. To study this, it is necessary to include Y into the set of the MSS parameters. For example, let us suppose that the number of macrosystem elements is a type of resource ($q_1 = Y$, $t_{1n} = 1$, $n \in \overline{1,m}$).

Consider a *macrosystem with the Boltzmann-states.* In this case, $H(N)$ is the generalized Boltzmann entropy. The set $\mathfrak{D}$ from (2) can be represented as

$$\mathfrak{D} = \left\{ N : \sum_{n=1}^{m} N_n = Y, \quad \sum_{n=1}^{m} t_{kn} N_n = q_k, \quad k \in \overline{2,r} \right\}. \tag{29}$$

According to (6.36),

$$N_n^* = a_n G_n \exp\left(-\sum_{j=1}^{r} \lambda_j t_{jn} \right), \quad n \in \overline{1,m}. \tag{30}$$

The Lagrange multipliers $\lambda_1, \ldots, \lambda_r$ are determined from the equations

$$\sum_{n=1}^{m} a_n G_n e^{-\lambda_1} \exp\left(-\sum_{j=2}^{r} \lambda_j t_{jn} \right) = Y, \tag{31}$$

$$\sum_{n=1}^{m} a_n G_n t_{kn} e^{-\lambda_1} \exp\left(-\sum_{j=2}^{r} \lambda_j t_{jn}\right) = q_k, \quad k \in \overline{2,r}. \tag{32}$$

Let us determine $e^{-\lambda_1}$ from (31) and substitute it in (30) and (32). We obtain

$$N_n^* = Y \frac{a_n G_n \exp\left(-\sum_{j=2}^{r} \lambda_j t_{jn}\right)}{\sum_{n=1}^{m} a_n G_n \exp\left(-\sum_{j=2}^{r} \lambda_j t_{jn}\right)}, \quad n \in \overline{1,m}, \tag{33}$$

$$\sum_{n=1}^{m} a_n G_n \exp\left(-\sum_{j=2}^{r} \lambda_j t_{jn}\right)(t_{kn} Y - q_k) = 0, \quad k \in \overline{2,r}. \tag{34}$$

Note that it follows from the asymptotical properties of B-model that the resource stores $q_2, \ldots, q_r$ should be changed in order for the set $\mathfrak{D}$ from (29) to be non-empty as $Y \to \infty$. In terms of (34), this means that $\min_n t_{kn} Y < q_k < \max_n t_{kn} Y$, $k \in \overline{2,r}$. In what follows, let us assume that q_k, $k \in \overline{2,r}$, increase according to the increase of Y.

Then the system (34) has the solution $\{\lambda_2(Y,q), \ldots, \lambda_r(Y,q)\} = \lambda(Y,q)$ for any $Y > 0$. According to (33),

$$N_n^* = Y \varphi_n(Y,q), \quad n \in \overline{1,m}, \tag{35}$$

where

$$\varphi_n(Y,q) = \frac{a_n G_n \exp\left(-\sum_{j=2}^{r} \lambda_j(Y,q) t_{jn}\right)}{\sum_{n=1}^{m} a_n G_n \exp\left(-\sum_{j=2}^{r} \lambda_j(Y,q) t_{jn}\right)}; \tag{35'}$$

$$0 < \varphi_n(Y,q) < 1 \quad \text{for any} \quad Y > 0 \quad \text{and} \quad n \in \overline{1,m}. \tag{35''}$$

We have from (13″)

$$\beta_n = -\frac{1}{N_n^*} = -\frac{1}{Y\varphi(Y,q)}, \quad n \in \overline{1,m}. \tag{36}$$

By substituting (36) in (25), (28) and (28′), we obtain

$$A^- = \alpha^-(Y)/Y; \quad A^+ = \alpha^+(Y)/Y,$$

where $\alpha^-(Y)$ and $\alpha^+(Y)$ are bounded functions of Y, namely

$$\alpha^-(Y) \geq \alpha^-; \quad \alpha^+(Y) \leq \alpha^+.$$

It is seen from (28, 28′) that the upper and lower estimates of the absolute fluctuation have the same asymptotics, namely they grow like $\sqrt{Y}$. We have

$$D_{\min} \geq \kappa^- \sqrt{Y}; \quad D_{\max} \leq \kappa^+ \sqrt{Y} = \widehat{D}_{\max}, \tag{37}$$

where

$$\kappa^+ = \sqrt{\frac{2k}{\alpha^-}}; \quad \kappa^- = \sqrt{\frac{2k}{\alpha^+}}. \tag{37'}$$

According to the definition (3.6), the guaranteed relative fluctuation is

$$\rho = \max_n \frac{\widehat{D}_{\max}}{N_n^*}.$$

It follows from (37, 37′) that

$$\rho \approx 1/\sqrt{Y}.$$

Thus $\rho \to 0$ as $Y \to \infty$.

Hence the generalized information entropy of Boltzmann has a "sharp" maximum for the deviations from the maximum point which belong to the set $\mathfrak{D}$ of the B-model's feasible macrostates. Note that this does not depend on the number of constraints, unless the set $\mathfrak{D}$ is empty as $r < m$ and $Y \to \infty$.

Thus it is shown that for B-model with a linear and complete resource consumption, the probabilities of feasible macrostates are significantly less than the maximum probability, with the difference increasing as the number of macrosystem elements grows.

Consider a ***macrosystem with the Fermi-states***. For this macrosystem, the occupation numbers $N_1, \ldots, N_m$ may be equal to any values in the intervals $\overline{0, G_n}$, $n \in \overline{1, m}$. In this case, the entropy $H(N)$ is the generalized information entropy of Fermi-Dirac (2.14). If such a macrosystem has a linear and complete resource consumption, its set of feasible states is given by the equation (2). The macrostate maximizing the entropy $H_F(N)$ (2.14) is characterized by the occupation numbers

$$N_n^* = \frac{G_n}{1 + b_n \exp\left(\sum_{j=1}^{r} \lambda_j t_{jn}\right)}, \quad n \in \overline{1,m}, \tag{38}$$

where the Lagrange multipliers $\lambda_1, \ldots, \lambda_r$ are obtained by solving the equations

$$\begin{aligned} &\sum_{n=1}^{m} \frac{G_n}{1 + b_n \exp\left(\sum_{j=1}^{r} \lambda_j t_{jn}\right)} = Y, \\ &\sum_{n=1}^{m} \frac{t_{kn} G_n}{1 + b_n \exp\left(\sum_{j=1}^{r} \lambda_j t_{jn}\right)} = q_k, \qquad k \in \overline{2,r}. \end{aligned} \tag{39}$$

Note that in studying the asymptotic properties of F-model when the number Y of elements increases, we assume that the capacities $G_1, \ldots, G_m$ and the resource stores $q_2, \ldots, q_r$ are changed correspondingly. The capacities $G_1, \ldots, G_m$ should be changed, because the occupation rule should be valid as Y grows, namely each state can be occupied by no more than one element. Therefore $Y \leq \sum_{n=1}^{m} G_n$. For example, we can assume that $G_n = \gamma_n Y$ $(n \in \overline{1,m})$ and $\sum_{n=1}^{m} \gamma_n = 1$.

The resource stores should change, in order for the set $\mathfrak{D}$ from (2) to be nonempty as $Y \to \infty$.

Now let us return to the equalities (38) and represent them as

$$N_n^* = Y \Psi_n(\lambda), \quad n \in \overline{1,m}; \tag{40}$$

where

$$\Psi_n(\lambda) = \frac{\gamma_n}{1 + b_n \exp\left(\sum_{j=1}^{r} \lambda_j t_{jn}\right)}, \tag{41}$$

$$\gamma_n = G_n / Y, \quad 0 < \Psi^- \leq \Psi_n \leq \Psi^+ < 1, \quad n \in \overline{1,m}. \tag{41'}$$

Let us determine β_n (13″) at the maximum point (40). We obtain

$$\beta_n = -\left(\frac{1}{N_n^*} + \frac{1}{G_n - N_n^*}\right) = \frac{1}{Y} \alpha_n(\lambda), \quad n \in \overline{1,m}; \tag{42}$$

where

$$\alpha_n(\lambda) = \frac{1}{\Psi_n(\lambda)[1 - \Psi_n(\lambda)]}. \tag{43}$$

By substituting β_n in (25), (28) and (28′), we obtain

$$\begin{aligned} A^- &= \alpha^-(Y)/Y, \\ A^+ &= \alpha^+(Y)/Y, \end{aligned} \tag{44}$$

where $\alpha^+(Y)$ and $\alpha^-(Y)$ are bounded functions, namely $\alpha^+(Y) \leq \alpha^+$; $\alpha^-(Y) \geq \alpha^-$.

It follows from (28) that the minimum values of the upper and lower estimates of the absolute fluctuation grow similarly, namely like $\sqrt{Y}$. They are determined by (37), (37′) and (44). According to (3.6), $\rho \to 0$ as $Y \to \infty$.

Thus the generalized information entropy for a macrosystem with the Fermi-states also have a "sharp" maximum even with the constrained resources, i.e. the maximum of $H_F(N)$ remains "sharp" in sense of (3.6) for the macrostates from $\mathfrak{D}$ (2).

Finally, we should only show that the property of maximum "sharpness" holds for homogeneous *macrosystems with the Einstein-states* and a linear and complete resource consumption. Recall that for such macrosystems, the occupation numbers $N_1, \ldots, N_m$ are not constrained by the capacities of the close state subsets. Therefore, in studying the morphological properties of the generalized information entropy of Einstein-Boze, the capacities $G_1, \ldots, G_m$ can be considered as fixed as $Y \to \infty$.

The macrostate maximizing the entropy $H_E(N)$ from (2.23) on the set $\mathfrak{D}$ is characterized by the occupation numbers

$$N_n^* = G_n \omega_n(\lambda), \quad n \in \overline{1,m}, \tag{45}$$

where

$$\omega_n(\lambda) = \frac{1}{c_n \exp\left(\sum_{j=1}^{r} \lambda_j t_{jn}\right) - 1}. \tag{45′}$$

The Lagrange multipliers $\lambda_1, \ldots, \lambda_r$ are determined from the system of equations

$$\sum_{n=1}^{m} G_n \omega_n(\lambda) = Y, \tag{46}$$

$$\sum_{n=1}^{m} t_{kn} G_n \omega_n(\lambda) = q_k, \quad k \in \overline{2,r}. \tag{46'}$$

We assume that the set $\mathfrak{D}$ remains nonempty as $Y \to \infty$, i.e. $q_2, \ldots, q_r$ increase correspondingly. The functions $\omega_n(\lambda)$ from (45′) are positive, as it follows from Lemma 2.

It follows from (46) that Y is a linear combination of ω_n. Hence it follows that $\omega_1, \ldots, \omega_m$ grow like Y, i.e.

$$\omega_n \approx p_n Y, \quad n \in \overline{1,m}, \tag{47}$$

where p_n is a coefficient.

Let us determine β_n (13″) at the maximum point (45, 45′). We obtain

$$\beta_n = -\frac{1}{G_n \omega_n(\lambda)[1 + \omega_n(\lambda)]}, \quad n \in \overline{1,m}.$$

Thus

$$A^- \geq (\pi^-/Y); \quad A^+ \leq (\pi^+/Y),$$

where π^- and π^+ are constant. The minimum values of the upper and lower estimates of the absolute fluctuation are

$$D_{\min} \geq \kappa^- \sqrt{Y}, \quad D_{\max} \leq \kappa^+ \sqrt{Y}; \tag{48}$$

where

$$\kappa^- = \sqrt{\frac{2k}{\pi^+}}, \quad \kappa^+ = \sqrt{\frac{2k}{\pi^-}}.$$

It follows from the definition of the guaranteed relative fluctuation that $\rho \approx (1/\sqrt{Y})$ and $\rho \to \infty$ as $Y \to \infty$.

Therefore for E-models the property of maximum "sharpness" holds for the generalized information entropy of Einstein-Boze. Hence the macrostates from the feasible set $\mathfrak{D}$ (2) can be realized with the probabilities significantly less than that of the most probable macrostate.

2.8 Bibliographical comments

One of the key problems is to explain in what way a macrosystem containing a large number of elements can reach a stationary state characterized by constant values of a fairly small number of macrovariables. The contradictions arising here are most clearly seen in the case of a conservative mechanical macrosystem. Recall that the Hamiltonian of such a macrosystem does not depend on time, and the macrosystem elements move on the energetic surface (on which the Hamiltonian is constant and equals the sum energy). In doing so, the motion is considered in a $6Y$-dimensional space, where Y is the number of elements, while the state of each element is characterized by three moments and three generalized coordinates.

The well-known Poincare theorem states that a conservative system in the $6Y$-dimensional space moves so that for any neighborhood of the initial point there exists a finite time after which the phase point comes back to the neighborhood. Therefore there are no trajectories leading to a stationary state in this space. It means that the Lyapunov function does not exist for the system.

But by observing real macrosystems, we see that this does not hold. The observed system parameters (not the coordinates of the phase point in the $6Y$-dimensional space, but the macroparameters) reach their stationary values after a while and do not change them arbitrarily long.

This fact appears to mean that we observe the changing macrosystem states in another space which logically arises from the different time scales for the motion of elements and that of the system.

While the space in which the elements move is defined uniquely (its basis is constructed by the coordinates and moments of the elements), there are many alternatives for introducing the space of macroobservations. Its basis is determined by a set of macroparameters which, in general, are invented, although, of course, their physical realizability, scheme of observations and so on are taken into account.

The first macrosystem concept is related to Boltzmann (L. Boltzmann, 1877), who proposed the probabilistic scheme of interactions between the particles forming a physical body.

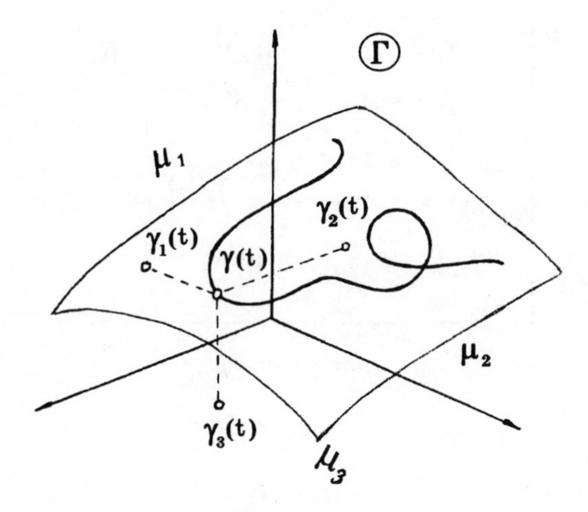

FIGURE 2.4.

Consider a 6-dimensional space μ in which a particle moves. Then the $6Y$-dimensional space is

$$\Gamma = \mu^1 \otimes \mu^2 \otimes \ldots \otimes \mu^Y,$$

where $\otimes$ is a direct product, i.e. μ^i $(i \in \overline{1,Y})$ are orthogonal subspaces in Γ. The upper index i means that the i-th particle moves in the space μ^i, but all spaces $\mu^1, \ldots, \mu^Y$ are the same.

Let us consider a point $\gamma(t)$ in Γ (fig. 2.4) and project it onto the spaces $\mu^1, \ldots \mu^Y$. In each μ^i we obtain the point $\gamma_i(t)$ characterizing the state of the i-th particle at the time t. Since all μ^i are similar, we may consider only one μ-space (fig. 2.5) and locate therein the points $\gamma_1(t), \ldots, \gamma_Y(t)$, each characterizing the states of a particle. As a result, the finite set $\mathfrak{M}_t$ of the phase points $\gamma_1(t), \ldots, \gamma_Y(t)$ arises in μ. The set $\mathfrak{M}_t$ can be characterized by the distribution density $f(p, q, t)$.

Thus the point $\gamma(t) \in \Gamma$ is mapped into the set $\mathcal{M}_t \in \mu$. For $t_1 > t$ the point $\gamma(t_1) \in \Gamma$ will be mapped into another set $\mathcal{M}_{t_1} \in \mu$. Hence when the system phase point moves in Γ-space, the set $\mathcal{M}_t$ containing Y phase points of distinct particles evolves in μ-space.

Maxwell showed that the distribution $f(p, q, t)$ of the phase points (or particles) in μ-space tends to the stationary distribution $f^\circ(p, q)$ as the time tends to the infinity. ($f^\circ(p, q)$ is the velocity distribution, since $p = mv$,

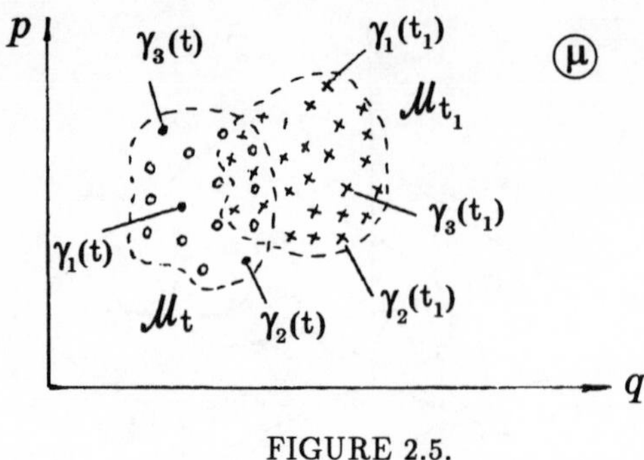

FIGURE 2.5.

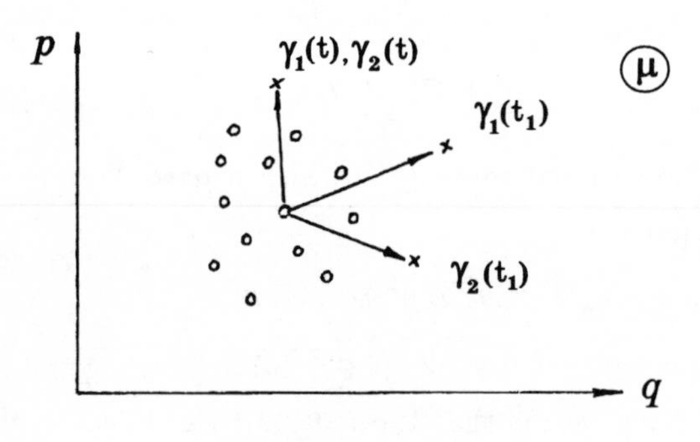

FIGURE 2.6.

and consequently $f^{\circ}(p,q) = f^{\circ}(v,q)$.) Therefore it is reasonable to consider the function $f(p,q,t)$ as a macrocharacteristic.

Since $f(p,q,t) \to f^{\circ}(p,q)$ as $t \to \infty$, the pass from $f(p,q,t)$ to $f^{\circ}(p,q)$ can be characterized by the Lyapunov function, which reaches its minimum at $f^{\circ}(p,q)$.

The function $f(p,q,t)$ evolves over time due to various factors related to the way in which the elements interact at the microlevel. The traditional scheme for describing this evolution belongs to Boltzmann. He proposed to consider temporary variations of the distribution function $f(p,q,t)$ as consisting of two components; the first one is determined by the flow of elements, and the second one does so by the elements collisions.

Boltzmann restricted himself to pairwise collisions. The point is that if we come back to the map of the trajectory points from Γ-space into μ-space, we can observe situations where pairs of projections are the same at t (for example, $\gamma_1(t) = \gamma_3(t)$; $\gamma_i(t) = \gamma_k(t);\ldots$) and different at $t_1 > t$. Such a situation is shown in fig. 2.6. The coincidence of the pairs means the collision of the phase point.

By accepting this hypothesis, Boltzmann derived the kinetic equation describing the way in which the distribution function evolves (Uhlenbeck, Ford 1963).

To investigate the properties of solutions of this equation, the function $H(t) = -\int f(p,q,t)\ln f(p,q,t)\,dpdq$ has been used. It was shown that $\frac{dH}{dt} > 0$ for the solutions of the Boltzmann kinetic equations, and $\frac{dH}{dt} = 0$ for $f(p,q,t) = f^{\circ}(p,q)$. Thus the function $H(t)$ is a Lyapunov function for the Boltzmann kinetic equation.

This is the well-known H-theorem of Boltzmann. It states that the stationary distribution of particles in μ-space is the distribution having the maximum entropy.

Sometimes it is sufficient to describe the distribution in μ-space by a discrete function. In this case, μ-space is divided into the cells $\omega_1,\ldots,\omega_m$, each characterized by the number N_i of particles occupying it. It is assumed that the apriori probabilities of particles are uniform. Then the distribution in μ-space will be characterized by the vector $N(t) = N_1(t),\ldots,N_m(t)$. According to the H-theorem, $N(t) \to N^{\circ}$ as $t \to \infty$ and N° maximizes the entropy $H = -\sum N_i \ln N_i$. If the apriori probabilities are $\nu_1,\ldots,\nu_m$ and are different then N° maximizes the entropy $H = -\sum N_i \ln \frac{N_i}{\nu_i}$.

Thus the probabilistic model of Boltzmann represents the first example of the macrosystem in which a random behaviour of particles is transformed into a nonrandom behaviour of the system consisting of a large number of particles.

Further, the Boltzmann macrosystem concept has been developed by the probabilistic schemes of Fermi-Dirac and Einstein-Boze (Landau, Lifshitz, 1964).

The Boltzmann scheme of equiprobable distribution was found to be the limit case of the Fermi-Dirac and the Einstein-Boze schemes.

Note that the Fermi-Dirac and the Einstein-Boze schemes are also based on the assumption that the particles are distributed among the groups of close states with equal probabilities; therefore these more general schemes do not include the Boltzmann probabilistic scheme with nonuniform distribution.

The assumption on equiprobable distribution of particles used in the Fermi-Dirac and Einstein-Boze schemes often appears to be too restrictive. In particular, such a situation arises in studying an ideal macrosystem consisting of N spins, with their magnetic moments directed either like the magnetic field or oppositely.

The magnetic moment of a spin orients with respect to the external field randomly and independently of the other spins. But we know the apriori probabilities p and q for a magnetic moment to be directed like the external field and in the opposite direction, respectively (Reif, 1977). If we consider two groups of states, with the first group containing the spins directed in the external field, while the second one containing the other spins, then we obtain the scheme of macrosystem with the Fermi-states, but provided with the apriori probabilities p and q.

It should be noted that, along with the physical postulates of probabilistic schemes, there exist abstract constructions of probability theory. One of them is the Bernoulli scheme of trials (Prochorov et al., 1967, Foeller, 1957). In particular, if we consider a scheme with two outcomes A and B occurring with the apriori probabilities p and $(1-p)$, such a scheme involves the macrosystem containing N spins with two possible orientations of the magnetic moment.

If we are interested in the number of the outcomes A preceding the first occurrence of the outcome B in the Bernoulli scheme then the corresponding probabilistic scheme coincides with the scheme of distributing particles among two groups of states, with the particles having the apriori probabilities p and $(1-p)$ of occupying the groups, respectively (see the Pascal distribution in Prochorov et al., 1967).

And, finally, the Bernoulli scheme with many outcomes, which is described by the polynomial distribution (Feller 1957), coincides with the Boltzmann probabilistic scheme with a non-uniform distribution of particles.

The macrosystem ideology proved to be very useful not only in studying thermodynamic systems. Recently, a number of papers dealing with the systems of entirely different nature (urban, transportation, regional and demographic) have appeared. This series starts with the paper by A. G. Wilson, 1967, who proposed to apply the equiprobable scheme of Boltzmann to the model of spatial interaction in a regional system. This approach proved to be so fruitful that the flow of papers developing, generalizing and implementing the Wilson models (see reviews in Wilson 1970, Popkov et al. 1984) is not stopped even now. Here the approach based on the phenomenological scheme of a stochastic communication system is the most universal. In such a scheme, the apriori probabilities of choosing the corresponding states take into account the characteristics of the communication network (Imelbayev, Shmulyan 1978). Models of this type are characterized by the generalized information entropy of Boltzmann, with its parameters being interpreted according to the problem to be solved.

The phenomenological scheme of a macrosystem with the Boltzmann states has been used for modeling stochastic hierarchy systems (Popkov, 1984). In such systems, the subsystems are randomly distributed among the fixed hierarchy levels.

And, finally, the phenomenological scheme of Boltzmann macrosystem proved to be useful in one more field, namely in studying the spatial migration of population (Rogers, Willekens, 1976). Various types of stochastic migration models were developed and intensively tested on extensive experimental data.

The phenomenological schemes of macrosystems with the Fermi- and and Einstein-states are used for modeling urban, transportation, regional and demographic systems more rarely. It is not because the appropriate problems are absent. To the contrary, the schemes of macrosystems with the Fermi-states should be tested first in all cases where the Boltzmann-type schemes are applied, since it is not known in advance, whether the assumption on the occupation numbers being small enough is valid.

But the schemes with the Fermi- and Einstein-states require many apriori data (the capacities $G_1, \dots, G_m$), and this requirement appears to restrain their practical use. For example, in modeling the distribution of passengers among the transport communications, we have to take into account the bounded capacity of the transport vehicles. It means that the capacities are finite and equal to the number of seats in the transport vehicles circulating on given communication. Since each seat can be occupied by only one passenger, we obtain the macrosystem model with the Fermi-states (Handjan, 1984).

The phenomenological schemes of macrosystems with the Fermi- and Einstein-states are used more often for modeling the human migration in a regional hierarchy. The two-level hierarchy which consists of a set of upper level regions, each containing a number of spatial cells, is the most common. The elements are distributed among the large upper level regions and then among the second level cells. The macrosystem schemes with either the Fermi-, or Boltzmann-, or Einstein-states arise in such schemes depending on the relations between the overall number of elements and capacities of both large regions and their spatial cells (Roy, 1987; O'Kelly, 1981; Roy, Lesse, 1981; Roy, Brotchie, 1984; Walsh, Gilbert, 1980).

CHAPTER THREE

PARAMETRIC PROPERTIES OF MSS

A model of stationary states (MSS) of a homogeneous macrosystem is described by the problem of maximizing the entropy on the set of feasible macrostates. For the macrosystems with complete consumption of resources, this set is given by the system of equalities. For the macrosystems with incomplete consumption of resources, the feasible set is given by the system of inequalities.

Parametric properties are the properties of macrostate under variations of the MSS parameters. The MSS parameters are divided into three groups. The first group consists of the parameters characterizing the stochastic mechanism of distributing the elements among the subsets of states. The second group includes the characteristics of the resource consumption. And the third group includes the parameters characterizing the resource stores.

In this chapter, we analyze the parametric properties of MSS, and apply the analysis to develop the quantitative estimates of parametric sensitivity of MSS.

3.1 Parameterization of MSS with complete consumption of resources

MSS with the complete consumption of resources are described by problems of constrained maximization of entropy functions depending on the parameters $a_1, \dots, a_m$ and $G_1, \dots, G_m$. The former group characterizes the apriori probabilities for a macrosystem element to occupy the corresponding subset of states, while the latter characterizes the capacities of the subsets. It is assumed that the feasible sets are also parameterized by a number of parameters. The first group includes parameters $q_1, \dots, q_r$. They characterize the stores of corresponding resources. The other group includes the parameters of functions $\varphi_k(N)$, which characterizes the resource consumption. Denote these parameters by $t_k = \{t_{k1}, \dots, t_{km_k}\}$. In general, their numbers m_k vary for each function φ_k.

MSS with the complete consumption of resources can be represented as

$$H(N, \alpha, G) \Rightarrow \max; \quad \varphi_k(N, t_k) = q_k, \quad k \in \overline{1, r}; \tag{1}$$

where

$N = \{N_1, \dots, N_m\}$	is a vector of system macrostate;
$a = \{a_1, \dots, a_m\}$	are the apriori probabilities for an element to occupy a state from the subsets $\mathfrak{S}_1, \dots, \mathfrak{S}_m$;
$G = \{G_1, \dots, G_m\}$	are the capacities of the subsets $\mathfrak{S}_1, \dots, \mathfrak{S}_m$;
$\varphi_k(N, t_k)$	are the functions characterizing the consumption of the resources of the k-th type ($k \in \overline{1, r}$);
$t_k = \{t_{k1}, \dots, t_{km_k}\}$	is the vector of parameters of the consumption functions and
q_k	are the stores of the k-th resource.

Here we assume that the functions φ_k are convex with respect to the variables N and

$$\mathfrak{D} = \{N : \varphi_k(N, t_k) = q_k, \quad k \in \overline{1, r}\} \neq \varnothing \tag{2}$$

for all t_k and q_k from the corresponding bounded sets.

The solution N^*, or a realizable macrostate, depends on all the parameters of the problem, namely

$$N^* = \Phi(\alpha, G, t, q). \tag{3}$$

The parametric properties of MSS are the parametric properties of this function. If we take into account that all the parameters in (3) are vectors, we can see that it is rather difficult to analyze the parametric properties, because of very high dimension of the argument in the vector-function (3). Although the difficulties are not principal, the constructiveness (and practical value) of parametric analysis in this situation seems rather doubtful.

Let us try to look at this problem from a slightly different point of view, namely let us ask of what are the reasons for the parameters to change? The question can be answered, at least, in two ways.

Assume that the MSS is adequate to the macrosystem. Since only isolated macrosystems are considered, the parameters cannot fluctuate. But if the fluctuations still exist then the system is not isolated, i.e. external forces influence its properties.

Another situation arises if the macrosystem and its MSS have the same structure, but the macrosystem parameters (and consequently the MSS parameters) are not known precisely. In this case, the MSS can be considered as an isolated macrosystem with the same structure and mechanism of functioning, but with the different parameters. Again, in order to pass from an isolated macrosystem with one set of parameters to an isolated system with another set of parameters, an external force should be applied to the system.

Note that in both situations, the changes in internal parameters of MSS are caused by external perturbations.

Let the external perturbations be simulated by a vector parameter $x = \{x_1, \ldots, x_s\}$, where $s < m$. Then $a = a(x)$, $G = G(x)$, $t = t(x)$ and $q = q(x)$, and MSS (1) can be written as

$$\begin{aligned} &H(N, x) = H(N, a(x),\ G(x)) \Rightarrow \max, \\ &\varphi_k(N, x) = \varphi_k(N, t_k(x)) = q_k(x), \quad k \in \overline{1, r}. \end{aligned} \tag{4}$$

The realizable macrostate determined by this model is

$$N^* = N^*(x) = (\alpha(x),\ G(x),\ t(x),\ q(x)) \tag{5}$$

The function $N^*(x)$ (5) is given implicitly by the solution of (4) and is defined on the space R^s, which has a dimension significantly lower than that of the space on which the function in (3) is defined. But this reduction needs additional data concerning the properties of the functions $a(x)$, $G(x)$, $t(x)$ and $q(x)$.

Consider the problem (4) and corresponding Lagrange function

$$L(N, \lambda, x) = H(N, x) + \sum_{k=1}^{r} \lambda_k (q_k(x) - \varphi_k(N, x)). \tag{6}$$

The stationarity conditions for L are

$$\frac{\partial L}{\partial N_n} = F_n(N, \lambda, x) = \frac{\partial H}{\partial N_n} - \sum_{k=1}^{r} \lambda_k \frac{\partial \varphi_k}{\partial N_n} = 0 \quad n \in \overline{1,m}; \tag{7}$$

$$\frac{\partial L}{\partial \lambda} = \Phi_k(N, \lambda, x) = q_k(x) - \varphi_k(N, x) = 0, \quad k \in \overline{1,r}. \tag{7'}$$

Let us consider a convex set $\mathfrak{S} \subset R^s$ and assume that $x \in \mathfrak{S}$. For example, $\mathfrak{S}$ may be a ball having the center at x^0 and the radius β, namely

$$\mathfrak{S}(x^0, \beta) = \{x : \|x - x^0\| \leqslant \beta\} \tag{8}$$

Assume that the system (7, 7′) has the solution N^0 for $x = x^0$, i.e.

$$F_n(N^0, \lambda^0, x^0) = 0, \quad n \in \overline{1,m}; \tag{9}$$

$$\Phi_k(N^0, \lambda^0, x^0) = 0, \quad k \in \overline{1,r}; \tag{10}$$

If for all x close to x^0 there exist solutions of (7, 7′) close to N^0, λ^0 then we say that this system determines the implicit function $N(x)$, $\lambda(x)$. The closeness of x and x^0 is characterized by the ε – neighborhood of x^0 contained in $\mathfrak{S}(x^0, \beta)$ $(\mathfrak{S}(x^0, \varepsilon) = \{x : \|x - x^0\| \leqslant \varepsilon\})$.

To illustrate the definitions, let us consider the MSS describing the distribution of passenger flows in a transportation system (Wilson, 1978)

$$H(N) = -\sum_{i,j=1}^{n,m} N_{ij} \ln \frac{N_{ij}}{e a_{ij}} \Rightarrow \max; \tag{11}$$

$$\sum_{i=1}^{n} N_{ij} = P_j; \quad j \in \overline{1,m}; \tag{12}$$

$$\sum_{j=1}^{m} N_{ij} = Q_i, \quad i \in \overline{1,n}; \tag{13}$$

$$\sum_{i,j}^{n,m} c_{ij} N_{ij} = T. \tag{14}$$

Here i and j denote the nodes of the transportation network; N_{ij} is the number of passengers travelling from i to j in a unit time; a_{ij} is a probability that a passenger chooses the pair (i, j); c_{ij} are unit travel costs for one passenger; P_j are the capacities of origin nodes (for example, the numbers of residence places); Q_i are the capacities of destination nodes (for example, the numbers of working places); and T is an average travel time.

In investigating the influence of new construction (housing places or working places) on the transport flows, the variables $P_1, \dots, P_m$ and $Q_1, \dots, Q_n$ are external parameters. Thus in this model the exogenous parameters are $x = (P, Q)$. The changes of P and Q should, generally speaking, influence the apriori probabilities, the unit costs and the average travel time. But multiple empirical studies (e.g. Goltz, 1981) show that a_{ij}, c_{ij} and T do not change significantly.

Thus the change of the correspondence matrix $N = [N_{ij}]$ is caused by the changes in the right-hand sides of the balance equations (12, 13), with all the other parameters of model (11–14) remaining constant.

In this case, the Lagrange function for (11–14) is

$$\begin{aligned} L(N, \lambda) &= \sum_{i,j=1}^{m,n} N_{ij} \ln \frac{N_{ij}}{e a_{ij}} + \sum_{j=1}^{m} \lambda_j \left(P_j - \sum_{i=1}^{n} N_{ij} \right) + \\ &+ \sum_{i=1}^{n} \lambda_{i+m} \left(Q_i - \sum_{j=1}^{m} N_{ij} \right) + \lambda_{n+m+1} \left(T - \sum_{i,j=1}^{n,m} c_j N_{ij} \right). \end{aligned}$$

Its stationarity conditions are

$$\begin{aligned} N_{ij}(P, Q) &= a_{ij} \exp(-\lambda_j(P, Q) - \lambda_{i+m}(P, Q) - \\ &\quad -\lambda_{n+m+1}(P, Q) c_{ij}); \quad i \in \overline{1, n}; \quad j \in \overline{1, m}; \end{aligned} \tag{15}$$

$$\begin{aligned} &\exp\left(-\lambda_j(P, Q)\right) \sum_{i=1}^{n} a_{ij} \exp\left(-\lambda_{i+m}(P, Q) - \right. \\ &\left. -\lambda_{n+m+1}(P, Q) c_{ij}\right) - P_j = 0, \quad j \in \overline{1, m}; \end{aligned} \tag{16}$$

$$\begin{aligned} &\exp\left(-\lambda_{i+m}(P, Q)\right) \sum_{j=1}^{m} a_{ij} \exp\left(-\lambda_j(P, Q) - \right. \\ &\left. -\lambda_{n+m+1}(P, Q) c_{ij}\right) - Q_i = 0, \quad i \in \overline{1, n}; \end{aligned} \tag{17}$$

$$\begin{aligned} &\sum_{i,j=1}^{n,m} c_{ij} a_{ij} \exp\left(-\lambda_j(P, Q) - \lambda_{i+m}(P, Q) - \right. \\ &\qquad \left. -\lambda_{n+m+1}(P, Q) c_{ij}\right) - T = 0. \end{aligned} \tag{18}$$

Equations (16) – (18) determine the way in which the Lagrange multipliers λ depend on the external parameters P and Q. The existence of implicit function $\lambda(P,Q)$ is related to the investigation of system (16–18).

3.2 Theorems on implicit functions

The equations (1.7) and (1.7′) determines the implicit functions $N(x)$ and $\lambda(x)$. Some general theorems on implicit functions are used in studying the properties of these functions. We formulate the theorems for unified form of (1.7) and (1.7′).

To obtain this form, we introduce the notations

$$z' = (N, \lambda); \quad W = \begin{pmatrix} F \\ \Phi \end{pmatrix}; \quad F = \{F_1, \dots, F_m\}; \quad \Phi = \{\Phi_1, \dots, \Phi_r\}, \tag{1}$$

where F_n and Φ_k are determined by (1.7) and (1.7′).

Then (1.7) and (1.7′) can be written as

$$W(z, x) = 0, \tag{2}$$

where

$$z \in R^{m+r}, \quad x \in R^s, \quad W : R^{m+r} \times R^s \Rightarrow R^{m+r}.$$

The matrix

$$J(z,x) = W'_z(z,x) = \left[\frac{\partial W_i}{\partial z_j}; \quad i, j \in 1, \overline{(m+r)}\right].$$

is important in studying the equation (2).

Local properties

We assume that there exists a point $(z^0, x^0) \in R^s \times R^{m+r}$ for which $W(z^0, x^0) = 0$. We also assume that the derivative $J(z,x)$ is continuous in a neighborhood of (z^0, x^0) and

$$\det J(z^0, x^0) \neq 0. \tag{3}$$

Denote

$$\mathfrak{S}(z^0, \rho) = \{z : \|z^0 - z\| \leq \rho\} \tag{4}$$

1. Existence and continuity. The conditions for a single-valued implicit function to exist in $\mathfrak{S}(z^0, x^0)$ are given by the following theorem.

Theorem 1 (Krasnoselsky et al, 1969). *Let the following conditions be satisfied*

(a) $\|W(z^1, x^0) - W(z^2, x^0) - J(z^0, x^0)(z^1 - z^2)\| \leq \omega(\rho)\|z^1 - z^2\|$,

$$\|[W(z^1, x) - W(z^2, x)] - [W(z^1, x^0)] - W(z^2, x^0)]\| \leq \omega_1(\beta, \rho)\|z^1 - z^2\|$$

for $x \in \mathfrak{S}(x^0, \beta)$ *and* $z^1, z^2 \in \mathfrak{S}(z^0, \rho)$;

(b) $c_0 = \lim_{\rho,\beta \to 0} \|J(z^0, x^0)\|(\omega(\rho) + \omega_1(\rho, \beta)) < 1$;

(c) *the function* $W(z^0, x)$ *is continuous at* x^0.

Then there exist $\varepsilon \leq \beta$ *and* $\eta \leq \rho$ *such that the system* (2) *has a unique solution* $z^*(x)$ *in the ball* $\mathfrak{S}(z^0, \eta)$ (4) *for* $x \in \mathfrak{S}(x^0, \varepsilon)$ *and*

$$\lim_{x \to x^0} \|z^*(x) - z^0\| = 0.$$

where $z^*(x^0) = z^0$.

The proof of this theorem is based on the analysis of an equation equivalent to (2), namely

$$z = z - J^{-1}(z^0, x^0)\, W(z, x) = T(z, x). \tag{5}$$

The first condition in (a) can be represented as

$$\|J(z^0, x^0)[T(z^2, x^0) - T(z^1, x^0)]\| \leq \omega(\rho)\|z^1 - z^2\|$$

and the second can be represented as

$$\|J(z^0, x^0)[T(z^1, x^0) - T(z^1, x) + T(z^2, x) - T(z^2, x^0)]\| \leq \omega_1(\rho, \beta)\|z^1 - z^2\|.$$

Theorem 1 gives the sufficient conditions for the implicit function $z^*(x)$ (solution of (2)) to exist. By narrowing the class of functions $W(z, x)$, we can not only establish the existence of the implicit function, but also assure some of its properties.

Theorem 2 (Lyusternik, Sobolev, 1965). *Let the conditions (a) and (b) of Theorem* 1 *be satisfied and let the function* $W(z, x)$ *be continuous with respect to* x. *Then the implicit function* $z^*(x)$ *exists and is continuous in the ball* $\mathfrak{S}(x^0, \varepsilon)$.

To prove the continuity, we consider the equation $z = T(z, x)$, where T is determined by (5). We obtain

$$\|z^*(x) - z^*(x^0)\| \leq \|T(z^*(x), x) - T(z^*(x^0), x)\| + \\ + \|T(z^*(x^0), x) - T(z^*(x^0), x^0)\|. \quad (6)$$

According to (5), the derivative is

$$\frac{\partial T}{\partial z} = T_z(z, x) = E - J^{-1}(z^0, x^0)\, J(z, x) = \\ = J^{-1}(z^0, x^0)(J(z^0, x^0) - J(z, x)).$$

Note that $T_z = 0$ at (z^0, x^0). From the assumption on the continuity of $T(z, x)$ in $\mathfrak{S}(x^0, \beta)$ and from the assumption on its nondegeneracy at (z^0, x^0) (3) it follows that $T_z(z, x)$ is continuous, i.e. there exists a domain in $\mathfrak{S}(z^0, \rho) \times \mathfrak{S}(x^0, \beta)$ for which

$$\|T_z'(z, x)\| \leq q(\rho, \beta), \quad \rho = \rho(\beta), \quad (7)$$

and

$$\rho \to 0, \quad q(\rho, \beta) \to 0 \text{ as } \beta \to 0. \quad (8)$$

Choose β and ρ so that $q(\rho, \beta) = q < 1$. Then

$$\|T(z^*(x), x) - T(z^*(x^0), x)\| \leq q\, \|z^*(x) - z^*(x^0)\|.$$

By substituting this estimate to (6), we obtain

$$\|z^*(x) - z^*(x^0)\| \leq \frac{1}{1-q} \|T(z^*(x^0), x) - T(z^*(x^0), x^0)\|.$$

It follows from (5) that

$$T(z^*(x^0), x) = z^*(x^0) - J^{-1}(z^0, x^0)\, W(z^*(x^0), x).$$

Thus

$$\|z^*(x) - z^*(x^0)\| \leq \frac{1}{1-q} \|J^{-1}(z^0, x^0)\|\, \|W(z^0, x)\| \quad (9)$$

The continuity of $z^*(x)$ at x^0 follows from this. The continuity at the other points of $\mathfrak{S}(x^0, \varepsilon)$ is proved similarly.

From (9) we can obtain a useful estimate of the norm of an implicit function. We have

$$\|W(z^0, x)\| = \|W(z^0, x) - W(z^0, x^0)\| \leq \|J(z^0, \xi)\| \, \|x - x^0\|, \tag{10}$$

where

$$\xi = \Theta x + (1 - \Theta) x^0, \quad 0 \leq \Theta \leq 1.$$

If

$$\|J(z, x)\| \leq \alpha \tag{11}$$

in $\mathfrak{S}(z^0, \rho) \times \mathfrak{S}(x^0, \varepsilon)$ then it follows from (9–11) that

$$\|z^*(x) - z^*(x^0)\| \leq \frac{\alpha}{1 - q} \, \|J^{-1}(z^0, x^0)\| \, \|x - x^0\| \tag{12}$$

2. *Homogeneous forms and posynoms.* For the further analysis of the properties of implicit functions, we recall a number of definitions related to homogeneous forms and posynoms.

Let us consider a collection of real variables $u_1, \ldots, u_s$ and produce all possible linearly independent products of k multiplies.

Denote

$$u_j^{(k)} = u_1^{p_1} \cdots u_s^{p_s}, \tag{13}$$

where $p_1, \ldots, p_s$ are integers;

$$j \to (p_1, \ldots, p_s); \quad j \in \overline{1, m};$$

$$p_i \geq 0 \quad i \in \overline{1, s}; \quad \sum_{i=1}^{s} p_i = k; \quad m = \binom{s + k - 1}{k}. \tag{14}$$

We introduce the vector

$$u^{(k)} = \{u_1^{(k)}, \ldots, u_m^{(k)}\}$$

with lexicographically-ordered coordinates $u_j^{(k)}$ (13) and an arbitrary real $r \times m$ matrix A_k. $V_k(u)$ is an r-dimensional function.

$$V_k(u) = A_k u^{(k)}, \tag{15}$$

is called a *homogeneous k-form.*

It follows from (15) that

$$\frac{\partial}{\partial u^{(k)}} V_k(u) = V_k^{(1)}(u) = A_k. \tag{16}$$

A k-linear form is a special case of k-forms. If $k < s$ then all the numbers $p_1, \ldots, p_s$ in (13) can be 0 or 1, and their sum should be equal to k. In this case, we can assume that all coordinates of the vector $u^{(k)}$ are zeros, except for the coordinates $u_{i_1} \cdots u_{i_k}$ $(i_1, \ldots, i_k \in \overline{1,s})$.

A sum of finite number of k-forms (15) is called a *posynom*

$$P_n(u) = \sum_{k=1}^{n} V_k(u), \tag{17}$$

where n is a posynom order. It is easy to see that

$$\left.\frac{\partial^k}{\partial u^{(k)}} P_n(u)\right|_{u^{(k)}=0} = A_k = P_n^{(k)}(0). \tag{18}$$

Therefore (17) can be rewritten as

$$P_n(u) = \sum_{k=1}^{n} P_n^{(k)}(0) u^{(k)}. \tag{19}$$

Consider the function $f(x)$ mapping R^m into R^r. Recall that $f(x)$ is differentiable if the increment is such that

$$f(x+v) - f(x) = f'(x)v + \alpha(x,v), \tag{20}$$

where

$f'(x)$ is a bounded function and

$\alpha(x,v)$ is a function such that

$$\lim_{\|v\| \to 0} \frac{\|\alpha(x,v)\|}{\|v\|} \to 0.$$

The function $f'(x)$ is called a *strong derivative* (or the *Frechet derivative* or simply *derivative*), while

$$f'(x)\, v = d f(x,v) \tag{21}$$

is called a *strong differential* of $f(x)$.

In defining the increment of $f(x)$, we considered arbitrary increments v (in other words, we considered the function behaviour in moving from x in any directions v). Now let us fix a direction v and consider the function behaviour in this direction. The limit

$$\lim_{t \to 0} \frac{f(x+tv) - f(x)}{t} = \frac{d}{dt} f(x+tv) \Big|_{t=0} = Df(x,v), \tag{22}$$

is called a *weak differential* of $f(x)$. Here t is a scalar parameter characterizing the motion from x along v. It is seen from this expression that the weak differential can be nonlinear with respect to v (see (20)).

If the linearity holds, we have

$$Df(x,v) = f'_v(x)v, \tag{23}$$

where $f'_v(x)$ is a v-directional derivative of $f(x)$ (a weak derivative).

If $f(x)$ has the strong derivative, it has the weak derivative as well, and $f'(x) = f'_v(x)$. The strong differentiability follows from the weak differentiability only if $f'_v(x)$ is continuous.

Assume that there exists a posynom $P_n(v)$ such that

$$f(x+v) - f(x) = P_n(v) + \omega_n(x,v), \tag{24}$$

where

$$\lim_{\|v\| \to 0} \frac{\|\omega_n(x,v)\|}{\|v\|^n} = 0.$$

Or, if taken (17) and (15) into account,

$$f(x+v) - f(x) = \sum_{k=1}^{n-1} A_k v^{(k)} + A_n v^{(n)} + \omega_n(x,v). \tag{25}$$

We determine the n-th strong differential of $f(x)$ as

$$d^n f(x,v) = n! A_n v^{(n)}. \tag{26}$$

In particular, for $n = 1$ we obtain $d^{(1)} f(x,v) = A_1 v^{(1)}$, where $A_1 = f'(x)$. In general,

$$n! A_n = f^{(n)}(x) \tag{27}$$

is the *n-th strong derivative* of $f(x)$. By substituting this expression into (25), we obtain the Taylor expansion for $f(x)$, namely

$$f(x+v) = f(x) \ + \ f'(x)v^{(1)} + \frac{1}{2}f''(x)v^{(2)} + \ldots + \frac{1}{n!}f^{(n)}(x)v^{(n)} + \\ + \ \ldots + \omega_n(x,v). \tag{28}$$

If $f(x)$ can be represented as in (25), any its strong differential (up to the n-th) is equal to the corresponding weak one, and the same holds for the strong and the weak derivatives.

Really,

$$f(x+tv) = f(x) + \sum_{k=1}^{n-1} t^k A_k v^{(k)} + t^n A_n v^{(n)} + \omega_n(x,v).$$

Hence

$$D^n f(x,v) \ = \ \frac{d^n}{dt^n} f(x+tv)\bigg|_{t=0} = n! A_n v^{(n)} = \\ = \ f_v^{(n)}(x) v^{(n)} = d^n f(x,v), \tag{29}$$

where $f_v^{(n)}(x)$ is the n-th v-directional derivative and equal to $f^{(n)}(x)$ (26).

The formula (29) is useful in practical constructing the Taylor expansion (28).

Below we will need the posynoms of two variables $u = \{u_1, \ldots, u_s\}$ and $v = \{v_1, \ldots, v_p\}$. Like in (13), we introduce the vector $(u,v)^{(k+l)}$, with its coordinates being linearly independent products

$$(u,v)_j^{(k+l)} = u_1^{p_1} \cdots u_s^{p_s} \cdot v_1^{q_1} \cdots v_p^{q_p},$$

where $p_1, \ldots, p_s$ and $q_1, \ldots, q_p$ are integers;

$$j \to (p_1, \ldots, p_s, \ q_1, \ldots, q_p), \quad j \in \overline{1,m};$$

$$\sum_{i=1}^{s} p_i + \sum_{i=1}^{p} q_i = k + l, \quad p_i \geq 0, \quad q_i \geq 0$$

and

$$m = \binom{s+p+k+l-1}{k+l}.$$

Then the homogeneous $(k+l)$-form becomes

$$V_{k+l}(u,v) = B_{(k+l)} u^{(k)} v^{(l)}, \tag{30}$$

where $B_{(k+l)}$ is a $(r \times m)$ matrix.

The sum of homogeneous $(k+l)$-forms produces the posynom of order n, namely

$$P_n(u,v) = \sum_{k+l=1}^{n} V_{k+l}(u,v). \tag{31}$$

Let us assume that there exists a posynom $P_n(u,v)$ (31) for which the function $f(x,y)$ of two variables can be represented as

$$f(x+u,\ y+v) = f(x,y) + P_n(u,v) + \Omega_n(x,y,u,v), \tag{32}$$

where

$$\lim_{\substack{\|u\|\to 0\\ \|v\|\to 0}} \frac{\|\Omega_n(x,y,u,v)\|}{\|u\|^n + \|v\|^n} = 0. \tag{32'}$$

Then, by following (26, 27), we can define the n-th strong differential

$$d^n f(x,y,u,v) = n!\ B_n\ (u,v)^{(n)}$$

and the n-th strong derivative

$$f^{(n)}(x,y) = n! B_n. \tag{33}$$

The Taylor expansion for a two-variable function follows from (32) and (33), namely

$$\begin{aligned} f(x+u,\ y+v) &= f(x,y) + f^{(1)}(x,y)(u,v)^{(1)} + \dots \\ &\dots + \frac{1}{n!} f^{(n)}(x,y)\ (u,v)^{(n)} + \Omega_n(x,y,u,v). \end{aligned} \tag{34}$$

Now let us return to the equation (2) determining the implicit function $z^*(x)$.

3. *Differentiability.* **Theorem 3.** *Let the conditions of Theorem* 1 *be satisfied and let the function $W(z,x)$ in* (2) *have all the derivatives of order up to m in $\mathfrak{S}(z^0,\rho) \times \mathfrak{S}(x^0,\varepsilon)$, with* $\det J(z,x) \neq 0$ *in $\mathfrak{S}(z^0,\rho) \times \mathfrak{S}(x^0,\varepsilon)$. Then the implicit function $z^*(x)$ has the derivatives $z^{(k)}(x)$ up to order m in $\mathfrak{S}\, x^0,\varepsilon)$.*

The theorem is proved by induction. Let the theorem hold for $k < s < m$, i.e. $z^*(x)$ has the $(s-1)$-th derivative in $\mathfrak{S}(x^0, \varepsilon)$. Then, by virtue of (28), in $\mathfrak{S}(x^0, \varepsilon)$ we have

$$z^*(x) = z^*(x^0) + z_{s-1}(x) + \omega_{s-1}(x - x^0, x^0), \tag{35}$$

where

$$z_{s-1}(x) = z^{(1)}(x^0)(x - x^0)^{(1)} + \ldots + \frac{1}{(s-1)!} z^{(s-1)}(x^0)(x - x^0)^{(s-1)}. \tag{35'}$$

and

$$\lim_{x \to x^0} \frac{\|\omega_{s-1}(x - x^0, x^0)\|}{\|x - x^0\|^{s-1}} = 0 \tag{36}$$

Let us show that the theorem holds for $k = s$. Following the idea proposed in (Krasnoselsky, 1969), we consider the auxiliary equation

$$G(u, x) = 0, \tag{37}$$

where

$$\begin{aligned} G(u, x) \quad = \quad & W\Big[z^*(x^0) + z^{(1)}(x^0)(x - x^0)^{(1)} + \ldots \\ & \ldots + \frac{1}{(s-1)!} z^{(s-1)}(x^0)(x - x^0)^{(s-1)} + u, x \Big]. \end{aligned} \tag{37'}$$

The equation (37, 37′) has the solution

$$u(x) = z^*(x) - z_{s-1}(x) - z^*(x^0), \tag{38}$$

which is $o\left[\|x - x^0\|^{s-1}\right]$ according to (36), i.e.

$$\lim_{x \to x^0} \frac{\|u(x)\|}{\|x - x^0\|^{s-1}} = 0. \tag{39}$$

It is easily seen from (37′) that $G(u, x)$ at $(0, x^0)$ coincides with $W(z^0, x^0)$. By the theorem conditions, the latter has $s \leq m$ derivatives in the neighborhood $\mathfrak{S}(z^0, \rho) \times \mathfrak{S}(x^0, \varepsilon)$. Hence, according to (34), $\mathfrak{S}(u, x)$ can be represented as

$$G(u, x) = P_s(u, x - x^0) + \Omega(u, x - x^0),$$

where

$P_s(u, x - x^0)$ is a posynom of two vector variables (u, x);
$\Omega(u, x - x^0) = o\left[\| u \|^s + \| x - x^0 \|^s\right]$.

Since the posynom $P_s(u, x-x^0)$ and the function $\Omega(u, x-x^0)$ satisfy the Lipschitz condition in $\mathfrak{S}(0,\rho) \times \mathfrak{S}(x^0,\varepsilon)$, we obtain

$$\begin{aligned} \|G(u(x),x) - G(0,x)\| &= L_{\varepsilon,\rho}\|u(x)\| \; \|W[z^*(x^0) + z_{s-1}(x)+ \\ &\qquad +u(x), x] - W[z^*(x^0) + z_{s-1}(x), x]\|. \end{aligned}$$

But, according to (39), we have

$$u(x) = o(\|x - x^0\|^{s-1}).$$

By taking into account that $G(u(x), x) = 0$, where $u(x)$ is given by (38), we obtain

$$G(0,x) = o(\|x - x^0\|^{s-1}).$$

It follows immediately from this that

$$G'_x(0,x^0) = G''_x(0,x^0) = \ldots = G_x^{(s-1)}(0,x^0) = 0 \tag{40}$$

at x^0. Since, by the theorem condition, $W(z,x)$ is differentiable, its m-weak differential exists. Thus,

$$\begin{aligned} d^{(k)}G(0,x^0) &= G_x^{(k)}(0,x^0)h^{(k)} = \\ &= \frac{d^{(k)}}{dt^k} W[z^*(x^0) + z^{(1)}(x^0)th^{(1)} + \ldots \\ &\quad \ldots + \frac{1}{(s-1)!} z^{(s-1)}(x^0)t^{s-1}h^{(s-1)},\; x^0 + \\ &\quad +th^{(1)}]\big|_{t=0}, \quad k \le m. \end{aligned} \tag{40'}$$

Now the Taylor expansion for $G(u,x)$ in $\mathfrak{S}(0,\rho) \times \mathfrak{S}(x^0,\varepsilon)$ is

$$G(u,x) = \frac{1}{s!} G_x^{(s)}(0,x^0)(x-x^0)^{(s)} + G_u^{(1)}(0,x^0)u + \widetilde{\Omega}(u, x-x^0), \tag{41}$$

where

$$\widetilde{\Omega}(u, x-x^0) = o(\|u\| + \|x - x^0\|^s). \tag{42}$$

Now let us proceed to (38), which determines the solution of (37, 37′). According to (39), the function $u(x)$ has the derivatives

$$u'(x^0) = u''(x^0) = \ldots = u^{(s-1)}(x^0) = 0$$

at x^0. Therefore, according to (38), we have

$$u(x) = \frac{1}{s!} u^{(s)}(x^0)(x - x^0) + \widetilde{\omega}_s(x - x^0), \tag{43}$$

where

$$\widetilde{\omega}_s(x - x^0) = o(\|x - x^0\|^s). \tag{43'}$$

Now we substitute this expression into the equation $G(u, x) = 0$. The latter should be an identity, because $u(x)$ is its solution. By taking (41) into account, we obtain

$$\frac{1}{s!} G_x^{(s)}(0, x^0)(x - x^0)^{(s)} + G_u^{(1)}(0, x^0)u + \widetilde{\Omega}(u, x - x^0) = 0 \tag{44}$$

By substituting $u(x)$ from (43) into (44), we obtain

$$\frac{1}{s!} G_x^{(s)}(0, x^0)(x - x^0) + G_u^{(1)}(0, x^0)[u^{(s)}(x^0)(x - x^0)^{(s)} +$$

$$+ \widetilde{\omega}_s(x - x^0)] + \widetilde{\Omega}(u, x - x^0) = 0. \tag{44'}$$

For (44′) to hold identically, it it necessary and sufficient that

$$\frac{1}{s!} G_x^{(s)}(0, x^0) + G_u^{(1)}(0, x^0)\, u^{(s)}(x^0) = 0,$$

$$G_u^{(1)}(0, x^0)\, \widetilde{\omega}_s(x - x^0)^{(s)} + \widetilde{\Omega}(u, x - x^0) = 0$$

for any $x \in \mathfrak{S}(x^0, \varepsilon)$.

By taking (37′) into account, we obtain from the first equation

$$u^{(s)}(x^0) = -\frac{1}{s!} J^{-1}(z^0, x^0) G_x^{(s)}(0, x^0). \tag{45}$$

According to (38), $u^{(s)}(x^0) = z^{(s)}(x^0)$. Since J^{-1} exists, the s-th derivative of the implicit function $z(x)$ also exists at x^0. Thus, by assuming that $z(x)$ has $(s - 1)$ derivatives, we proved the existence of the s-th derivative. Hence, by induction, all the m derivatives of $z(x)$ exist at x^0 .

Now let us consider a point $x^1 \in \mathfrak{S}(x^0, \varepsilon)$ such that $z(x^1) \in \mathfrak{S}(z^0, \rho)$. The pair $z^1 = z(x^1)$ and x^1 solve the equation (2), i.e. $W(z^1, x^1) = 0$. By replacing x^1 and z^1 with x^0 and z^0, respectively, we obtain the existence of m derivatives of the implicit function $z(x)$ at $x^1 \in \mathfrak{S}(x^0, \varepsilon)$. Therefore the implicit function $z(x)$ given by (2) has m derivatives in the ε-neighborhood of the point x^0.

Thus (45) can be written as

$$z^{(s)}(x) = -\frac{1}{s!} J^{-1}(z, x) G_x^{(s)}(0, x),$$

where

$$x \in \mathfrak{S}(x^0, \varepsilon); \quad z \in \mathfrak{S}(z^0, \rho).$$

We can use this equality for the sequential computation of derivatives of $z(x)$ if, by taking into account $(40')$, we represent it as

$$d^{(s)}z(x) = z^{(s)}(x)h^{(s)} =$$

$$= -\frac{1}{s!}J^{-1}(z,x)\frac{d^s}{dt^s}W\left[z(x) + z^{(1)}(x)th^{(1)} + \dots\right.$$

$$\left.\left.\dots + \frac{1}{(s-1)!}z^{s-1}(x)t^{s-1}h^{s-1}, \ x + th^{(1)}\right]\right|_{t=0} \tag{46}$$

Let us write down the expressions for the first and second differentials

$$z^{(1)}(x)h^{(1)} = -J^{-1}(z,x)\ W_x^{(1)}(z,x)h^{(1)};$$

$$z^{(2)}(x)h^{(2)} = -\frac{1}{2}J^{-1}(z,x)[W_{zz}^{(2)}(z,x)z^{(1)}(x)z^{(1)}(x) + (W_{xz}^{(2)}(z,x) +$$

$$+W_{zx}^{(2)}(z,x))z^{(1)}(x) + W_{xx}^{(2)}(z,x)]\,h^{(1)}h^{(1)}. \tag{47}$$

Theorems 1–3 determine the local properties of an implicit function in a neighborhood of the solution z^0, x^0.

Global properties

For some classes of functions $W(z,x)$ from (2), we can find existence conditions and a smoothness order of an implicit function $z(x)$ for $x \in R^s$.

1. Existence. **Theorem 4.** *Let the function $W(z,x)$ from (2) be continuous with respect to its variables and let the following conditions hold for any fixed $x \in R^s$:*

(a) det $J(z,x) \not\equiv 0$ *as* $z \in R^n$;

(b) $\lim_{\|z\| \to \infty} W(z,x) = \infty$.

Then there exists the unique implicit function $z(x)$ defined on R^s.

Proof. The function $W(z,x)$ generates the vector field

$$\Phi_x(z) = W(z,x) \tag{48}$$

for any fixed $x \in R^s$. The field is continuous by the theorem conditions.

Let us introduce the vector field

$$\Pi_y(z) = \Phi_x(z) - y, \tag{49}$$

where $y \in R^s$ is a fixed vector.

It is clear that, according to (b), for the fixed y the vector field $\Pi_y(z)$ has no zeros on the spheres $||z|| = r$ if r are large enough.

Therefore the rotation of $\Pi_y(z)$ is defined on the spheres $||z|| = r$ with sufficiently large r (see Krasnoselsky, Zabreyko, 1975).

Let us consider two vectors fields produced in (49) by the fixed vectors y, namely

$$\Pi_{y_1}(z) = \Phi_x(z) - y_1; \quad \Pi_{y_2}(z) = \Phi_x(z) - y_2 \tag{50}$$

The vector fields are homotopic on spheres with sufficiently large radii, i.e. the field

$$K(z) = \lambda \Pi_{y_1}(z) + (1-\lambda)\Pi_{y_2}(z) = \Phi_x(z) - (\lambda y_1 + (1-\lambda)\, y_2)$$

has no zeros on such spheres for all $\lambda \in [0,1]$. Homotopic vector fields have the same rotations (Krasnoselsky, Zabreyko, 1975), namely

$$\gamma(\Pi_{y_1}) = \gamma(\Pi_{y_2}). \tag{51}$$

The vector fields $\Pi_{y_1}(z)$ and $\Pi_{y_2}(z)$ are nondegenerate on the spheres with large radii, but each of them can have a number of singular points in the ball $||z|| \leq r_1 < r$. Let $\kappa(y_1)$ and $\kappa(y_2)$ be the numbers of singular points for the fields Π_{y_1} and Π_{y_2}, respectively. Since the fields Π_{y_1} and Π_{y_2} are homotopic, we have

$$\kappa(y_1) = \kappa(y_2) = \kappa \tag{52}$$

Let the vector field $\Pi_y(z)$ have $\kappa(y)$ singular points in the ball $||z|| \leq r_1 \leq r$. These points are isolated by condition (a) of the theorem.

Recall (Krasnoselsky, Zabreyko 1975) that the index of the singular point z^0 is

$$\text{ind}(z^0) = (-1)^{\beta(z^0)}, \tag{53}$$

where $\beta(z^0)$ is the number of eigenvalues of matrix $\Pi'_y(z^0) = J(z^0, x)$ which have a negative real parts.

The definition shows that the index value, namely $+1$ or -1, depends on the evenness of $\beta(z^0)$, not on its absolute value.

The evenness of $\beta(z^0)$ turns out to be the same for all the singular points. This follows from condition (a) of the theorem. Really, since det $J(z,x) \not\equiv 0$, for any $x \in R^s$ the eigenvalues of $J(z,x)$ can pass from the left half-plane to the right one only by pairs, i.e. real eigenvalues are transformed into pairs of conjugates and the latter then intersect the imaginary axis.

By taking into account this fact and (52, 53), we obtain that the rotations of homotopic fields (50) are

$$\gamma(\Pi_y) = \kappa(-1)^\beta, \tag{54}$$

where β is the number of eigenvalues of matrix $\Pi'_y(z)$ which have a negative real part for some z.

Now let us show that the vector field $\Pi_y(z)$ has a unique singular point in the ball $\|z\| \leq r_1 < r$. Consider the equation

$$\Pi_y(z) = \Phi_x(z) - y = 0, \tag{55}$$

Let the equation have κ singular points, i.e. κ functions

$$z_1(y), \ldots, z_\kappa(y)$$

for any fixed y. Then (55) determines a multivalued function $z(y)$, with κ its branches being isolated (the latter is because the singular points are isolated). Each branch $z_i(y)$ determines an open subset $\mathfrak{Z}_i$ (by condition (b) of the theorem), with

$$\bigcup_{i=1}^{\kappa} z_i(y) = R^n.$$

This is possible only if $\kappa = 1$. Hence the rotation of $\Pi_y(z)$ equals $(-1)^\beta$ and, by the homotopy, the rotation of $\Pi_0(z) = \Phi_x(z)$ equals $(-1)^\beta$ too.

Hence for any $x \in R^s$ there exists the unique $z(x)$ for which the function $W(z,x)$ is zero.

Theorem 5. *Let the function $W(z,x)$ in (2) be continuous with respect to all variables and let the following conditions hold for any fixed $x \in R^s$:*

(a) det $J(z,x) \not\equiv 0$ for $z \in R^n$.

(b) $W(z,x) \geq c$ for all $z \in R^n$ (c is a fixed positive vector from R^n).

(c) The matrix $J(z,x)$ has semi-sign-preserving elements for all $x \in R^s$ and $z \in R^n$.

Then there exists the implicit function $z(x)$ defined on R^s.

Before proving the theorem, we note that the condition (c) should be understood as either $\dfrac{\partial W_i}{\partial z_j} \leq 0$ or $\dfrac{\partial W_i}{\partial z_j} \geq 0$ for all $i, j \in \overline{1,n}$ and $z \in R^n$, $x \in R^s$.

The theorem can be proved like Theorem 4. Only some stages are different, and we consider them in detail.

In contrast to (49), we define the field $\Pi_y(z)$ as

$$\Pi_y(z) = \Phi_x(z) - c\,(e^{-y} - 1),$$

where $y \in R^n$ is a fixed vector, e^{-y} is a vector with the coordinates e^{-y_i}, and ce^{-y} is a vector with the coordinates $c_i e^{-y_i}$. Note that

$$\Pi_0(z) = \Phi_x(z)$$

Consider the field $\Pi_y(z)$ on the sphere with the radius $r(\,\|z\| = r)$. From the definition of $\Pi_y(z)$ and condition (c) of Theorem 5, it is clear that for any fixed $y \in R^n$ we can choose a sufficiently large radius r such that the field $\Pi_y(z)$ has no singular points on the sphere $\|z\| = r$. Therefore the rotation of $\Pi_y(z)$ is defined on the spheres $\|z\| = r$ with sufficiently large r.

Consider the vector fields

$$\Pi_{y_1}(z) = \Phi_x(z) - c\,(e^{-y_1} - 1), \quad \Pi_{y_2}(z) = \Phi_x(z) - c(e^{-y_2} - 1)$$

and the vector field

$$K(z) = \lambda \Pi_{y_1}(z) + (1-\lambda)\,\Pi_{y_2}(z) = \Phi_x(z) - c\left(\lambda e^{-y_1} + (1-\lambda)\,e^{-y_2}\right),$$

where $0 \leq \lambda \leq 1$.

The field $K(z)$ has no zeros on the spheres with sufficiently large radii by condition (c) of Theorem 5. Therefore the vector fields Π_{y_1} and Π_{y_1} are homotopic and consequently have the same rotations.

The other stages of the proof coincide with those of Theorem 4.

2. *Differentiability.* **Theorem 6.** *Let the conditions of Theorems 3 or 5 be satisfied and let the function $W(z,x)$ have all derivatives up to order ρ with respect to both arguments at each point $(z(x), x)$, where $x \in R^s$ and $z \in R^n$. Then the implicit function $z(x)$ has the derivatives up to order ρ.* The derivatives can be computed by using the recurrent formula (47). The theorem proof follows from local Theorem 3 and from the existence of an implicit function on R^s.

3.3 Parametric properties of B-model with the complete consumption of resources

B – **model with the linear resource consumption.** Consider the B-model of stationary states in the following form

$$H_B(N) = -\sum_{n-1}^{m} N_n \ln \frac{N_n}{e\tilde{a}_n(x)} \Rightarrow \max, \tag{1}$$

$$\sum_{n=1}^{m} t_{kn}(x)\, N_n = q_k(x), \quad k \in \overline{1,r}; \quad r < m, \tag{2}$$

$$x \in \mathfrak{X} \subset R^s,$$

where

$$\tilde{a}_n(x) = a_n(x) G_n(x) \geq 0, \quad q_k(x) \geq \varepsilon_k > 0, \tag{3}$$

$$t_{kn}(x) \geq 0, \quad k \in \overline{1,r}; \quad n \in \overline{1,m}.$$

The matrix $T(x) = [t_{kn}(x)]$ has the full rank r for any $x \in \mathfrak{X} \subset R^s$.

Recall that in B-model considered, we assume its (internal) parameters to depend on the external parameters $x \in R$. The interpretation of the external parameters depends on the topic for which the formal construction is used. In particular, if the model (1) is implemented to simulate a stationary distribution of passengers in a transportation network, the parameters x may characterize the investments in transportation infrastructure development. It is clear that the investments influence the behaviour of passengers in using the transport ($a(x)$), the capacities of the transport modes ($G(x)$), and the costs ($t(x)$ and $q(x)$).

Recall that the set

$$\mathfrak{D}(x) = \left\{ N : \sum_{n=1}^{m} t_{kn}(x) N_n = q_k(x), \quad k \in \overline{1,r} \right\} \tag{4}$$

is feasible for the macrosystem considered, i.e. $\mathfrak{D} \cap \mathfrak{N} \neq \varnothing$, $\mathfrak{N} \in R_+^m$.

Let us proceed to the problem (1). In this case, the solution of the problem (1) is

$$N_n = \tilde{a}_n(x) \exp\left(-\sum_{j=1}^{r} \lambda_j\, t_{jn}(x) \right), \quad n \in \overline{1,m}; \tag{5}$$

$$\sum_{n=1}^{m} t_{kn}(x)\tilde{a}_n(x) \exp\left(-\sum_{j=1}^{r} \lambda_j t_{jn}(x)\right) - q_k(x) = 0, \quad k \in \overline{1,r}; \tag{6}$$

where $\lambda_1, \ldots, \lambda_r$ are the Lagrange multipliers.

It is seen from the system that subsystem (6) consists of the equations determining the Lagrange multipliers as functions of x, while subsystem (5) consists of the equations connecting the macrostate N and the Lagrange multipliers. Denote

$$\Phi_k(\lambda, x) = \sum_{n=1}^{m} t_{kn}(x)\tilde{a}_n(x) \exp\left(-\sum_{j=1}^{r} \lambda_j t_{jn}(x)\right) - q_k(x), \quad k \in \overline{1,r}. \tag{7}$$

Let us consider the matrix

$$\Phi_\lambda(\lambda, x) = [\Phi'_{kl}(\lambda, x)] =$$

$$= \left[-\sum_{n=1}^{m} t_{kn}(x)t_{ln}(x)\tilde{a}_n(x) \exp\left(-\sum_{j=1}^{r} \lambda_j t_{jn}(x)\right)\right], \tag{8}$$

which is important in investigating the properties of the implicit function $\lambda(x)$.

It is easy to see that $\Phi_\lambda(\lambda, x)$ is a symmetrical matrix for any (λ, x). We will show that it is negative-definite. For this purpose, we consider the quadratic form $G = \langle \Phi_\lambda z, z \rangle$ for $z \in R^r$, namely

$$G = -\sum_{k,s=1}^{r} \sum_{n=1}^{m} \beta_n(\lambda, x) t_{kn}(x) t_{sn}(x) z_k z_s, \tag{9}$$

where

$$\beta_n(\lambda, x) = \tilde{a}_n(x) \exp\left(-\sum_{j=1}^{r} \lambda_j t_{jn}(x)\right) \geq 0. \tag{9'}$$

By changing the order of summing up in (9), we obtain

$$G = -\sum_{n=1}^{m} \beta_n(x, \lambda) \left[\sum_{k=1}^{r} t_{kn}(x) z_k\right]^2 \leq 0. \tag{10}$$

The form G is zero on the vectors z satisfying the condition

$$\sum_{k=1}^{k} t_{kn}(x) z_k = \sum_{k=1}^{k} t'_{nk}(x) z_k = 0, \tag{11}$$

where t'_{nk} are elements of the transpose $T'(x) = [t_{kn}(x)]'$.

Since the matrix $T(x) = [t_{kn}(x)]$ has the full rank r for all $x \in \mathfrak{X}$, the equation in (11) has the unique solution $z_k \equiv 0$, $k \in \overline{1,r}$. Hence the quadratic form G is zero only on zero vectors, and consequently the matrix $\Phi_\lambda(\lambda, x)$ is negative-definite for all $\lambda \in R^r$ and $x \in \mathfrak{X}$.

In this case, the matrix $\Phi_\lambda(\lambda, x)$ has r real negative eigenvalues $\mu_k(\lambda, x)$. Indeed, by transforming the quadratic form G (10), we obtain

$$G = \sum_{k=1}^{r} \mu_k(\lambda, x) u_k^2, \tag{12}$$

where $u_1, \ldots, u_r$ are connected with $z_1, \ldots, z_r$ by the equation

$$z = Bu; \quad B^{-1}\Phi_\lambda(\lambda, x)B = \text{diag}[\mu_k(\lambda, x);\ k \in \overline{1,r}].$$

Since u and z are related by a nondegenerate linear transformation (rank of B equals r), we see that $G = 0$ for $u \equiv 0$. $G < 0$ for all $u \not\equiv 0$. It follows from (12) that $\mu_k(\lambda, x)$ are negative reals.

Consequently

$$\det \Phi_\lambda(\lambda, x) \not\equiv 0 \tag{13}$$

for $\lambda \in R^r$ and $x \in \mathfrak{X}$. Now we proceed to the system (6) and represent it as

$$\Phi_k(\lambda, x) = 0 \quad k \in \overline{1,r}. \tag{14}$$

Theorem 1. *Let the functions $t_{kn}(x)$, $\widetilde{a}_n(x)$ and $q_k(x)$ be continuous and let the matrix $T(x) = [t_{kn}(x)]$ have a full rank for $x \in \mathfrak{X}$. Then the system* (14, 6) *determines the unique continuous implicit function $\lambda(x)$ for $x \in \mathfrak{X}$.*

The proof is based on checking the conditions of Theorem 2.5. Condition (a) is satisfied because of (13). It follows from (8) that

$$\Phi'_{kl}(\lambda, x) \le 0, \quad k, l \in \overline{1,r}; \quad \lambda \in R^r; \quad x \in \mathfrak{X}.$$

Therefore condition (c) of Theorem 2.5 is also satisfied. It follows from (8) that the elements of the derivative matrix are zero only if $\sum_{j=1}^{r} \widetilde{\lambda}_j t_{jn}(x) \to \infty$ for all $x \in \mathfrak{X}$. $\Phi'_{kl}(\lambda, x) < 0$ at all the other points $\lambda \in R^r$. Thus the points $\widetilde{\lambda}$ minimize the functions $\Phi_k(\lambda, x)$ from (7), namely

$$\min_{\lambda} \Phi_k(\lambda, x) = -q_k(x) \quad \text{for} \quad x \in \mathfrak{X}.$$

Hence

$$\Phi_k(\lambda, x) \geq -q_k(x) \geq -\varepsilon_k$$

and condition (*b*) of Theorem 2.5 is also satisfied.

Theorem 2. *Let the functions $t_{kn}(x)$, $q_k(x)$ and $\widetilde{a}_n(x)$ have the derivatives up to order p in $\mathfrak{X}$. Then there exists the implicit function $\lambda(x)$ in $\mathfrak{X}$, it is determined by* (14) *and has the derivatives up to order p.*

The statement of the theorem follows from Theorems 2.6 and 2.3, since $\Phi_k(\lambda, x)$ (7) is infinitely differentiable with respect to λ.

Corollary 1. *If the functions $t_{kn}(x)$, $q_k(x)$ and $\widetilde{a}_n(x)$ are analytical in $\mathfrak{X}$ then the implicit function $\lambda(x)$* (14) *is also analytical in $\mathfrak{X}$.*

Now let us consider practical methods for approximate reconstruction of the implicit function $N(x)$ determined by system (5, 6).

Assume that the functions $t_{kn}(x)$, $\widetilde{a}_n(x)$ and $q_k(x)$, which characterize the resource consumption, the apriori occupation numbers for the subsets $\mathfrak{S}_n$ and the resource stores, respectively, are differentiable in $\mathfrak{X}$.

Then, according to Theorem 2, the relation between the realizable macrostate in B-model and the parameters x can be represented as

$$N(x) = N(x^0) + N_x(x^0)(x - x^0) + \Omega(x - x^0), \qquad (15)$$

where

$$N(x) = \{N_1(x), \ldots, N_m(x)\};$$

$$N_x(x^0) = [N_{ni}(x^0)] = \left[\frac{\partial N_n}{\partial x_i}\right]_{x^0} \quad \text{is a } (m \times s) \text{ matrix.} \qquad (15')$$

Here $\Omega(x - x^0) = o(\|x - x^0\|)$.

To determine the elements of N_x, we use the equalities (5). We obtain from (5)

$$\frac{\partial N_n}{\partial x_i} = \beta_n(x^0)\left[a_{ni}(x^0) - \widetilde{a}_n(x^0)\sum_{p=1}^{r}\left[\lambda_p(x^0)t_{pni}(x^0) + \right.\right.$$

$$\left.\left. + t_{pn}(x^0)\frac{\partial \lambda_p}{\partial x_i}\bigg|_{x^0}\right]\right]; \quad n \in \overline{1,m}; \quad i \in \overline{1,s}; \qquad (16)$$

where

$$\beta_n(x^0) = \exp\left(-\sum_{j=1}^{r} \lambda_j(x^0) t_{jn}(x^0)\right),$$

$$a_{ni}(x^0) = \left.\frac{\partial a_n}{\partial x_i}\right|_{x^0}; \quad t_{pni}(x^0) = \left.\frac{\partial t_{pn}}{\partial x_i}\right|_{x^0}. \tag{16'}$$

Let us introduce the following notations

$P(x^0)$ is a $(s \times m)$ matrix with the elements

$$P_{ni}(x^0) = \beta_n(x^0)\left[a_{ni}(x^0) - \tilde{a}_n(x^0) \sum_{p=1}^{r} \lambda_p(x^0) t_{pni}(x^0)\right]; \tag{17}$$

$\widehat{T}(x^0)$ is a $(r \times m)$ matrix with the elements

$$\widehat{t}_{pn}(x^0) = \beta_n(x^0)\, \tilde{a}_n(x^0)\, t_{pn}(x^0); \tag{18}$$

and $\Lambda_x(x^0)$ is a $(r \times s)$ matrix with the elements

$$\lambda_{pi}(x^0) = \left.\frac{\partial \lambda_p}{\partial x_i}\right|_{x^0}. \tag{19}$$

Then, by taking into account (16) and the notations, we have

$$N_x(x^0) = P(x^0) + \widehat{T}'(x^0)\Lambda_x(x^0). \tag{20}$$

The expression shows that in order to determine the matrix $N_x(x^0)$, we need the derivatives of the implicit functions $\lambda_1(x), \ldots, \lambda_r(x)$. These can be obtained from the equations (6).

By virtue of Theorem 2.3, $\lambda'(x)$ exist in $\mathfrak{X}$, and, according to (2.47), we have

$$\Lambda_x(x^0) = -\Phi_\lambda^{-1}(x^0)\, \Phi_x(x^0), \tag{21}$$

where

$$\Phi_\lambda^{-1}(x^0) = \Phi_\lambda^{-1}(\lambda(x^0), x^0);$$

$\Phi_\lambda(\lambda, x)$ is given by (8);

$\Phi_x(x)$ is a $(r \times s)$ matrix with the elements

$$\Phi_{ki} = \frac{\partial \Phi_k}{\partial x_i} = \sum_{n=1}^{m} \beta_n(x^0)\Big[\tilde{a}_n(x^0)\Big(t_{kni}(x^0) - t_{kn}(x^0) \sum_{j=1}^{r} (\lambda_j(x^0)\, t_{jni}(x^0) +$$

$$+ \lambda_{ji}(x^0)\, t_{jn}(x^0))\Big) + t_{kn}(x^0)\, a_{ni}(x^0)\Big] - q_{ki}(x^0); \tag{21'}$$

and $q_{ki}(x^0) = \left.\dfrac{\partial q_k}{\partial x_i}\right|_{x^0}$.

By substituting (21) into (20), we obtain

$$N(x) = N(x^0) + \left[P(x^0) - \widehat{T}(x^0)\Phi_\lambda^{-1}(x^0)\Phi_x(x_0)\right](x - x^0) +$$

$$+ \Omega(x - x^0). \tag{22}$$

We call the value

$$\delta = \lim_{x \to x^0} \frac{\|N(x) - N(x^0)\|}{\|x - x^0\|} \tag{23}$$

a *parametric sensitivity of B-model.* The sensitivity can be estimated by using (23). Taking into account (15′), we have

$$\delta \leq \|P(x^0)\| + \|\widehat{T}(x^0)\| \, \|\Phi_\lambda^{-1}(x^0)\| \, \|\Phi_x(x^0)\|, \tag{24}$$

where $\|\circ\|$ denotes the norms of corresponding matrices.

Consider the simplest version of model (1) in which the feasible set is described by one constraint ($r = 1$) and the parameters are $a_n(x) = a_n$ and $t_{kn}(x) = t_{kn}$. x is assumed to be a scalar. In this case, $P(x^0) = 0$,

$$\Phi_x(x^0) = \left. \frac{\partial q_1(x_1)}{\partial x_1} \right|_{x^0};$$

$$\Phi_\lambda(x^0) = \sum_{n=1}^{m} t_{1n}\widetilde{a}_n \exp(-\lambda_1(x^0)t_{1n}),$$

$$\widehat{t}_{1n}(x^0) = \widetilde{a}_n t_{1n} \exp(-\lambda_1(x^0)t_{1n}), \quad n \in \overline{1,m}.$$

Take the maximum element of the matrix as its norm. Then it follows from (24) that

$$\delta \leq \max \frac{\widetilde{a}_n t_{1n} \exp(-\lambda_1(x^0)t_{1n})}{\sum_{n=1}^{m} t_{1n}^2 \widetilde{a}_n \exp(-\lambda_1(x^0)t_{1n})} \max_x \left| \frac{\partial q_1}{\partial x_1} \right|. \tag{25}$$

It follows from (7) that the function $q_1(x) + \Phi_1(\lambda, x)$ monotonically decreases, is positive and $\Phi_1(-\infty, x) + q_1(x) = 0$. A solution of (5) is monotonically decreasing function of q_1, since $\dfrac{\partial \lambda_1}{\partial q_1} < 0$. Denote $q_1^{\max} = \sum_{n=1}^{m} t_{1n}\widetilde{a}_n$ and assume that $\varepsilon_1 \leq q_1(x) \leq q_1^{\max}$. Then the solution $\lambda_1(x^0)$ of (6) belongs to the interval $0 \leq \lambda_1(x^0) \leq \lambda_{1,\varepsilon}$, where $\lambda_{1,\varepsilon}$ is a solution of (6) with $q_1(x^0) = \varepsilon_1$.

In this case, the estimate

$$\delta \leq \frac{\max_n \tilde{a}_{n_0} t_{1n_0}}{\sum_{n=1}^{m} t_{1n}^2 \tilde{a}_n \exp\left(-\lambda_{1,\varepsilon} t_{1n}\right)} q_{11}, \tag{26}$$

holds, where

$$q_{11} = \max_x \left| \frac{\partial q_1}{\partial x_1}\bigg|_{x^0} \right|.$$

Thus, under constant resource consumption and apriori probabilities of occupation, the estimate for the parametric sensitivity of B-model is proportional to the maximum speed of change of the resource stores.

B-model with the nonlinear resource consumption. Consider the following reduced form of B-model.

$$H_B(N) = -\sum_{n=1}^{m} N_n \ln \frac{N_n}{e\tilde{a}_n(x)} \Rightarrow \max, \tag{27}$$

where

$$\mathfrak{D}(x) = \left\{N : \varphi_k(N, t_k(x)) = q_k(x), \quad k \in \overline{1,r}\right\}, \tag{28}$$

$$x \in \mathfrak{X} \subset R^s, \quad r < m;$$

$$\mathfrak{D}(x) \cap \mathfrak{N} \neq \varnothing; \quad \mathfrak{N} = R_+^m,$$

$$t_k(x) = [t_{k1}(x), \ldots, t_{km_k}(x)] \geq 0$$

and $q_k(x) \geq \varepsilon_k > 0$; $\tilde{a}_n(x) \geq 0$.

Recall that in this model, the function $\varphi_k(\bullet, \bullet)$ characterizes the consumption of the k-th resource in distributing the macrosystem elements among the states; and $t_k(x)$ is a vector of internal macrosystem parameters depending on the external (exogenous) parameters x.

The functions $\varphi_k(\bullet, \bullet)$ have some logical properties as characteristics of the consumption mechanism. For a macrostate $N \in \mathfrak{N}$, the consumption functions are

(a) nonnegative

$$\varphi_k(N, t_k(x)) \geq 0, \quad k \in \overline{1,r}; \tag{29}$$

and (b) monotone increasing

$$\frac{\partial \varphi_k(N, t_k(x))}{\partial N_n} \geq 0, \quad k \in \overline{1, r}; \quad n \in \overline{1, m} \tag{30}$$

for all $x \in \mathfrak{X}$

The nonnegativeness follows from that of the resource itself, while the monotonicity arises because the consumption grows proportionally to the occupation numbers.

We also assume the consumption functions to have some additional (formal) properties. The latter, on the one hand, do not restrict the applications, but, on the other hand, help to investigate the model (27, 28) constructively. Assume that the consumption functions $\varphi(N, t_k(x))$ are

(i) twice differentiable with respect to all variables;

(ii) strictly convex with respect to N, namely

$$\varphi_k(N + \Delta N, t_k(x)) > \varphi_k(N) + \langle \nabla_N \varphi_k(N, t_k(x)), \ \Delta N \rangle;$$

$$\Gamma_k = \left[\frac{\partial \varphi_k}{\partial N_n \partial N_p} \right] > 0; \tag{31}$$

(iii) regular, i.e. for all $x \in \mathfrak{X}$ there exists a subset $\mathfrak{M} \subset \mathfrak{D}(x)$ in which

$$\text{rank } F(N, t_k(x)) = r, \tag{32}$$

where

$$F(N, t_k(x)) = \left[\frac{\partial \varphi_k(N, t_k(x))}{\partial N_n}, \quad k \in \overline{1, r}; \quad n \in \overline{1, m} \right]. \tag{33}$$

Now let us consider all vectors $h \in R^m$ for which $Fh = 0$. Since we assume the matrix F to be of full rank, the set of these vectors, called the matrix kernel ($\ker F$), contain more than one element.

To formulate the optimality conditions, we need the Lagrange function. The latter is

$$L(N, \lambda, x) = H_B(N, x) + \sum_{k=1}^{r} \lambda_k (q_k(x) - \varphi_k(N, t_k(x))). \tag{34}$$

The Lagrange function also depends on the exogenous parameters x, and, by the assumptions on $\widetilde{a}_n(x)$, $q_k(x)$ and $t_k(x)$, is differentiable with respect to all variables.

Now we assume the vector of exogenous parameters x to be fixed.

The optimality conditions for (27, 28) are

$$\Psi_n(N,\lambda,x)=\frac{\partial L}{\partial N_n}=\ln\frac{\tilde{a}_n(x)}{N_n}-\sum_{k=1}^{r}\lambda_k\frac{\partial\varphi_k}{\partial N_n}=0,\quad n\in\overline{1,m}; \tag{35}$$

$$\Phi_k(N,\lambda,x)=\frac{\partial L}{\partial\lambda_k}=q_k(x)-\varphi_k(N,t_k(x))=0,\quad k\in\overline{1,r}. \tag{36}$$

These equations are the stationarity conditions for the Lagrange function $L(N,\lambda,x)$, i.e. if N^0 is a local extremum point in (27–28) then there exist the Lagrange multiplies $\lambda_1^0,\ldots,\lambda_r^0$ such that

$$\Psi_n(N^0,\lambda^0,x)=0;\qquad \Phi_k(N^0,\lambda^0,x)=0; \tag{37}$$

$$n\in\overline{1,m};\qquad\qquad k\in\overline{1,r}.$$

Note that the equations are only necessary extremum conditions, i.e. they say nothing about its type (maximum or minimum). Since the Lagrange function is differentiable with respect to all variables, the necessary second-order conditions can be used in order to define the local maximum conditions. The latter will be useful, because they involve the properties of Hessian of the Lagrange function at the local maximum point. In turn, the Hessian determines the existence conditions for the implicit function given by (35, 36).

We reproduce these conditions in formulation of (Alekseyev, Tikhomirov, Fomin, 1979).

Theorem 3. *Let N^0 be a local maximum point for problem* (27–28) *and let $N^0\in\mathfrak{M}$ be a regularity set for the matrix $F(N^0,x)$* (33). *Then there exist the Lagrange multipliers $\lambda_1^0,\ldots,\lambda_r^0$ for which the conditions* (37) *are satisfied and*

$$K(N^0,\lambda^0,x)=\langle L_{NN}(N^0,\lambda^0,x)\,h,h\rangle\le 0 \tag{38}$$

for all $h\in\ker F(N^0,x)$, where

$$L_{NN}(N,\lambda,x)=\left[\frac{\partial^2 L}{\partial N_p\,\partial N_n}\right]=\begin{cases}\dfrac{1}{N_n}-\displaystyle\sum_{k=1}^{r}\lambda_k\frac{\partial\varphi_k}{\partial N_n^2} & (n=p);\\[2ex] -\displaystyle\sum_{k=1}^{r}\lambda_k\frac{\partial\varphi_k}{\partial N_p\,\partial N_n} & (n\ne p).\end{cases} \tag{39}$$

According to the theorem, the Hessian of the Lagrange function is nonpositive-definite with respect to the basic variables at the local maximum point N^0 of problem (27–28), with this property being valid for all vectors belonging to the space tangential to the set of gradients of the functions $\varphi_k(N, t_k(x))$ at N^0. The conditions (37) and (38) are the criteria of the local maximum point of problem (27–28) for any fixed x.

Investigation of the parametric properties of B-model is reduced to the analysis of how the solution of (35, 36) depends on the exogenous parameters x. The Jacobian of (35, 36) is important in this analysis:

$$G(N,\lambda,x) = \begin{bmatrix} A(N,\lambda,x) & C(N,\lambda,x) \\ B(N,\lambda,x) & D(N,\lambda,x) \end{bmatrix}, \tag{40}$$

where

$$A(N,\lambda,x) = \left[\frac{\partial \Psi_n}{\partial N_p}\right], \qquad C(N,\lambda,x) = \left[\frac{\partial \Psi_n}{\partial \lambda_i}\right],$$

$$B(N,\lambda,x) = \left[\frac{\partial \Phi_k}{\partial N_p}\right], \qquad D(N,\lambda,x) = \left[\frac{\partial \Phi_k}{\partial \lambda_i}\right].$$

Comparing these expressions with (35, 36, 39), we see that

$$A(N,\lambda,x) = L_{NN}(N,\lambda,x) \text{ and is a } (m \times m) \text{ matrix;}$$

$$\begin{aligned} C(N,\lambda,x) = [c_{ni}] = \left[-\frac{\partial \varphi_i}{\partial N_n}\right] \quad & \text{and is a } (m \times r) \text{ matrix;} \\ & n \in \overline{1,m}; \quad i \in \overline{1,r}; \\ B(N,\lambda,x) = [b_{kp}] = \left[-\frac{\partial \varphi_k}{\partial N_p}\right] \quad & \text{and is a } (r \times m) \text{ matrix;} \\ & k \in \overline{1,r}; \quad p \in \overline{1,m}; \end{aligned} \tag{41}$$

$D(N,\lambda,x) = 0$ and is a $(r \times r)$ matrix.

It can be seen from this that $C = B'$ and the Jacobian G is a block matrix

$$G = \begin{bmatrix} A & B' \\ B & 0 \end{bmatrix}. \tag{42}$$

Matrices with this structure have a very useful property formulated by the following lemma.

Lemma 1. *Let A be a $(m \times m)$ matrix, let it be positive- (negative-) definite and let the $(r \times m)$ matrix B have the full rank r. Then G is a nondegenerate matrix* (*i.e.* det $G \neq 0$).

To prove the lemma, let us assume the opposite, i.e. let det $G = 0$. Then there exists an eigenvector $z = \begin{pmatrix} u \\ v \end{pmatrix} \neq 0$ which corresponds to the zero eigenvalue and is such that

$$Gz = 0. \tag{43}$$

Or, taking into account the structure of G (42), we obtain

$$Au + B'v = 0, \quad Bu = 0. \tag{44}$$

By multiplying the first equality by u' from the left, we obtain

$$u'Au + u'B'v = 0$$

or

$$u'Au + v'Bu = 0.$$

But, from the second equality in (44) we have $v'Bu = 0$ and $u'Au = 0$.

Since the matrix A is strictly sign-preserving, the corresponding quadratic form is zero only on zero vectors ($u \equiv 0$). Return to the first equality in (44). It implies that $Bv = 0$ for $u \equiv 0$. Thus $z \equiv 0$ and this finishes the proof.

Referring to (41), note that the matrix B coincides with F (33), which, by the assumption, has the full rank r in the set $\mathfrak{M}$.

Thus the non-degeneracy of G is linked with the sign-preserving property of $L_{NN}(N, \lambda, x)$. Note that the necessary 2-nd order optimality conditions (38) assures that the latter matrix is negative-definite.

Consider the quadratic form K from (38). According to (39), it is transformed as

$$K = -\sum_{n=1}^{m} \frac{1}{N_n} h_n^2 - \sum_{k=1}^{r} \lambda_k \left(\sum_{n,p=1}^{m} \frac{\partial^2 \varphi_k}{\partial N_n \, \partial N_p} h_p h_n \right). \tag{45}$$

The first term in the expression is negative and is zero only for $h \equiv 0$. The functions φ_k, which characterize the resource consumption in distributing the macrosystem elements among the states, were assumed to be strictly convex, i.e.

$$\Gamma_k(h) = \sum_{n,p=1}^{m} \frac{\partial^2 \varphi_k}{\partial N_n \, \partial N_p} h_n h_p > 0, \tag{46}$$

with equality being only for $h \equiv 0$.

Let us return to the system (35, 36) and rewrite it by using the special form of Boltzmann entropy in B-model.

From (35) we obtain

$$N_n - \tilde{a}_n(x) \exp\left(-\sum_{j=1}^{r} \lambda_j \beta_{jn}(N)\right) = 0, \quad n \in \overline{1,m};$$

where

$$\beta_{jn}(N) = \frac{\partial \varphi_j}{\partial N_n}.$$

Let us introduce the variables

$$z_j = e^{-\lambda_j} \geq 0$$

and rewrite the system (35, 36) in the form

$$\Psi_n(N, z, x) = N_n - \tilde{a}_n(x) \prod_{j=1}^{r} z_j^{\beta_{jn}(N)} = 0, \quad n \in \overline{1,m}; \tag{47}$$

$$\Phi_k(N, z, x) = q_k(x) - \varphi_k(N, t_k(x)) = 0, \quad k \in \overline{1,r}. \tag{48}$$

We show that under auxiliary conditions upon the consumption functions $\varphi_k(N, t_k(x))$, the system in (47, 48) has the solution belonging to the rectangle $0 \leq z_j \leq 1\,(j \in \overline{1,r})$.

Consider the equations (47) and investigate the behaviour of the functions Ψ_n in the rectangle $0 \leq z_j \leq 1$ for all $N_n \geq 0$, $n \in \overline{1,m}$. From (47) we obtain

$$\Psi_n(N; z_1, \ldots, z_{j-1},\ 0,\ z_{j+1}, \ldots, z_r;\ x) = N_n > 0, \quad j \in \overline{1,r}; \quad n \in \overline{1,m};$$

$$\Psi_n(N, 1; x) = N_n - \tilde{a}_n(x).$$

It follows from the first group of the equalities, that the solution z^* of (47, 48) cannot have zero components, since $N_n^* > 0$; $n \in \overline{1,m}$; ($z_k^* > 0$, $k \in \overline{1,r}$).

We again return to the functions $\Psi_n(N, z, x)$ (47) and consider a fixed vector $z \geq 0$ and the direction $p = \{p_k \geq 0,\ k \in \overline{1,r}\}$. Let us determine the derivative of Ψ_n at the point z in the direction p. We obtain

$$\Psi_n(N, z + \delta p, x) = N_n - \tilde{a}_n(x) \prod_{j=1}^{r} (z_j + \delta p_j)^{\beta_{jn}(N,x)}, \quad \delta > 0;$$

and

$$\left.\frac{d\Psi_n}{d\delta}\right|_{\delta=0} = -a_n(x)\sum_{\rho=1}^{r}\prod_{j=1}^{r}{}^{(\rho)}(z_j+\delta p_j)^{\beta_{jn}(N,x)}\times$$

$$\times\beta_{\rho n}(N,x)p_\rho(z_\rho+\delta p_\rho)^{\beta_{\rho n}(N,x)-1}<0. \qquad (48')$$

It is clear that the derivative sign does not depend on the choice of z or p. Hence Ψ_n decreases strictly monotonously.

Assume that the consumption functions satisfy the following conditions

$$\varphi_k(N_1,\dots,N_{j-1},\ \tilde{a}_n(x),\ N_{j+1},\dots,N_m;\ t_k(x)) - q_k(x) \ge 0, \qquad (49)$$

$$j\in\overline{1,m};\quad k\in\overline{1,r};$$

$$\varphi_k(0,t_k(x)) - q_k(x) < 0 \qquad (49')$$

The condition (49′) is obvious. It means that if all subsets of the states are empty, no resource is consumed.

On the contrary, (49) is rather nontrivial. First, note that the absolute maximum of the generalized information entropy of Boltzmann (see MSS (2.3.17)) is attained at the macrostate $N_n^0 = \tilde{a}_n(x)$. Therefore (49) means that, under the constraints (the constrained maximum), the resource consumption exceeds the store in realizing any macrostate such that at least one its coordinate is equal to that of the macrostate with the absolute maximum of entropy.

Together with the monotonicity (30), the conditions (49, 49′) assure that the solution to (47, 48) satisfies the conditions

$$0 < N_n^* \le \tilde{a}_n(x),\quad n\in\overline{1,m}.$$

Thus

$$\Psi_n(N^*,z_1,\dots,z_{j-1},0,\ z_{j+1},\dots,z_r,x) > 0 \quad j\in\overline{1,r};$$

$$\Psi_n(N^*,1,x)\le 0;$$

and Ψ_n is a strictly monotone-decreasing function. Hence the system (47, 48) has the solution

$$0 < z_j^* \le 1,\quad j\in\overline{1,r}.$$

Therefore the solution of (35, 36) satisfies the following inequalities

$$0\le\lambda_j^* < \infty,\quad j\in\overline{1,r}. \qquad (50)$$

Now let us proceed to the quadratic form K (45). By taking into account (46) and (50), we can conclude that K (45) is negative for $h \in R^m$ and is zero only for $h \equiv 0$.

Hence the matrix $L_{NN}(N, \lambda)$ is negative-definite and, by Lemma 1, the Jacobian $G(N, \lambda)$ of the system (35, 36) is nondegenerate.

Thus we prove the following theorem.

Theorem 4. *Let B-model have the following properties for the fixed* $x \in \mathfrak{X}$

(a) *the functions* $\varphi_1(N, t_1(x)), \ldots, \varphi_r(N, t_r(x))$ *are twice continuously differentiable.*

(b) $\dfrac{\partial \varphi_k}{\partial N_n} > 0;$

$$\Gamma_k = \left[\frac{\partial^2 \varphi_k}{\partial N_n \, \partial N_p}\right] \geq 0 \quad \text{is positive-definite}, \quad k \in \overline{1, r}; \quad n \in \overline{1, m};$$

(c) *there exists a set* $\mathfrak{M} \subset R^m$ *on which the matrix* $F = \left[\dfrac{\partial \varphi_k}{\partial N_n}\right]$ *is regular (has the full rank* r*);*

(d) $\varphi_k(0, t_k(x)) - q_k(x) < 0,$

$$\varphi_k(N_1, \ldots, N_{j-1}, \ \tilde{a}_j(x), \ N_{j+1}, \ldots, N_m, \ t_k(x)) - q_k(x) \geq 0.$$

Then the Hessian of the Lagrange function in problem (27, 28) *is negative defined, i.e.*

$$L_{NN}(N, \lambda, x) = \left[\frac{\partial^2 L}{\partial N_n \, \partial N_p}\right] < 0. \tag{51}$$

Let us consider the system (35, 36) and represent it as

$$W(y, x) = 0, \tag{52}$$

where $y = \{N, \lambda\}$ is a $(m + r)$ vector;

x is a vector of the exogenous parameters, $x \in \mathfrak{X} \subset R^s$;

$W(y, x) = \begin{bmatrix} \Psi(y, x) \\ \Phi(y, x) \end{bmatrix}$ is a $(m + r)$ vector function with the components

$$\Psi_i(y, x) = \ln \frac{\tilde{a}_i(x)}{y_i} - \sum_{k=1}^{r} y_{k+m} \frac{\partial \varphi_k(y_1, \ldots, y_m, t_k(x))}{\partial y_i}; \quad i \in \overline{1, m}; \tag{53}$$

$$\Phi_{i-m}(y, x) = q_{i-m}(x) - \varphi_{i-m}(y_1, \ldots, y_m, \ t_{i-m}(x));$$

$$i \in \overline{(m + 1), \ (m + r)}.$$

The parametric properties of the considered B-model are characterized by the implicit function $y^*(x)$ which is determined by (52). From (53) it is seen that $W(y,x)$ is differentiable.

Let us consider a point $x^0 \in \mathfrak{X}$ and its neighborhood $\mathfrak{S}(x^0, \rho)$, and let all the conditions of Theorem 4 be satisfied. Then there exists locally the implicit function $y^*(x)$, with its smoothness order determined by the minimum smoothness order of the functions $\tilde{a}_n(x)$, $q_k(x)$ and $t_k(x)$.

The global parametric characteristics of B-model significantly depend on the global properties of the functions $\varphi_1, \ldots, \varphi_k$.

Let

(a) the conditions of Theorem 4 be satisfied and let

(b) $\lim_{\|y\| \to \infty} \varphi_k(y_1, \ldots, y_n, t_k(x)) = +\infty$

for all $x \in \mathfrak{X}$. Then, as is easily checked, the function $W(y,x)$ in (52) will satisfy the conditions of Theorem 2.4, and thus assure the existence of the implicit function $y^*(x)$ for $x \in \mathfrak{X}$.

3.4 Parametric properties of F-models with the complete consumption of resources

F-models with the linear consumption of resources.

Consider the following F-model

$$H_F(N) = -\sum_{n=1}^{m} N_n \ln \frac{N_n}{\tilde{a}_n(x)} + (G_n(x) - N_n)\ln(G_n(x) - N_n) \Rightarrow \max$$

$$\mathfrak{D}(x) = \left\{ N : \sum_{n=1}^{m} t_{kn}(x) N_n = q_k(x), \quad k \in \overline{1,r} \right\}; \tag{1}$$

$$\mathfrak{D}(x) \cap \mathfrak{N} \neq \varnothing;$$

$$\mathfrak{N} = \{N : 0 \leq N_n \leq G_n, \quad n \in \overline{1,m}\};$$

$$\tilde{a}_n(x) \geq 0, \quad G_n(x) > 0, \quad t_{kn}(x) \geq 0, \quad q_k(x) \geq \varepsilon > 0;$$

$$n \in \overline{1,m}; \quad k \in \overline{1,r}; \quad r < m; \quad x \in \mathfrak{X} \subset R^s.$$

The parameters $\widetilde{a}_n(x)$, $G_n(x)$, $t_{kn}(x)$ and $q_k(x)$ of this model have the same meaning as in B-model considered in Section 3.

The solution of the problem (1) is

$$N = \frac{G_n(x)}{1 + b_n(x)\exp\left(\sum_{j=1}^{r} \lambda_j t_{jn}(x)\right)}; \quad n \in \overline{1,m}; \tag{2}$$

$$\sum_{n=1}^{m} \frac{t_{kn}(x) G_n(x)}{1 + b_n(x)\exp\left(\sum_{j=1}^{r} \lambda_j t_{jn}(x)\right)} - q_k(x) = 0, \quad k \in \overline{1,r}; \tag{3}$$

where $b_n(x) = 1/\widetilde{a}_n(x)$.

For any $x \in \mathfrak{X}$, this system of equations determines the pair N^*, λ^* which characterizes the macrostate with maximum entropy. In this system, the variables N are expressed via the Lagrange multipliers analytically (2), while the Lagrange multipliers λ are given by the equations (3).

We represent these equations as

$$A_k(\lambda, x) = 0; \quad k \in \overline{1,r}; \quad x \in \mathfrak{X}, \tag{4}$$

where

$$A_k(\lambda, x) = \sum_{n=1}^{m} \frac{t_{kn}(x) G_n(x)}{1 + b_n(x)\exp\left(\sum_{j=1}^{r} \lambda_j t_{jn}(x)\right)} - q_k(x). \tag{5}$$

The matrix $A_\lambda(\lambda, x)$ is important in investigating the existence, continuity and smoothness of the implicit function $\lambda^*(x)$ determined by (4, 5). The matrix has the elements

$$\frac{\partial A_k}{\partial \lambda_l} = -\sum_{n=1}^{m} \frac{t_{kn}(x) t_{ln}(x) G_n(x) \exp\left(\sum_{j=1}^{r} \lambda_j t_{jn}(x)\right)}{\left[1 + b_n(x)\exp\left(\sum_{j=1}^{r} \lambda_j t_{jn}(x)\right)\right]^2};$$

$$l, k \in \overline{1,r}. \tag{6}$$

Theorem 1. *Let the matrix* $T(x) = [t_{kn}(x);\ k \in \overline{1,r};\ n \in \overline{1,m}]$ *have a full rank for all* $x \in \mathfrak{X}$. *Then*

$$\det\ A_\lambda(\lambda, x) \neq 0 \tag{7}$$

for all $x \in \mathfrak{X} \subset R^s$ *and* $\lambda \in R^r$.

To prove the theorem, we consider the quadratic form (since $A_\lambda(\lambda, x)$ is a symmetrical matrix) $G = \langle A_\lambda z, z\rangle$, $z \in R^r$, such that

$$G = -\sum_{k,l=1}^{r} \sum_{n=1}^{m} t_{kn}(x)\, t_{ln}(x) \beta_n(\lambda, x) z_k z_l, \tag{8}$$

where

$$\beta_n(\lambda, x) - \frac{G_n(x) \exp\left(\sum_{j=1}^{r} \lambda_j t_{jn}(x)\right)}{\left[1 + b_n(x) \exp\left(\sum_{j=1}^{r} \lambda_j t_{jn}(x)\right)\right]^2} \geq 0. \tag{8'}$$

Changing the summing order in (8), we obtain

$$G = -\sum_{n=1}^{m} \beta_n(\lambda, x) \left(\sum_{k=1}^{r} t_{kn}(x) z_k\right)^2 \leq 0. \tag{9}$$

Since the matrix $T(x)$ has the full rank r for all $x \in \mathfrak{X}$, equality is obtained in (9) only for $z \equiv 0$. Hence the matrix $A_\lambda(\lambda, x)$ is negative-definite. It is easy to show that the matrix $A_\lambda(\lambda, x)$ has real, negative and non-zero eigenvalues for all $x \in \mathfrak{X}$ and $\lambda \in R^r$. A theorem follows from this.

Theorem 2. *Let the functions* $\tilde{a}_n(x)$, $t_{kn}(x)$, $G_n(x)$ *and* $q_k(x)$ *be continuous and let the matrix* $T(x)$ *have the full rank* r *for all* $x \in \mathfrak{X}$. *Then the equations* (3) *determine the unique continuous implicit function* $\lambda(x)$ *for* $x \in \mathfrak{X}$.

The proof is based on checking the conditions of Theorem 2.2.

Note that under the conditions of Theorem 2, the smoothness order (maximum order of function $\lambda^*(x)$ derivative) will be determined by the minimum smoothness order of the functions $\tilde{a}_n(x)$, $G_n(x)$, $t_{kn}(x)$ and $q_k(x)$.

Having these functions the first derivatives, the way in which the realizable macrostate $N^*(x)$ depends on the exogenous parameters x can be described as in (3.15). According to (2), the elements of matrix $N_x(x^0)$ are represented by the expressions

$$\frac{\partial N_n}{\partial x_i} = \left. \frac{g_{ni}(x)c_n(x) - G_n(x)\beta_n(x)\left[b_{ni}(x) + b_n(x)\sum_{p=1}^{r}(\lambda_p t_{pni}(x) + t_{pn}(x)\lambda_{pi}(x))\right]}{c_n^2(x)} \right|_{x^0}, \tag{10}$$

where

$$c_n(x) = 1 + b_n(x)\beta_n(x),$$

$$\beta_n(x) = \exp\left(\sum_{j=1}^{r} \lambda_j t_{jn}(x)\right); \quad \lambda_{pi}(x) = \frac{\partial \lambda_p(x)}{\partial x_i};$$

$$g_{ni}(x) = \frac{\partial G_n(x)}{\partial x_i}; \quad b_{ni}(x) = \frac{\partial b_n(x)}{\partial x_i}; \quad t_{pni}(x) = \frac{\partial t_{pn}(x)}{\partial x_i}. \tag{10'}$$

Under such a notation, the matrices $P(x^0)$ and $\widehat{T}(x^0)$ from (3.20) are like

$$P_{ni}(x^0) = \widetilde{g}_{ni}(x^0) - a_n(x^0)\left[b_{ni}(x^0) + b_n(x^0)\sum_{p=1}^{r}\lambda_p(x^0)t_{pni}(x^0)\right],$$

$$\widehat{t}_{pn}(x^0) = \frac{G_n(x^0)\beta_n(x^0)b_n(x^0)}{c_n^2(x^0)} t_{pn}(x^0), \tag{11}$$

where

$$\widetilde{g}_{ni}(x) = \frac{g_{ni}(x^0)}{c_n^2(x^0)}; \quad a_n(x) = \frac{G_n(x^0)\beta_n(x^0)}{c_n^2(x^0)}; \tag{11'}$$

To determine the matrix $\Lambda_x(x^0)$ from (3.20), let us use the equations (3), namely

$$\Lambda_x(x^0) = -A_\lambda^{-1}(x^0)A_x(x^0), \tag{12}$$

where

$$A_\lambda(x^0) = A_\lambda(\lambda(x^0),\ x^0) \quad \text{is determined by (6);}$$

$A_x(x^0)$ is a matrix with the elements

$$A_{ki}(x^0) = \frac{\partial A_k}{\partial x_i} = \sum_{n=1}^{m}\left[t_{kni}^0(x^0)N_n(x^0) + t_{kn}(x^0)N_{ni}(x^0)\right] - q_{ki}(x^0); \tag{12'}$$

$$N_{ni}(x^0) = \left.\frac{\partial N_n}{\partial x_i}\right|_{x^0}; \quad q_{ki}(x^0) = \left.\frac{\partial q_k}{\partial x_i}\right|_{x^0}.$$

Thus if the relation between the endogenous and exogenous parameters is differentiable in F-model then the estimate (3.24) can be used to characterize the parametric sensitivity. The matrices involved in (3.24) are determined in this case by (11–12).

***F*-model with the nonlinear consumption of resources.** Consider F-model

$$H_F(N, x) = -\sum_{n=1}^{m} N_n \ln \frac{N_n}{\tilde{a}_n(x)} + (G_n(x) - N_n)\ln(G_n(x) - N_n) \Rightarrow \max,$$

$$\mathfrak{D}(x) = \left\{N : \varphi_k(N, t_k(x)) = q_k(x), \quad k \in \overline{1,r}\right\}, \tag{13}$$

where

$$\mathfrak{D}(x) \cap \mathfrak{N} \neq \varnothing; \quad \mathfrak{N} = \left\{N : 0 \leq N_n \leq G_n, \; n \in \overline{1,m}\right\},$$

$$G_n(x) \geq 0, \quad \tilde{a}_n(x) \geq 0, \quad q_k(x) \geq \varepsilon > 0,$$

$$t_k(x) = \{t_{k1}(x), \ldots, t_{km_k}(x)\} \geq 0; \quad r < m.$$

We assume that the functions $\tilde{a}_n(x)$, $q_k(x)$, $t_k(x)$ and $\varphi_k(N, t_k(x))$ ($n \in \overline{1,m}$; $k \in \overline{1,r}$) have all the properties described in Section 3. The Lagrange function for this problem is

$$L(N, \lambda, x) = H_F(N, x) + \sum_{k=1}^{r} \lambda_k (q_k(x) - \varphi_k(N, t_k(x))), \tag{14}$$

and for any fixed $x \in \mathfrak{X}$ its stationarity conditions are determined by the solvability of the system of equations

$$B_n(N, \lambda) = \frac{\partial L}{\partial N_n} = -\ln \frac{N_n}{\tilde{a}_n(x)(G_n(x) - N_n)} -$$

$$-\sum_{k=1}^{r} \lambda_k \frac{\partial \varphi_k}{\partial N_n} = 0; \quad n \in \overline{1,m}; \tag{15}$$

$$A_k(N, \lambda, x) = \frac{\partial L}{\partial \lambda_k} = q_k(x) - \varphi_k(N, t_k(x)) = 0, \quad k \in \overline{1,r}.$$

Let us assume that the conditions of regularity of matrix $F = \left[\frac{\partial \varphi_k}{\partial N_n}\right]$ for the problem (13) are satisfied. Therefore the stationary point of the Lagrange function $L(N, \lambda, x)$ is the point of its local maximum, i.e. the conditions (3.38) are satisfied, where

$$L_{NN}(N,\lambda,x) = \begin{cases} -\dfrac{1}{N_n} - \dfrac{1}{G_n(x) - N_n} - \displaystyle\sum_{k=1}^{r} \lambda_k \frac{\partial^2 \varphi_k}{\partial N_n^2} & (n = p); \\ -\displaystyle\sum_{k=1}^{r} \lambda_k \frac{\partial^2 \varphi_k}{\partial N_n \, \partial N_n} & (n \neq p). \end{cases} \tag{16}$$

The equations (15) are similar to those in (3.35) and (3.36). Therefore the nondegeneracy of the Jacobian of (15) is, by Lemma 3.1, determined by the non-degeneracy of the matrix $L_{NN}(N,\lambda,x)$ (16).

Let us consider a special case of F-model (13) in which the functions $\varphi_k(N, t_k(x))$ are separable with respect to the variables $N_1, \ldots, N_m$; for example, they are as

$$\varphi_k(N, t_k(x)) = \sum_{l=1}^{m} \varphi_{lk}(N_l, t_k(x)). \tag{17}$$

This structure of consumption functions is quite logical. It arises, if the consumption of the k-th resource is a linear function of the consumptions of resources in occupying each subset of the close states by the macrosystem elements. The consumption function for each subset depends only on the number of elements occupying this subset. In this case,

$$B_n(N_n,\lambda) = -\ln \frac{N_n}{\tilde{a}_n(x)(G_n(x) - N_n)} - \sum_{k=1}^{r} \lambda_k \frac{\partial \varphi_{nk}}{\partial N_n};$$

$$n \in \overline{1,m}; \tag{18}$$

and the matrix in (16) is diagonal

$$L_{NN}(N,\lambda,x) = \operatorname{diag} L_{NN}^{(n)}(N,\lambda,x) =$$

$$= \left[-\frac{1}{N_n} - \frac{1}{G_n(x) - N_n} - \sum_{k=1}^{r} \lambda_k \frac{\partial^2 \varphi_{nk}}{\partial N_n^2} \right]. \tag{19}$$

We introduce the following notations.

$D(N)$ is a m-vector with the components

$$d_n(N_n, x) = -\ln \frac{N_n}{\tilde{a}_n(x)(G_n(x) - N_n)}. \tag{20}$$

$\widehat{\Phi}(N, x)$ is a $(m \times r)$ matrix with the elements

$$\widehat{\varphi}_{nk}(N_n, x) = \frac{\partial \varphi_{nk}}{\partial N_n}. \tag{21}$$

$C(N, x)$ is a m-vector with the components

$$c_n(N_n, x) = -\frac{1}{N_n} - \frac{1}{G_n(x) - N_n}. \tag{22}$$

$\widehat{\Psi}(N, x)$ is a $(m \times r)$ matrix with the elements

$$\widehat{\psi}_{nk}(N_n, x) = \frac{\partial^2 \varphi_{nk}}{\partial N_n^2}. \tag{23}$$

$M(N, x)$ is a $[2m \times (r+1)]$ matrix of the form

$$M(N, x) = \begin{bmatrix} \widehat{\Phi}\,(N, x) & D\,(N, x) \\ \widehat{\Psi}\,(N, x) & C\,(N, x) \end{bmatrix}. \tag{24}$$

Theorem 3. *Let all the $(r+1)$-order minors of the matrix $M(N, x)$ be nonzero at the local minimum point N^0 of problem (13), i.e. let*

$$\det{}_{(r+1)} M(N^0, x) \neq 0. \tag{25}$$

Then the matrix $L_{NN}(N^0, \lambda^0, x)$ (19) is nonsingular.

To prove the theorem, we consider

$$\det\ L_{NN}(N, \lambda, x) = \prod_{n=1}^{m} \beta_n(N, \lambda, x)\,(-1)^n, \tag{26}$$

where

$$\beta_n(N, \lambda, x) = -c_n(N_n, x) - \langle \widehat{\Psi}_n(N_n, x), \lambda \rangle,$$

λ is a vector with the components $\lambda_1, \ldots, \lambda_r$, and
$\widehat{\Psi}_n(N_n, x)$ is a vector row with the components $\widehat{\psi}_{nk}(N_n, x)$ (23).

The determinant (26) is zero if at least one of the multipliers is zero, i.e.

$$c_{n_i}(N_{n_i}, x) - \langle \widehat{\Psi}_{n_i}(N_{n_i}, x),\ \lambda \rangle = 0.$$

According to the necessary optimality conditions, at the local minimum point there exist the Lagrange multipliers λ^0 such that

$$\widehat{\Phi}(N^0, x)\lambda^0 = D\,(N^0, x).$$

Let us consider the extended system

$$\widehat{\Phi}(N^0, x)\lambda^0 = D(N^0, x),$$
$$\widehat{\Psi}_{n_i}(N^0_{n_i}, x)\lambda^0 = c_{n_i}(N^0, x). \tag{27}$$

The extended matrix of this system

$$\begin{bmatrix} \widehat{\Phi}(N^0, x) & D(N^0, x) \\ \widehat{\Psi}_{n_i}(N^0, x) & c_{n_i}(N^0, x) \end{bmatrix}$$

has the dimension $(m+1) \times (r+1)$. By (25), all its $(r+1) \times (r+1)$ – minors are nonzero. Hence λ^0 does not solve the system, i.e. none of the multipliers $\beta_n(N, \lambda)$ is zero in (26).

This allows us to use the global theorems on implicit functions (Theorems 2.4 and 2.5) in analyzing the parametric properties of F-model. The theorems are used like for B-model (Section 3).

3.5 Parametric properties of E-models with the complete consumption of resources

E-models with the linear consumption of resources.

Consider the following E-model

$$H_E(N, x) = -\sum_{n=1}^{m} N_n \ln \frac{N_n}{a_n(x)} - (G_n(x) + N_n) \ln (G_n(x) + N_n) \Rightarrow \max,$$

$$\mathfrak{D}(x) = \left\{ N : \sum_{n=1}^{m} t_{kn}(x) N_n = q_k(x), \quad k \in \overline{1, r} \right\}, \tag{1}$$

$$\mathfrak{D}(x) \cap \mathfrak{N} \neq \varnothing, \quad \mathfrak{N} \in R^m_+;$$

$$a_n(x) \geq 0, \quad G_n(x) > 0 \quad t_{kn}(x) \geq 0;$$

$$q_k(x) \geq \varepsilon > 0, \quad k \in \overline{1, r}; \quad n \in \overline{1, m}; \quad r < m; \quad x \in \mathfrak{X} \subset R^s.$$

The solution of the problem (1) is

$$N_n = \frac{G_n(x)}{b_n(x) \exp\left(\sum_{j=1}^{r} \lambda_j t_{jn}(x) \right) - 1}, \quad n \in \overline{1, m}; \tag{2}$$

$$\Omega_k(\lambda, x) = \sum_{n=1}^{m} \frac{t_{kn}(x)\, G_n(x)}{b_n(x) \exp\left(\sum_{j=1}^{r} \lambda_j t_{jn}(x)\right) - 1} - q_k(x) = 0,$$
$$k \in \overline{1,r}; \tag{3}$$

where $b_n(x) = 1/a_n(x)$.

Note that, unlike B- and F-models, in which the similar optimality conditions cause immediately the nonnegativeness of primal variables, equations (2) mean that $N_n^* \geq 0$ if

$$\exp\left(\sum_{j=1}^{r} \lambda_j t_{jn}(x)\right) > a_n(x) \tag{4}$$

for all $x \in \mathfrak{X}$.

The solution of (1) with respect to the dual variables turns out to have such a property under auxiliary assumptions on the feasible set of the problem.

Theorem 1. *Let the following conditions be satisfied for a fixed $x \in \mathfrak{X}$.*

(a) *The (r, m)-matrix $T(x) = [t_{kn}(x)]$ has the full rank r in* (1).

(b) *The set $\mathfrak{D}(x)$ from* (1) *contains more than one point.*

Then the system in (3) *has the unique solution*

$$\lambda^* \in \operatorname{int} \widetilde{\Lambda}, \tag{5}$$

where

$$\widetilde{\Lambda} = \left\{ \lambda : \sum_{j=1}^{r} \lambda_j t_{jn}(x) \geq \ln a_n(x), \quad n \in \overline{1,m} \right\}. \tag{5'}$$

Proof. We start with investigating the uniqueness of the solution of (3). The Lagrange function for (1) is

$$L(N, \lambda, x) = H_E(N, x) + \sum_{k=1}^{r} \lambda_k \left(q_k(x) - \sum_{n=1}^{m} t_{kn}(x) N_n \right). \tag{6}$$

Consider the dual function for the fixed x

$$\widetilde{L}(\lambda) = L(N^*(\lambda), \lambda, x) =$$
$$= H_E(N^*(\lambda), x) + \sum_{k=1}^{r} \lambda_k \left(q_k(x) - \sum_{k=1}^{r} t_{kn}(x) N_n^*(\lambda) \right), \tag{7}$$

where $N^*(\lambda)$ is determined by (2)

It follows from (6) and (2) that

$$\frac{\partial \widetilde{L}}{\partial \lambda_k} = -\Omega_k(\lambda, x) = -\sum_{n=1}^{m} \frac{a_n(x) G_n(x) t_{kn}(x)}{\exp\left(\sum_{j=1}^{r} \lambda_j t_{jn}(x)\right) - a_n(x)} \tag{8}$$

$$\frac{\partial^2 \widetilde{L}}{\partial \lambda_i \, \partial \lambda_k} = -\frac{\partial \Omega_k}{\partial \lambda_i} = \sum_{n=1}^{m} t_{kn}(x) \, t_{in}(x) \, \omega_n(\lambda), \tag{9}$$

$$k, i \in \overline{1, r};$$

where

$$\omega_n(\lambda) = \frac{a_n(x) G_n(x) \exp\left(\sum_{j=1}^{r} \lambda_j t_{jn}(x)\right)}{\left[\exp\left(\sum_{j=1}^{r} \lambda_j t_{jn}(x)\right) - a_n(x)\right]^2}. \tag{10}$$

Let us consider the quadratic form

$$G = \sum_{k,i=1}^{r} \sum_{n=1}^{m} t_{kn}(x) \, t_{in}(x) \, \omega_n(\lambda) \, z_i z_k,$$

where $z \in R^r$. Since T has the full rank, we obtain

$$G = \sum_{n=1}^{r} \omega_n(\lambda) \left(\sum_{k=1}^{r} t_{kn} z_k\right)^2 \geq 0, \tag{11}$$

with equality only for $z \equiv 0$. It follows from this that $\widetilde{L}(\lambda)$ is a strictly convex function.

Let λ^* be a solution of (3). Note that

$$\lambda^* \notin \Lambda = \left\{\lambda : \sum_{k=1}^{r} \lambda_k t_{kn}(x) - \ln a_n(x) = 0, \quad n \in \overline{1, m}\right\}. \tag{12}$$

Let us assume the contrary. Then it follows immediately from (2) that $N_n = \infty$, $n \in 1, m$. But this is contrary to the constraints in (1).

Let us consider the behaviour of the dual function $\widetilde{L}(\lambda, x)$ (7) near the point λ^*. From $\Omega_k(\lambda^*, x) = 0$ we obtain

$$\widetilde{L}(\lambda, x) = \widetilde{L}(\lambda^*, x) + \frac{1}{2}(\lambda - \lambda^*)' \nabla^2 \widetilde{L}(\xi, x)(\lambda - \lambda^*),$$

where

$$\xi = (1-\Theta)\,\lambda + \Theta\,\lambda^*, \quad 0 < \Theta < 1,$$

$$\nabla^2 \widetilde{L} = \left[\frac{\partial^2 \widetilde{L}}{\partial \lambda_i \, \partial \lambda_j} \right].$$

According to (11), we obtain the estimate

$$\widetilde{L}\,(\lambda, x) \geq \widetilde{L}\,(\lambda^*, x).$$

The function $\widetilde{L}\,(\lambda, x)$ is continuous on the set $\widetilde{\Lambda} = R^r \setminus \Lambda$; therefore if $\lambda^* \neq \pm\infty$, then $\widetilde{L}\,(\lambda^*, x) > -\infty$, i.e. $\widetilde{L}\,(\lambda, x)$ is bounded from below.

Thus $\widetilde{L}\,(\lambda, x)$ has the unique minimum point λ^*, and the system (3), which determines the gradient $\widetilde{L}\,(\lambda)$ being zero, has the unique solution λ^*.

Now we prove that the solutions of (3) belong to the set (5). Let us consider the behaviour of $\widetilde{L}\,(\lambda, x)$ on the set $\widetilde{\Lambda}$ from (5′). We choose two points, namely z and u, such that

$$\sum_{j=1}^{r} z_j t_{j\gamma}(x) = \ln a_\gamma(x);$$

$$\sum_{j=1}^{r} z_j t_{jn}(x) > \ln a_n(x); \quad n \in \overline{1,m}; \quad n \neq \gamma; \tag{13}$$

$$\sum_{j=1}^{r} u_j t_{jn}(x) > \ln a_n(x); \quad n \in \overline{1,m}.$$

Let us join them with a segment and consider the point

$$y\,(\beta) = (1-\beta)\,z + \beta u, \quad 0 \leq \beta \leq 1.$$

Then

$$\widetilde{L}\,(y(\beta)) = \widetilde{L}\,(\beta) = H\,(N^*(y(\beta))) + \sum_{k=1}^{r} y_k(\beta) \left(q_k(x) - \sum_{n=1}^{m} t_{kn}(x) N_n^*(y) \right).$$

By virtue of (13), we have

$$\left. \frac{\partial \widetilde{L}}{\partial \beta} \right|_{\beta=0} = -\infty.$$

Therefore for any point from $\widetilde{\Lambda}$ there exists a neighborhood in the interior of Λ (5′) such that the values of $\widetilde{L}$ in it are less than at the boundary point. Hence the unique solution λ^* of (3) satisfies (5, 5′).

Note that since the matrix $T(x)$ has the full rank for the fixed $x \in \mathfrak{X}$, the Jacobian of (3) is positive-definite (cf. (9) and (11)).

Theorem 2. *Let the functions $a_n(x)$, $t_{kn}(x)$, $q_k(x)$ and $G_n(x)$ be continuous and let the matrix $T(x)$ have the full rank r. Then the system* (3) *determines the unique continuous implicit function $\lambda^*(x)$ for $x \in \mathfrak{X} \subset R^s$.*
The *proof* follows from checking the conditions of Theorem 2.5.

If the functions $a_n(x)$, $t_{kn}(x)$ and $G_n(x)$ are smooth then $\lambda^*(x)$ is also smooth and its smoothness order is determined by the minimum smoothness order of these functions.

If only the first derivatives of these functions exist the first-order approximation for the implicit function $N^*(x)$ can be constructed by using (3.15).

Here the elements of $N_x(x^0)$ are determined like for the F-model (4.10–4.12′), except for the expression for $c_n(x)$ in (4.10). The latter in this case is

$$c_n(x) = b_n(x)\,\beta_n(x) - 1, \quad n \in \overline{1,m}. \tag{14}$$

***E*-models with non-linear consumption of resources.**

Consider the E-model

$$\begin{gathered} H_E(H) = -\sum_{n=1}^{m} N_n \ln \frac{N_n}{a_n(x)} - (G_n(x) + N_n) \ln (G_n(x) + N_n)\,, \\ \mathfrak{D}(x) = \{N : \ \varphi_k(N, t_k(x)) = q_k(x)\}\,, \quad k \in \overline{1,r}; \end{gathered} \tag{15}$$

where

$$\mathfrak{D}(x) \cap \mathfrak{N} \neq \varnothing; \quad \mathfrak{N} \subset R_+^m$$

$$G_n(x) \geq 0, \quad a_n(x) \geq 0, \quad q_k(x) \geq \varepsilon > 0,$$

$$t_k(x) = \{t_{k1}(x), \ldots, t_{km_k}(x)\} \geq 0$$

We assume that the functions $a_n(x)$, $q_k(x)$, $G_n(x)$, $t_k(x)$ and $\varphi_k(N, t(x))$ have the properties mentioned in Section 3. The Lagrange function for this problem is

$$L(N, \lambda, x) = H_E(N, x) + \sum_{k=1}^{r} \lambda_k \left(q_k(x) - \varphi_k(N, t_k(x))\right). \tag{16}$$

The stationarity conditions for each fixed $x \in \mathfrak{X}$ give the following system of equations

$$\begin{aligned} S_n(N, \lambda, x) &= \frac{\partial L}{\partial N_n} = \\ &= -\ln \frac{N_n}{a_n(x)(G_n(x) - N_n)} - \sum_{k=1}^{r} \lambda_k \frac{\partial \varphi_k}{\partial N_n} = 0, \ n \in \overline{1, m}; \\ Q_k(N, \lambda, x) &= \frac{\partial L}{\partial \lambda_k} = \\ &= q_k(x) - \varphi_k(N, t_k(x)) = 0, \quad k \in \overline{1, r}. \end{aligned} \tag{17}$$

For the considered E-model, the regularity conditions for $F = \dfrac{\partial \varphi_k}{\partial N_n}$ are satisfied by the assumptions; therefore the stationary point of the Lagrange function with respect to the variables N is the point of its maximum, i.e. there exist the Lagrange multipliers λ^* for which the Hessian of L (16) is nonpositive-definite. The Hessian elements are

$$L_{NN}(N, \lambda, x) = \begin{cases} -\dfrac{1}{N_n} + \dfrac{1}{G_n(x) + N_n} - \displaystyle\sum_{k=1}^{r} \lambda_k \frac{\partial^2 \varphi_k}{\partial N_n^2}, & (n = p); \\ -\displaystyle\sum_{k=1}^{r} \lambda_k \frac{\partial^2 \varphi_k}{\partial N_p \partial N_n}, & (n \neq p; \quad n, p \in \overline{1, m}). \end{cases} \tag{18}$$

In studying the parametric properties of E-models, the modifications with negative-definite Hessian (18) are the most interesting. In such cases, according to Lemma 3.1, the system (17) will have non-degenerate Hessian (this is one of the conditions for the implicit function $N^*(x)$ to exist).

Consider the E-model (15) with the resource consumption functions from (4.17). In this case,

$$S_n(N_n, \lambda, x) = -\ln \frac{N_n}{a_n(x)(G_n(x) - N_n)} - \sum_{k=1}^{r} \lambda_k \frac{\partial \varphi_{nk}(N_n)}{\partial N_n},$$
$$n \in \overline{1, m};$$

and the matrix L_{NN} (18) has a diagonal structure

$$L_{NN}(N, \lambda, x) = \left[-\frac{1}{N_n} + \frac{1}{(G_n(x) + N_n)} - \sum_{k=1}^{r} \lambda_k \frac{\partial^2 \varphi_{nk}}{\partial N_n^2} \right]. \tag{19}$$

In order to formulate the conditions for the matrix (19) to be negative-definite, it is convenient to use the notation similar to (4.20–4.24), namely

$$
\begin{aligned}
d_n(N_n, x) &= -\ln \frac{N_n}{a_n(x)(G_n(x) + N_n))}, \\
c_n(N_n, x) &= -\frac{1}{N_n} + \frac{1}{G_n(x) + N_n}, \quad n \in \overline{1, m}.
\end{aligned} \tag{20}
$$

The other notation ($\widehat{\Phi}(N)$, $\widehat{\Psi}(N)$ and $M(N, x)$) is the same (4.21), (4.23) and (4.24).

Under such a notation, Theorem 4.3 holds for the E-model considered.

3.6 Parameterization and qualitative analysis of MSS with the incomplete resources consumotion

Models discussed in this section are applied to the cases where the resources in the system are consumed incompletely. Recall that in the model, this is taken into account by describing the feasible set with a system of inequalities.

The stationary macrostate determined by this model depends on the same parameters as for the MSS with complete consumption of the resources considered in Sections 3–5. Recall that the parameters are divided into two groups. One of them includes the parameters characterizing the random way in which the macrosystem elements are distributed among the states or the subsets of close states ($a_1, \dots, a_m$ are the apriori probabilities for elements to occupy the subsets $\mathfrak{S}_1, \dots, \mathfrak{S}_m$ and $G_1, \dots, G_m$ are the capacities of the subsets). The second group includes the parameters characterizing the consumptions and stores of r resources, namely $t_{11}, \dots, t_{1m_1}; \dots; t_{r1}, \dots, t_{rm_r}$ are the consumption parameters; and $q_1, \dots, q_r$ are the resource stores.

These parameters are endogenous for a macrosystem. Earlier we assumed that the parameters are changed because of the external perturbation of a macrosystem (the exogenous parameters x), namely

$$
a = a(x), \quad G = G(x), \quad t_k = t_k(x), \quad q_k = q_k(x), \quad k \in \overline{1, r}. \tag{1}
$$

Taken into account these comments, the MSS with incomplete consumption of resources can be represented as

$$H(N, a(x),\ G(x)) \Rightarrow \max,$$
$$\varphi_k(N, t_k(x)) \leq q_k(x), \quad k \in \overline{1,r}; \tag{2}$$
$$x \in \mathfrak{X} \subset R^s,$$

where

H is the macrosystem entropy determined by the type of statistics characterizing its elements;

φ_k is a consumption function for the k-th resource and

$\mathfrak{X}$ is a feasible set for the exogenous parameters.

For any fixed realization of the exogenous parameters $x \in \mathfrak{X}$, the MSS (2) determines the stationary macrostate N^*. Enumerating all the realizations of exogenous parameters $x \in \mathfrak{X}$, we obtain the way in which the stationary macrostate depends on x, i.e. we obtain the function $N^*(x)$. Investigation of MSS (2) parametric properties is reduced to studying the properties of this function.

MSS can be rewritten as

$$H(N, x) \Rightarrow \max,$$
$$\mathfrak{D}(x) = \{N : \varphi_k(N, x) \leq q_k(x), \quad k \in \overline{1,r}\} \subset \mathfrak{N}, \tag{3}$$
$$x \in \mathfrak{X} \subset R^s.$$

where

$$\mathfrak{N} = \begin{cases} R_+^m & (B\text{- and } E\text{–MSS}); \\ \{N : 0 \leq N_n \leq G_n, \quad n \in \overline{1,m}\} & (F\text{-MSS}); \end{cases} \tag{3'}$$

$$\mathfrak{D}(x) \neq \varnothing.$$

In such a form, it is more convenient to study the model as a mathematical object; if any special properties of H or φ_k as functions of x are necessary, they will be formulated taking into account the properties of the functions in (1), which characterize the relations between the endo- and exogenous parameters.

Problem (3) belongs to the class of perturbed mathematical programming problems. But it has specific features which generates some "good" properties of its solution.

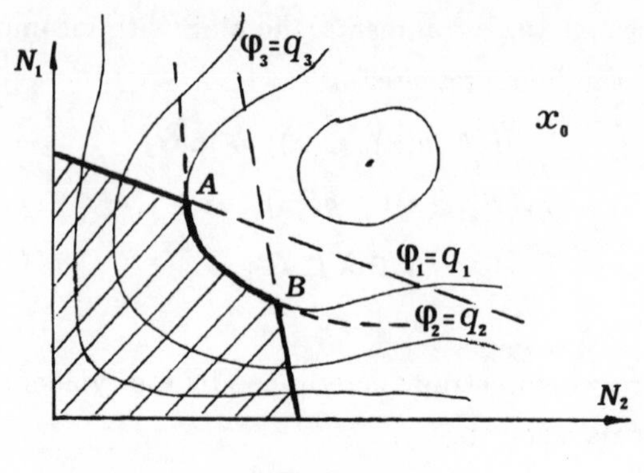

FIGURE 3.1.

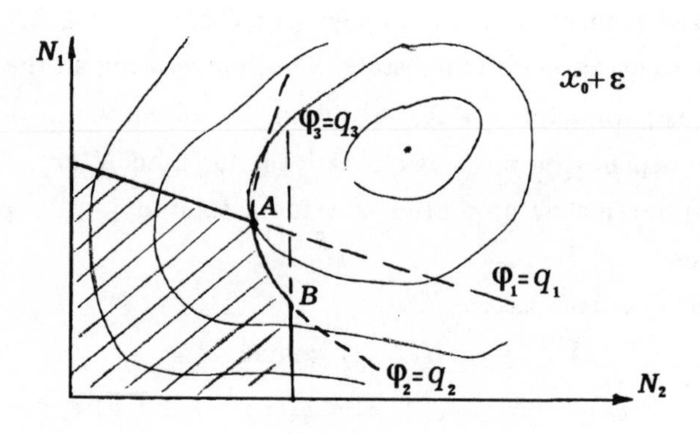

FIGURE 3.2.

Recall that the objective functions in (3), i.e. the entropy $H(N, x)$, is a strictly concave function for any $x \in \mathfrak{X}$. The functions φ_k, in general, can be arbitrary continuous functions. In what follows we consider φ_k as convex or monotone convex differentiable functions, which often occurs in applied problems.

The mathematical analysis of problem (3) is not simple, therefore it seems useful to begin with a qualitative investigation of the behaviour of its solution under the parameter changes.

Let $\varphi_1, \ldots, \varphi_r$ be continuous functions. Then the feasible set in (3) has an arbitrary structure (it is non-convex). Assume that for some parameter x_0 we observe the situation shown in fig. 3.1. Thin lines show the level lines of

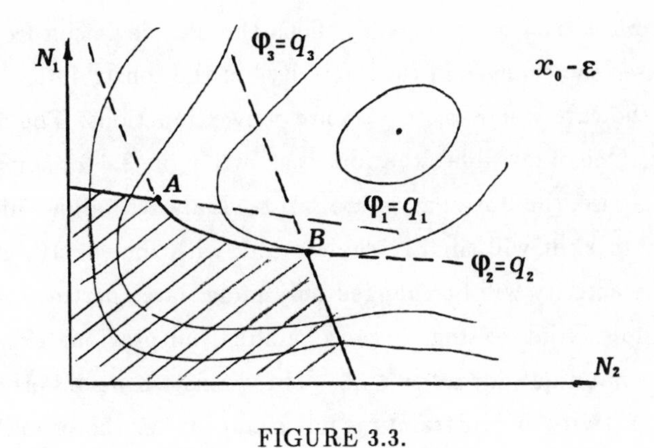

FIGURE 3.3.

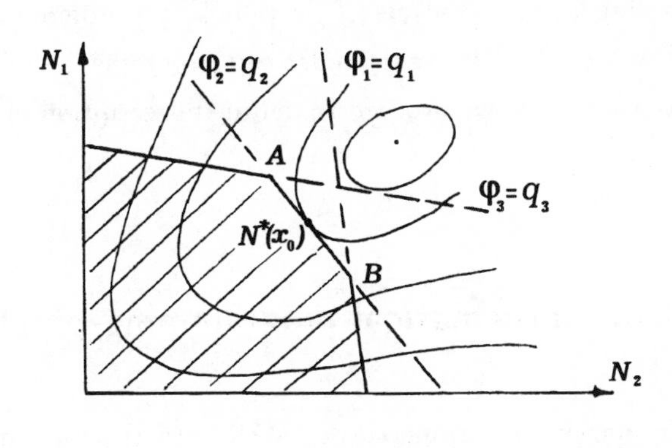

FIGURE 3.4.

$H(N, x_0)$. The feasible set is dashed. The picture shows that all the points on the segment AB of the boundary of the feasible set solve (3). Hence the point x_0 from the set $\mathfrak{X}$ is corresponded by a bounded set $N^*(x_0)$ of solutions of (3). Any small deviations from x_0 lead to the situations shown in fig. 3.2 and 3.3. In both cases the set $N^*(x_0)$ is "contracted" into one point, namely either into A for $x_0 + \varepsilon$ or into B for $x_0 - \varepsilon$.

For this example, consider the behaviour of the inequalities $\varphi_k(N, x) \leq q_k(x)$. If the solution of the problem falls in A, both the first and second inequality are active, i.e. are equalities. This situation occurs if the exogenous parameters are equal to x_0 or $x_0 + \varepsilon$. If the solution falls on the line AB, except for its endpoints, the second constraint is active. Here $x = x_0$. At last, if the solution falls in B, the second and third constraints are active.

This is possible if $x = x_0$ or $x_0 - \varepsilon$. Thus the $\pm\varepsilon$ -deviation from a fixed point x_0 causes the changes in the "activity" of the constraints.

Consider the case where $\varphi_1, \ldots, \varphi_r$ are convex functions. The admissible set is convex. One of possible situations is shown in fig. 3.4 for some parameter x_0. In this case, the stationary state $N^*(x_0)$ is always unique; under small deviations from x_0 it will slightly deviate too, with the constraints remain "active". The activity will be changed only under large perturbations.

This example is interesting, because, unlike the previous example, the solution here is unique for all $x \in \mathfrak{X}$. There exists a neighborhood of x_0 in which the activity of constraints is not changed, i.e. the problem having three inequalities is transformed into a problem having one equality (namely the 2nd). According to the analysis of Section 2, the differentiability of $N^*(x)$ should be expected in this case, under certain properties of $\varphi_k(N, x)$.

After qualitative analysis, we proceed to detailed description of the properties of $N^*(x)$.

3.7 Perturbed mathematical programming problems

Analysis of the parametric properties of MSS with the incomplete consumption is based on certain general results characterizing the properties of perturbed mathematical programming problems.

Consider the problem

$$\begin{gathered} f(x, \varepsilon) \Rightarrow \max; \quad g(x, \varepsilon) \leq 0; \\ x \in \mathfrak{X} \subset R^n; \quad \varepsilon \in \mathfrak{E} \subset R^s; \quad g: \mathfrak{X} \times \mathfrak{E} \to R^r; \\ \mathfrak{X} \text{ and } \mathfrak{E} \text{ are closed sets.} \end{gathered} \tag{1}$$

Here ε is a parameter (generally speaking, a vector parameter) from the set $\mathfrak{E}$.

We introduce some notation and concepts which are necessary in formulating the main results.

Problem (1) for $\varepsilon = \varepsilon_0$ is said to be unperturbed and its solution is denoted by $x^*(\varepsilon_0) = x_0^*$. This solution is not necessary unique, i.e. there can be a set $\widetilde{\mathfrak{X}}(\varepsilon_0)$ of solutions of the unperturbed problem.

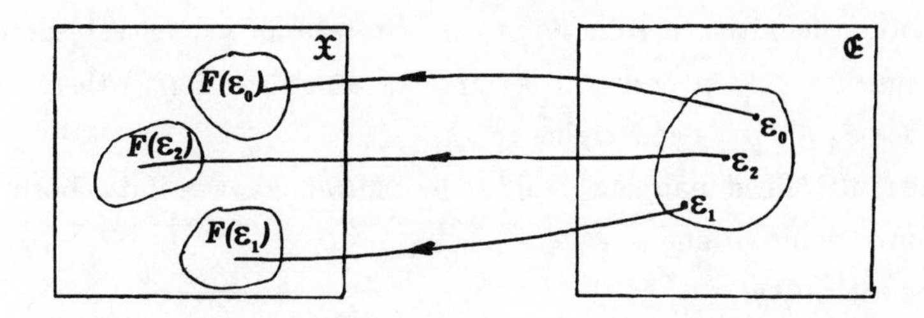

FIGURE 3.5.

Let us consider the subset $\mathfrak{V} \subseteq \mathfrak{E}$ containing ε_0. For each $\varepsilon \in \mathfrak{V}$ we introduce the following notation

$$\widetilde{\mathfrak{X}}(\varepsilon) = \{x^* : x^* = \arg\max[f(x,\varepsilon) \mid g(x,\varepsilon) \leq 0, \quad x \in \mathfrak{X}]\} \tag{2}$$

is a set of optimal solutions of the perturbed problem;

$$\mathfrak{G}(\varepsilon) = \{x : g(x,\varepsilon) \leq 0, \quad x \in \mathfrak{X}\} \tag{3}$$

is a set of feasible solutions of the perturbed problem;

$$\mathfrak{J}(\varepsilon) = \max[f(x,\varepsilon) \mid g(x,\varepsilon) \leq 0, \quad x \in \mathfrak{X}] \tag{4}$$

is a perturbation function.

The equalities (2) and (3) define the mappings $\mathfrak{V} \to \mathfrak{X}$ such that a point $\varepsilon \in \mathfrak{V}$ is transformed into the sets $\mathfrak{X}(\varepsilon)$ or $\mathfrak{G}(\varepsilon)$ from $\mathfrak{X}$. The mappings are called *multivalued* (fig. 3.5).

The multivalued mapping $F(\varepsilon)$ is said to be *upper-semicontinuous at the point* ε_0 if for any neighborhood $\mathfrak{O}_x(F(\varepsilon_0))$ of the set $F(\varepsilon_0)$ there exists a neighborhood $\mathfrak{O}_\varepsilon$ of ε such that

$$F(\varepsilon) \subset \mathfrak{O}_x(F(\varepsilon_0)) \tag{5}$$

for all $\varepsilon \in \mathfrak{O}_\varepsilon$.

The multivalued mapping $F(\varepsilon)$ is said to be *lower-semicontinuous at the point* ε_0 if for any $x_0 \in F(\varepsilon_0)$ and any neighborhood $\mathfrak{O}_x(x_0)$ there exists a neighborhood $\mathfrak{O}_\varepsilon$ of ε such that

$$F(\varepsilon) \cap \mathfrak{O}_x(x_0) \neq \varnothing \tag{6}$$

for all $\varepsilon \in \mathfrak{O}_\varepsilon$.

Another definition is sometimes used (in term of sequences), namely for any sequence ε_n converging to ε_0 and for any $x_0 \in F(\varepsilon_0)$ there exists a sequence $x_n \in F(\varepsilon_n)$ converging to x_0.

The multivalued mapping is said to be *continuous* at ε_0 if it is both upper- and lower-semicontinuous at ε_0.

Semicontinuity.

Theorem 1. *Let*

(1) *$g(x,\varepsilon)$ and $f(x,\varepsilon)$ in* (1) *be continuous with respect to all variables;*

(2) *there exist a neighborhood $\mathfrak{O}_\varepsilon$ of ε_0 such that $\bigcup_{\varepsilon \in \mathfrak{O}_\varepsilon} \widetilde{\mathfrak{X}}(\varepsilon)$ is a bounded set, where $\widetilde{\mathfrak{X}}(\varepsilon)$ is given by* (2);

(3) *there exists a sequence $y_s \to x^*$ such that $g(y_s, \varepsilon_0) < 0$ for some $x^* \in \widetilde{\mathfrak{X}}(\varepsilon_0)$.*

Then the mapping $\widetilde{\mathfrak{X}}(\varepsilon)$ is upper-semicontinuous.

We will prove the theorem by contradiction. Assume that it is wrong. It means that there is a neighborhood $\mathfrak{O}_x(\widetilde{\mathfrak{X}}(\varepsilon_0))$ of $\widetilde{\mathfrak{X}}(\varepsilon_0)$ in $\mathfrak{X}$ such that

$$\widetilde{\mathfrak{X}}(\varepsilon) \notin \mathfrak{O}_x(\widetilde{\mathfrak{X}}(\varepsilon_0)), \quad \varepsilon \in \mathfrak{O}_\varepsilon \tag{7}$$

for any neighborhood $\mathfrak{O}_\varepsilon$ of ε_0. ($\mathfrak{R}(\mathfrak{Y}(y))$ means $\mathfrak{R}(y)$ for all $y \in \mathfrak{Y}$, where $\mathfrak{R}$ and $\mathfrak{Y}$ are sets.)

Let us consider a sequence ε_k which converges to ε_0. By using (2), we construct the sequence $x_k \in \widetilde{\mathfrak{X}}(\varepsilon_k)$. By (7), $x_k \notin \mathfrak{O}_x$.

By condition (2) of the theorem, the sequence x_k is bounded. Therefore a converging subsequence may be chosen from it. Denote its limit point by $x_* \notin \mathfrak{O}_x$.

It follows from the continuity of functions $g(x,\varepsilon)$ that

$$g(x_*, \varepsilon_0) = \lim_{k\to\infty} g(x_k, \varepsilon_k) \leq 0,$$

Hence x_* is an admissible point, but it is not optimal by virtue of (7), since it is the limit point of the sequence out of $\mathfrak{O}_x$. Such a situation is possible only if

$$f(x_*, \varepsilon_0) \leq \mathfrak{J}(\varepsilon_0). \tag{8}$$

But it follows from the continuity of f that

$$f(x_*, \varepsilon_0) = \lim_{k\to\infty} f(x_k, \varepsilon_k) = \lim_{k\to\infty} \mathfrak{J}(\varepsilon_k).$$

By substituting this result into (8), we obtain

$$\lim_{k\to\infty} \mathfrak{J}(\varepsilon_k) \le \mathfrak{J}(\varepsilon_0). \tag{9}$$

Let us return to the condition (3) of the theorem and consider the sequence $y_s \to x^* \in \widetilde{\mathfrak{X}}(\varepsilon_0)$. Then, as a consequence of the continuity of g, we have

$$g(y_s, \varepsilon_k) \le 0 \tag{10}$$

for any sequences $\varepsilon_k \to \varepsilon_0$ and y_s for sufficiently large $k(k \ge K(s))$. Hence y_s is feasible for (1); therefore

$$f(y_s, \varepsilon_k) \le \mathfrak{J}(\varepsilon_k) \quad \text{as} \quad k \ge K(s). \tag{11}$$

Since f is continuous, by passing to the limit with respect to k and then with respect to s in (11), we obtain

$$\lim_{k\to\infty} f(y_s, \varepsilon_k) \le \lim_{k\to\infty} \mathfrak{J}(\varepsilon_k), \tag{12}$$

$$\lim_{s\to\infty} f(y_s, \varepsilon_0) = f(x_*, \varepsilon_0) = \mathfrak{J}(\varepsilon_0) \le \lim_{k\to\infty} \mathfrak{J}(\varepsilon_k),$$

which contradicts (9).

In many cases it is difficult to check condition (2) of the theorem. Therefore the sufficient conditions for boundedness of $\bigcup_{\varepsilon\in\mathfrak{O}_\varepsilon} \widetilde{\mathfrak{X}}(\varepsilon) = \mathfrak{X}(\mathfrak{O}_\varepsilon)$ can be useful.

Let us consider a scalar function $\mu(x, \varepsilon)$ defined on $\mathfrak{X} \times \mathfrak{E}$ and determine

$$\mathfrak{M}(\varepsilon, c) = \{x : \ \mu(x, \varepsilon) \le c, \quad x \in \mathfrak{X}\}. \tag{13}$$

Lemma 1. *Let*

(1) *the function $\mu(x, \varepsilon)$ be x-uniformly continuous with respect to ε on $\mathfrak{X} \times \mathfrak{E}$ and let*

(2) *$\mathfrak{M}(\varepsilon_0, c)$ be bounded for $c > c_0$.*

Then there exists a neighborhood $\mathfrak{O}_\varepsilon$ of ε_0 for which $\bigcup_{\varepsilon\in\mathfrak{O}_\varepsilon} \mathfrak{M}(\varepsilon, c_0) = \mathfrak{M}(\mathfrak{O}_\varepsilon, c_0)$ is bounded.

Proof. It follows from the condition (1) that

$$\mu(x, \varepsilon_0) = \mu(x, \varepsilon) + [\mu(x, \varepsilon_0) - \mu(x, \varepsilon)] \le c_0 + \omega(\varepsilon - \varepsilon_0),$$

as $x \in \mathfrak{M}(\varepsilon, c_0)$, where $\omega(\varepsilon)$ is a module of continuity of the function μ with respect to ε. It means that

$$\mathfrak{M}(\varepsilon, c_0) \subset \mathfrak{M}(\varepsilon_0, c_0 + \omega(\varepsilon - \varepsilon_0)).$$

As a consequence of the monotonicity (with respect to the inclusion) of $\mathfrak{M}(c)$, we have $\mathfrak{M}_0(\mathfrak{O}_\varepsilon, c_0) \in \mathfrak{M}(\varepsilon_0, c_0 + \sup_{\varepsilon \in \mathfrak{O}_\varepsilon} \omega(\varepsilon - \varepsilon_0))$ for any neighborhood $\mathfrak{O}_\varepsilon$ of ε_0, and the latter set is bounded by the condition (1).

By using Lemma 1, the sufficient conditions can be formulated for the set $\widetilde{\mathfrak{X}}(\mathfrak{O}_\varepsilon) = \bigcup_{\varepsilon \in \mathfrak{O}_\varepsilon} \widetilde{\mathfrak{X}}(\varepsilon)$ to be bounded.

Let us consider

$$\mu(x, \varepsilon) = -f(x, \varepsilon). \tag{14}$$

Theorem 2. *Let*

(1) *there exist a sequence $y \to x^*$ such that $g_i(y_s, \varepsilon_0) < 0$, $i \in \overline{1, r}$ for some $x^* \in \widetilde{\mathfrak{X}}(\varepsilon_0)$ and let*

(2) *the assumptions of Lemma* 1 *be satisfied for $\mu(x, \varepsilon)$* (14) *for $c_0 = \mathfrak{J}(\varepsilon_0)$. Then the set $\widetilde{\mathfrak{X}}(\mathfrak{O}_\varepsilon)$ is bounded.*

The proof is obvious if we note that the upper-semicontinuity of $\mathfrak{J}(\varepsilon)$ (4) at ε_0 follows from (1), while $\mathfrak{X}(\varepsilon) \in \mathfrak{M}(\varepsilon, \mathfrak{J}(\varepsilon))$, with the latter being bounded, by Lemma 1. Thus $\widetilde{\mathfrak{X}}(\mathfrak{O}_\varepsilon)$ is bounded.

Consider

$$\mu(x, \varepsilon) = \max_i g_i(x, \varepsilon). \tag{15}$$

Theorem 3. *Let the conditions of Lemma* 1 *be satisfied for $\mu(x, \varepsilon)$ for $c_0 = 0$. Then $\widetilde{\mathfrak{X}}(\mathfrak{O}_\varepsilon)$ is bounded.*

The proof follows from the inclusion $\widetilde{\mathfrak{X}}(\mathfrak{O}_\varepsilon) \subset \mathfrak{G}(\mathfrak{O}_\varepsilon) \in \mathfrak{M}(\mathfrak{O}_\varepsilon, 0)$. The properties of the set $\widetilde{\mathfrak{X}}(\varepsilon)$ in a neighborhood of ε_0 are determined by the properties of $\mathfrak{G}(\varepsilon)$ (3). In this situation, the following criteria of upper-semicontinuity of $\mathfrak{G}(\varepsilon)$ at ε_0 can be useful.

Theorem 4. *Let*

(1) $\mu(x, \varepsilon)$ (15) *be x - uniformly continuous with respect to ε on $\mathfrak{X} \times \mathfrak{E}$, let*

(2) $\mathfrak{M}(\varepsilon_0, c)$ (13) *be bounded for some $c > 0$ and let $\mathfrak{M}(\varepsilon_0, 0) \neq \varnothing$, and finally let*

(3) *$g(x,\varepsilon)$ be continuous with respect to x and ε.*

Then $\mathfrak{G}(\varepsilon)$ is upper-semicontinuous at ε_0.

Proof. Let the theorem do not hold. Then there exist a neighborhood $\mathfrak{O}_x$ of $\mathfrak{G}(\varepsilon_0)$ and the sequences $\varepsilon_n \to \varepsilon_0$, $x_n \in \mathfrak{G}(\varepsilon_n)$ such that $x_n \in \mathfrak{O}_x$. By Lemma 1, (see conditions (1) and (2) of this theorem) the sequence x_n is bounded and, without loss of generality, it can be assumed to converge to x_0. Since the set $\mathfrak{X}$ is assumed to be closed in problem (1), we see that $x_0 \in \mathfrak{X}$. It follows from (3) that

$$g(x_0,\varepsilon_0) = \lim_{n\to\infty} g(x_n,\varepsilon_n) \leq 0,$$

which contradicts the initial assumption on $x_n \notin \mathfrak{O}_x$.

Theorem 5. *Let*

(1) *$\mathfrak{G}(\varepsilon_0) \neq \varnothing$ and be bounded, let*

(2) *$\mathfrak{X}$ be a convex set, and let*

(3) *$g(x,\varepsilon)$ be ε-continuous with respect to $x \in \mathfrak{X}$ and be x-convex with respect to x, for $\varepsilon \in \mathfrak{E}$.*

Then $\mathfrak{G}(\varepsilon)$ is upper-semicontinuous at ε_0.

The theorem will be proved, as before, by contradiction. Let the statement do not hold, i.e. there exist a neighborhood $\mathfrak{O}_x$ and the sequences ε_n and x_n mentioned in the proof of Theorem 4.

If the sequence x_n is bounded then, as in Theorem 4, the contradiction follows immediately.

If the sequence x_n is not bounded, it can be considered as divergent ($\|x_n\| \to \infty$). Consider a sphere $\mathfrak{S}_r$ and a ball $\mathfrak{B}_r$ centered at the point $x_0 \in \mathfrak{G}(\varepsilon_0)$. Choose its radius r sufficiently large so that $\mathfrak{S}_r \cap \mathfrak{G}(\varepsilon_0) = \varnothing$ and $\mathfrak{B}_r \supset \mathfrak{G}(\varepsilon_0)$. At the sphere $\mathfrak{S}_r$, consider the points $y_n = (1-a_n)x_0 + a_n x_n$; $y_n \in \mathfrak{X}$; $a_n = r/\|x_0 - x_n\|$. The sequence y_k converges to $y_* \in \mathfrak{S}_r$. By condition (3) of the theorem and because of $g(x_0,\varepsilon_0) \leq 0$ and $\lim_{n\to 0} a_n = 0$, we obtain

$$g(y_*,\varepsilon_0) = \lim_{n\to\infty} g[(1-a_n)x_0 + a_n x_n, \varepsilon_0] \leq$$

$$\leq \lim_{n\to\infty} [(1-a_n)\, g(x_0,\varepsilon_0) + a_n g(x_n,\varepsilon_0)] = g(x_0,\varepsilon_0) \leq 0.$$

It follows from this that $y_* \in \mathfrak{G}(\varepsilon_0)$ and, at the same time, $y_* \in \mathfrak{S}_r$, which contradicts $\mathfrak{S}_r \cap \mathfrak{G}(\varepsilon_0) = \varnothing$.

Continuity

The upper-semicontinuity of corresponding sets means that, under perturbations, their configuration can change abruptly, but the perturbed sets belong to a neighborhood of the unperturbed ones.

In studying the parametric properties of macrosystem MSS, the case where the perturbed sets are continuous is interesting.

Theorem 6. *Let the conditions of Theorem* 5 *be satisfied and let the unperturbed problem* (1) *satisfy the Slater condition, namely there exists* $x_0 \in \mathfrak{G}(\varepsilon_0)$ *such that* $g(x_0, \varepsilon_0) < 0$. Then $\mathfrak{G}(\varepsilon)$ *is continuous at* ε.

Proof. Taken into account the statement of Theorem 5, it is sufficient to prove the lower semicontinuity of $\mathfrak{G}(\varepsilon)$ at ε_0, i.e. to prove that for any sequence $\varepsilon_n \to \varepsilon_0$ and $y \in \mathfrak{G}(\varepsilon_0)$ there exists a sequence $x_n(\mathfrak{G}(\varepsilon_n))$ which converges to y_0.

Consider the points $x_k = \left(1 - \frac{1}{k}\right) y + \frac{1}{k} x_0 \in \mathfrak{X}$. From the x-convexity of g we obtain

$$g(x_k, \varepsilon_0) < \left(1 - \frac{1}{k}\right) g(y, \varepsilon_0) + \frac{1}{k} g(x_0, \varepsilon_0) < 0.$$

It follows from the ε-continuity of g that

$$g(x_k, \varepsilon_n) < 0$$

for sufficiently large n. Hence $x_k \in \mathfrak{G}(\varepsilon_n)$ for sufficiently large $n \leq N(k)$.

Consider the sequence $k(n) \to \infty$ as $n \to \infty$ so that $n \geq N(k(n))$. Then $x_{k(n)} \in \mathfrak{G}(\varepsilon_n)$ and it follows from the definition of x_k that $x_{k(n)} \to y$.

Differentiability.

Let us consider the problem (1) and assume that the set of solutions of the unperturbed problem $\mathfrak{X}(\varepsilon_0)$ (2) consists of one element, i.e. it has the unique solution $x^*(\varepsilon_0)$.

Let us recall the Kuhn-Tucker conditions, which should be satisfied by the solution. For this purpose we introduce the Lagrange function

$$L(x, \lambda, \varepsilon_0) = f(x, \varepsilon_0) - \sum_{k=1}^{r} \lambda_k g_k(x, \varepsilon_0). \tag{16}$$

Assume that the set $\mathfrak{G}(\varepsilon_0)$ (3) satisfies the Slater condition, namely there exists a point $\overline{x}$ for which $g_k(\overline{x}, \varepsilon_0) < 0$, $k \in \overline{1, r}$. The necessary optimality conditions are

$$\frac{\partial L}{\partial x_i} = \left(\frac{\partial f}{\partial x_i} - \sum_{k=1}^{r} \lambda_k \frac{\partial g_k}{\partial x_i} \right) \Bigg|_{x^*, \lambda^*} = 0, \tag{17}$$

$$\lambda_k^* \geq 0,$$

$$\lambda_k^* g_k(x^*, \varepsilon_0) = 0, \tag{18}$$

$$g_k(x^*, \varepsilon_0) \leq 0, \quad k \in \overline{1, r}.$$

At the point $x^*(\varepsilon_0)$, some constraints are transformed into equalities (the active constraints). The set of such constraints is denoted by $K_A^*(\varepsilon_0)$, i.e.

$$g_k(x^*, \varepsilon_0) = 0, \quad k \in K_A(\varepsilon_0).$$

It follows immediately from (18) that

$$\lambda_k \geq 0, \quad k \in K_A(\varepsilon_0); \quad \lambda_k \equiv 0, \quad k \notin K_A(\varepsilon_0);$$

and (17, 18) can be written as

$$\nabla_x L(x^*, \lambda^*, \varepsilon_0) = \nabla_x f(x^*, \varepsilon_0) - \sum_{k \in K_A(\varepsilon_0)} \lambda_k^* \nabla_x g(x^*, \varepsilon_0) = 0 \tag{19}$$

$$\nabla_\lambda L(x^*, \lambda, \varepsilon_0) = \{g_k(x^*, \varepsilon_0) = 0, \quad k \in K_A(\varepsilon_0)\},$$

$$\lambda_k^* \geq 0, \quad k \in K_A(\varepsilon_0).$$

Here $\nabla_x L$ is a vector of the Lagrange function derivatives with respect to the variables x; $\nabla_\lambda L$ is a vector of the Lagrange function derivatives with respect to λ corresponding to the active constraints.

Theorem 7. *In problem* (1), *let the functions* $f(x, \varepsilon), g_1(x, \varepsilon), \ldots, g_r(x, \varepsilon)$ *be continuously differentiable* m *times with respect to both arguments, and let the following conditions be satisfied for the unperturbed problem* ($\varepsilon = \varepsilon_0$).

(a) *There exists the unique solution* $x^*(\varepsilon_0)$.

(b) *The Lagrange multipliers corresponding to the active constraints at* $x^*(\varepsilon_0)$ *are such that*

$$\lambda_k^*(\varepsilon_0) > 0, \quad k \in K_A(\varepsilon_0).$$

(c) *The gradients of the active constraints are linearly independent at* $x^*(\varepsilon_0)$.

(d) *The matrix*

$$\nabla_x^2 L(x^*, \lambda^*, \varepsilon_0) = \nabla_x^2 f(x^*, \varepsilon_0) - \sum_{k \in K_A(\varepsilon_0)} \lambda_k^*(\varepsilon_0) \nabla_x^2 g_k(x^*, \varepsilon_0)$$

is nondegenerate, i.e.

$$\det \nabla_x^2 L(x^*, \lambda, \varepsilon_0) \neq 0.$$

Then there exists a neighborhood $\mathfrak{O}_\varepsilon$ of the point ε_0 such that for all $\varepsilon \in \mathfrak{O}_\varepsilon$ the solution $x^*(\varepsilon)$, $\lambda^*(\varepsilon)$ of (4) *is m times differentiable.* (This theorem differs from Theorem 3.2 in Pervozvansky, Graytzgory, 1979 by the condition (*d*)).

The proof is based on Theorem 2.3 on existence of the differentiable implicit function.

It follows from the differentiability of functions in (1) and from the conditions (b) and (c) that there exists a set $\mathfrak{O}_\varepsilon$ such that the following conditions of type (19) should hold for all $\varepsilon \in \mathfrak{O}_\varepsilon$:

$$\lambda_k^* > 0, \quad k \in K_A(\varepsilon_0):$$

$$\nabla_x L(x^*, \lambda^*, \varepsilon) = \nabla_x f(x^*, \varepsilon) - \sum_{k \in K_A(\varepsilon)} \lambda_k^* \nabla g_x(x^*, \varepsilon) = 0, \tag{20}$$

$$\nabla_\lambda L(x^*, \lambda^*, \varepsilon) = \{g_k(x^*, \varepsilon)\} = 0.$$

According to Theorem 2.3, this system of equations determines m-times-differentiable implicit functions $x^*(\varepsilon)$ and $\lambda^*(\varepsilon)$ if the matrix

$$\nabla_{x,\lambda}^2 L = \begin{bmatrix} \nabla_x^2 L(x^*, \lambda^*, \varepsilon_0) & \nabla_{x\lambda} L(x^*, \lambda^*, \varepsilon_0) \\ \nabla_{\lambda x} L(x^*, \lambda^*, \varepsilon_0) & \nabla_\lambda^2 L(x^*, \lambda^*, \varepsilon_0) \end{bmatrix} \tag{21}$$

is non-degenerate.

It follows from (20) that

$$\nabla_{x,\lambda}^2 L = \left[\begin{array}{c|c} \nabla^2 f_x(x^*, \varepsilon_0) - \sum_{k \in K_A(\varepsilon_0)} \lambda_k^* \nabla_k^2 g(x^*, \varepsilon_0) & -\nabla_x g(x^*, \varepsilon_0) \\ \hline \nabla_x g(x_x^*, \varepsilon_0) & 0 \end{array}\right], \tag{22}$$

where

$$\nabla_x g(x^*, \varepsilon_0) = [\nabla_x g_k(x^*, \varepsilon_0); \quad k \in K_A(\varepsilon_0)]$$

is a $(\rho_A \times n)$ matrix of gradients of active constraints, ρ_A is the number of elements in the index set $K_A(\varepsilon_0)$ and $\rho_A < n$.

The matrices like (22) have already arisen (see 3.42). It follows from conditions (c) and (d) that

$$\det \nabla^2_{x,\lambda} L(x^*, \lambda^*, \varepsilon_0) \neq 0$$

Hence the matrix $\nabla^2_{x,\lambda} L$ is non-degenerate, and this finishes the proof of the theorem.

The differential properties of $x^*(\varepsilon)$ and $\lambda^*(\varepsilon)$ can be represented by the series

$$x^*(\varepsilon) = x^*(\varepsilon_0) + x^{(1)}(\varepsilon_0)(\varepsilon - \varepsilon_0) + \ldots + \frac{1}{m!} x^{(m)}(\varepsilon_0)(\varepsilon - \varepsilon_0)^{(m)};$$

$$\lambda^*(\varepsilon) = \lambda^*(\varepsilon_0) + \lambda^{(1)}(\varepsilon_0)(\varepsilon - \varepsilon_0) + \ldots + \frac{1}{m!} \lambda_0^{(m)}(\varepsilon_0)(\varepsilon - \varepsilon_0)^{(m)},$$

where

$x^{(k)}(\varepsilon_0)(\varepsilon - \varepsilon_0)^{(k)}, \lambda^{(k)}(\varepsilon_0)(\varepsilon - \varepsilon_0)^{(k)}$ are homogeneous k-forms of the variables $\varepsilon_1, \ldots, \varepsilon_s$. These forms are given by the recurrent expressions (2.27).

3.8 Parametric properties of MSS with the incomplete consumption of resources

The analysis of perturbed mathematical programming problems given in the previous section allow us to investigate the parametric properties of macrosystem MSS with the incomplete consumption of resources.

Consider the following MSS

$$H(N, x) \Rightarrow \max_{x \in \mathfrak{X} \subset R^s}; \quad \varphi_k(N, x) \leq q_k(x), \quad k \in \overline{1, r}; \tag{1}$$

Unlike the general problem considered in Section 7, this problem has certain specific features. The objective function $H(N, x)$ in (1) is strictly concave for all $x \in \mathfrak{X}$.

The functions $\varphi_k(N, x)$ in (1) have the following properties

$$\varphi_k(N, x) \geq 0, \quad k \in \overline{1, r}; \tag{2}$$

$$\frac{\partial \varphi_k}{\partial N_n} \geq 0; \quad k \in \overline{1,r}, \quad n \in \overline{1,m}; \tag{3}$$

$$\Gamma_k = \left[\frac{\partial^2 \varphi_k}{\partial N_p \, \partial N_n}\right] > 0 \tag{4}$$

for $x \in \mathfrak{X}$ and $N \in R^m_+$.

These properties of φ_k are caused by the qualitative features of the process of distributing the macrosystem elements among the states discussed in Section 3 (B-model with the nonlinear resource consumption).

The Lagrange function for this problem is

$$L(N, \lambda, x) = H(N, x) + \sum_{k=1}^{r} \lambda_k (q_k(x) - \varphi_k(N, x)). \tag{5}$$

Here we assume that the matrix $F(N, x) = \left[\frac{\partial \varphi_k}{\partial N_n}\right]$ is regular (has the full rank r) for all $N_n \geq 0$ and $x \in \mathfrak{X}$.

Let us consider the unperturbed problem corresponding to the exogenous parameter $x^0 \in \mathfrak{X}$. As a consequence of the strict concavity of the entropy function H, strict convexity (4) of the functions φ_k and regularity of the matrix F, problem (1) has the unique solution with respect to both the primal variables $N^*(x^0)$ and the dual variables $\lambda^*_k(x^0) = 0$ for $k \notin K_A(x^0)$.

We show that here the dual variables $\lambda^*_k(x^0)$ corresponding to the active constraints have a stronger property, namely $\lambda^*_k(x^0) > 0$, $k \in K_A(x^0)$.

For $\{N^*(x^0), \lambda^*(x^0)\}$, we have

$$L(N^*, \lambda^*, x^0) = H(N^*, x^0) + \sum_{k \in K_A(x^0)} \lambda^*_k (q_k(x^0) - \varphi_k(N^*, x^0)) \tag{6}$$

and

$$\frac{\partial L}{\partial N_n} = \frac{\partial H}{\partial N_n} + \sum_{k \in K_A(x^0)} \lambda_k \frac{\partial \varphi_k}{\partial N_n} = 0; \tag{7}$$

$$\begin{gathered} \frac{\partial L}{\partial \lambda_k} = q_k(x^0) - \varphi_k(N^*, x^0) = 0, \quad k \in K_A(x^0); \\ \lambda^*_k \geq 0, \quad k \in K_A(x^0). \end{gathered} \tag{8}$$

To analyze the solutions $\{N^*, \lambda\}$ of (7) and (8), the properties of entropy functions $H(N, x)$ should be used.

For B-model of stationary states, (7) is

$$N_n^* - \widetilde{a}_n(x^0) \exp\left(- \sum_{j \in K_A(x^0)} \lambda_k^* a_{jn}(N^*, x^0) \right) = 0, \quad n \in \overline{1,m}, \tag{9}$$

where

$$a_{jn}(N^*, x^0) = \left. \frac{\partial \varphi_j}{\partial N_n} \right|_{N^*, x^0}, \quad \widetilde{a}_n(x^0) = a_n(x^0)\, G_n(x^0). \tag{9'}$$

Let us transform the system (7, 8) by introducing new variables $z_j = e^{-\lambda_j}$. Then

$$\Psi_n(N^*, z^*, x^0) = N_n^* - \widetilde{a}_n(x^0) \prod_{j \in K_A(x^0)} (z_j^*)^{a_{jn}(N^*, x^0)} = 0,$$

$$n \in \overline{1,m}; \tag{10}$$

$$\Phi_k(N^*, z^*, x^0) = q_k(x^0) - \varphi_k(N^*, x^0) = 0, \quad k \in K_A(x^0). \tag{10'}$$

Let us consider the functions $\Psi_n(N^*, z^*, x^0)$ and assume that N^* is fixed and $0 < z_j^* < 1$ $(j \in \overline{1,r})$. Then we obtain

$$\Psi_n(N^*, z^{(s)}, x^0) = N_n^*, \tag{11}$$

$$\Psi_n(N^*, 1, x^0) = N_n^* - \widetilde{a}_n(x^0), \tag{11'}$$

where $z^{(s)}$ is a vector with the components z_j $(j \in K_A(x^0))$ such that $z_s = 0$ $(s \in K_A(x^0))$. It can be shown (cf. (3.46)) that $\dfrac{\partial \Psi_n}{\partial N_p} < 0$; $n, p \in \overline{1,m}$.

Let the consumption functions satisfy (3.49–3.49′), namely

$$\begin{gathered} \varphi_k(N_1^*, \ldots, N_{j-1}^*, \quad \widetilde{a}_j(x^0), \quad N_{j+1}^*, \ldots, N_m^* x^0) > q_k(x^0), \\ \varphi_k(0, x^0) < q_k(x^0), \quad j \in \overline{1,m}. \end{gathered} \tag{12}$$

Taken into account (9), it follows from this that

$$N_n^* < \widetilde{a}_n(x^0), \quad n \in \overline{1,m}.$$

Thus

$$\Psi_n(N^*, z^{(s)}, x^0) > 0; \quad \Psi_n(N^*, 1) < 0; \quad n \in \overline{1,m}.$$

Since the functions $\Psi_n(N^*, z^*, x^0)$ are strictly monotonous with respect to z^* for any fixed $N^* > 0$, we see that $0 < z_j^* < 1$, $j \in K_A(x^0)$. Thus $\lambda_k^* > 0$, $k \in K_A(x^0)$, under the conditions of the Lagrange function stationarity (7, 8).

Let us consider the matrix of second derivatives of the Lagrange function with respect to the primal variables at the point $(N^*, 1)$. According to (7), for B-model this is

$$\nabla_N^2 L(N^*, \lambda^*, x^0) = \begin{cases} -\dfrac{1}{N_n} - \displaystyle\sum_{k \in K_A(x^0)} \lambda_k^* \dfrac{\partial^2 \varphi_k}{\partial N_n^2} & (n = p); \\ - \displaystyle\sum_{k \in K_A(x^0)} \lambda_k^* \dfrac{\partial^2 \varphi_k}{\partial N_n \partial N_p} & n \neq p. \end{cases}$$

It can be shown (see Section 3) that $\nabla_N^2 L(N^*, \lambda^*, x^0)$ is negative-definite, hence $\det \nabla_N^2 L(N^*, \lambda^*, x^0) \neq 0$ and all the conditions of Theorem 7.7 are satisfied.

Thus the theorem is proved.

Theorem 1. *Let B-model* (9) *have the following properties.*

(1) *The functions $G_n(x)$, $a_n(x)$, $q_k(x)$ and $\varphi_k(N, t_k(x))$ are m-times differentiable with respect to x.*

(2) *The functions $\varphi_k(N, x)$ are*

(a) *twice differentiable with respect to N,*

(b) *non-negative, i.e. $\varphi_k(N, t_k(x)) \geq 0$, $k \in \overline{1, r}$,*

(c) *monotone increasing, i.e.*

$$\frac{\partial \varphi_k}{\partial N_n} \geq 0, \quad x \in \mathfrak{X}, \quad N \in R_+^m,$$

(d) *and satisfy the conditions*

$$\varphi_k(N_1, \ldots, N_{j-1}, \tilde{a}_n(x^0), \quad N_{j+1}, \ldots, N_n, x^0) > q_k(x^0),$$

$$\varphi_k(0, x^0) < q_k(x^0), \quad j \in \overline{1, m}; \quad k \in \overline{1, r}.$$

(3) *The matrix*

$$F = \left[\frac{\partial \varphi_k}{\partial N_n}\right]_{N^*, x^0}$$

is regular (its rank equals r).

(4) *The Hessian is*

$$\Gamma_k = \left[\frac{\partial^2 \varphi_k}{\partial N_n \partial N_p}\right] \geq 0, \quad x \in \mathfrak{X}, \quad N \in R_+^m.$$

Then there exists a neighborhood $\mathfrak{O}_x$ of x^0 in $\mathfrak{X}$ such that for all $x \in \mathfrak{O}_x$ the solution $N^*(x)$, $\lambda^*(x)$ of B-model (1) *is m times differentiable.*

Similar statements hold for F- and E- models.

CHAPTER

FOUR

COMPUTATIONAL METHODS FOR DETERMINING THE STATIONARY STATES OF HOMOGENEOUS MACROSYSTEMS

Practical use of models of stationary states (MSS) is associated with computational methods for solving the corresponding mathematical programming problems (or conditional extremum problems). This branch of applied mathematics is well developed and contains many efficient methods.

However, an MSS, as a mathematical object, has special features associated with objective functions in the models. These helps to construct simple and efficient computational schemes for computer implementation of MSS. In this chapter, these schemes are called multiplicative. Convergence of multiplicative algorithms for solving systems of nonlinear equations and convex programming problems is studied. Convergence theorems are used in investigating multiplicative algorithms for determining stationary states of macrosystems with complete and incomplete consumption of resources. The properties of multiplicative algorithms are illustrated by the results of computer experiments.

4.1 Classification of computational methods

To determine a macrosystem stationary state is to solve nonlinear equations obtained from the MSS optimality conditions. Combining the results of Chapter 2, we conclude that for an MSS with complete consumption of resources these equations are represented as

$$f(x) = 1, \quad x \in R^m \tag{1}$$

where $f(x)$ is a continuous function which maps $R^m \to R^m$.
Assume that the equation (1) has a unique solution x^*. Because of this and the continuity of $f(x)$, there exists a set $\mathcal{P} \subset R^m$ such that

$$x^* \in \mathcal{P}, \tag{2}$$

$f(x) > 0$ for all $x \in \mathcal{P}$.
Sometimes, determination of a stationary macrostate can be reduced to the equation

$$x \otimes \Psi(x) = 0, \quad x \in R^m_+, \tag{3}$$

where $\otimes$ is the coordinate-wise product.

The equation (3) can be considered as a modification of (1) if we put

$$\Psi(x) = f(x) - 1. \tag{4}$$

It is obvious that (3) may have both the trivial solution $x^* = 0$ and any other solution with a number of non-zero components.

Equations of these classes are solved by iterative processes. The latter are given by the *scheme of organization* determining the structure of current approximation and by the *operator* transforming the previous solutions to the next ones.

Only one-step iterative processes are considered below. In this class we distinguish iterative processes with parallel schemes where the previous approximation contains only the components of this approximation and iterative processes with coordinate-wise scheme where the previous approximation with respect to a group of variables can contain components of the subsequent approximation with respect to another group of variables.

If s denotes the previous approximation, while $s+1$ denotes the subsequent approximation, then for the *iterative process with the parallel scheme* we obtain

$$\{x_1^s, \ldots, x_m^s\} \to x_1^{s+1}, \ldots, \{x_1^s, \ldots, x_m^s\} \to x_m^{s+1}, \tag{5}$$

and for the *iterative process with the coordinate-wise scheme* we obtain

$$\{x_1^s, \ldots, x_m^s\} \to x_1^{s+1}, \ldots, \{x_1^s, \ldots, x_m^s\} \to x_{r_1}^{s+1};$$
$$\{x_1^{s+1}, \ldots, x_{r_1}^{s+1}; x_{r_1+1}^s, \ldots, x_m^s\} \to x_{r_1+1}^{s+1}, \ldots, \tag{6}$$
$$\cdots$$
$$\{x_1^{s+1}, \ldots, x_{r_1}^{s+1}; \ldots; x_{r_1+\ldots+r_{k-2}}^{s+1}, \ldots, x_{r_1+\ldots+r_{k-2}+r_{k-1}}^{s+1};$$
$$x_{r_1+\ldots+r_{k-1}+1}^s, \ldots, x_m^s\} \to x_{r_1+\ldots+r_{k-1}+1}^{s+1};$$
$$\cdots$$
$$\{x_1^{s+1}, \ldots, x_{r_1}^{s+1}; \ldots; x_{r_1+\ldots+r_{k-2}}^{s+1}, \ldots, x_{r_1+\ldots+r_{k-2}+r_{k-1}}^{s+1};$$
$$x_{r_1+\ldots+r_{k-1}+1}^s, \ldots, x_m^s\} \to x_m^{s+1},$$

where k is the number of groups of variables in the vector x; r_j is the number of variables in the j-th group and $\sum_{j=1}^{k} r_j = m$.

Let us denote the operator of iterative process by $B = \{B_1, \ldots, B_m\}$ and consider the equation

$$x = B(x). \tag{7}$$

Then the one-step iterative process may be represented as

$$x_i^{s+1} = B_i\left(X_i\left(s, s+1\right)\right), \quad i \in \overline{1, m}, \tag{8}$$

where $X_i(s, s+1)$ is the set of variables of the previous approximation and is given either by (5) for the parallel scheme or by (6) for the coordinate-wise one.

How various the iterative methods are, there exists a general scheme for constructing the operator B, namely

$$B(x) = x + \gamma p\left(x\right), \tag{9}$$

where $p(x)$ is the correction operator and γ is a vector parameter. Such a scheme will be called ***additive***.

Below we consider another, ***multiplicative***, scheme (see Dubov, Imelbayev, Popkov, 1983):

$$B(x) = x \otimes p^{\gamma}(x). \tag{10}$$

Multiplicative scheme (10) has appeared as a method for solving problems of entropy maximization (see Wilson, 1967; Imelbayev, Shmulyan, 1978).

But the subsequent researches showed that it is universal and, in some cases, is more efficient than the additive scheme.

Besides, the equations generated by the multiplicative scheme are somewhat similar to those of mathematical models of biological processes (the Lotka-Volterra equations, see Volterra, 1976). Studies of the evolution of biological communities confirm the adequacy of the mathematical models used. It is quite possible that the multiplicativity (as in 10) is a natural feature of biological processes, and its useful properties are worth to be used in other areas, for example, in developing efficient computational methods.

To obtain the correction operator $p(x)$ in (9) and (10), we may use various information on functions $f(x)$ and $\Psi(x)$ from (1, 3).

Only information on functions $f(x)$ or $\Psi(x)$ (1) and (3) is used in *zero-order iterative methods* for constructing $p(x)$. In *iterative methods of the* 1*st and higher orders*, $p(x)$ includes information on both $f(x)$ or $\Psi(x)$ and their derivatives.

Consider a group of *zero-order methods with multiplicative scheme*. The simplest version of this construction for the equation (1) is

$$p(x) = f(x). \tag{11}$$

Substituting (11) into (7) and (10), we obtain

$$x_i = x_i f_i^{\gamma}(x), \quad i \in \overline{1, m}. \tag{12}$$

By comparing (1) and (10), we can see that the latter can be obtained from the former by multiplying each component of both its sides by x.

The above approach for constructing the operator B in multiplicative scheme (10) is oriented on the equation (1). For the equations (3, 4), whose left-hand side has already the multiplicative structure, it is more convenient to use the traditional approach of the additive scheme. By adding x to both sides of (3, 4), we obtain

$$x = x \otimes (1 + \alpha \Psi(x)). \tag{13}$$

Here the operator B (13) gives the multiplicative scheme (10) with the correction function

$$p(x) = 1 + \alpha \Psi(x)$$

and $\gamma = 1$.

It should be noted that there is a relation between the additive and multiplicative schemes.

Thus, if (13) uses the correction function

$$p(x) = x \otimes \Psi(x),$$

it should be classified as additive scheme (9).

If $x > 0$ and $x \in \mathcal{P}$ in (12), new variables $y_i = \ln x_i$ may be introduced. Then (12) gives

$$y_i = y_i + \gamma \ln F_i(y), \quad i \in \overline{1,m}, \tag{14}$$

where $F_i(y_1, \ldots, y_m) = f_i(e^{y_1}, \ldots, e^{y_m})$.

In these equations, the correction function $p(y) = \ln F(y)$ and the operator B generated by (14) corresponds to the additive structure.

Thus the *multiplicative algorithm with the parallel scheme of iterations* can be written as follows

$$x_i^{s+1} = x_i^s p_i(x_1^s, \ldots, x_m^s), \quad i \in \overline{1,m}. \tag{15}$$

The *multiplicative algorithm with the coordinate-wise scheme of iterations* is of the form

$$\begin{aligned} x_{r_1+\ldots+r_{n-1}+i}^{s+1} = {} & x_{r_1+\ldots+r_{n-1}+i}^{s} p_{r_1+\ldots+r_{n-1}+i} \left(x_1^{s+1}, \ldots, x_{r_1}^{s+1}; \ldots;\right. \\ & x_{r_1+\ldots+r_{n-2}+1}^{s+1}, \ldots, x_{r_1+\ldots+r_{n-2}+r_{n-1}}^{s+1}; \\ & \left. x_{r_1+\ldots+r_{n-1}+1}^{s}, \ldots, x_{r_1+\ldots+r_{n-1}+i}^{s}, \ldots, x_m^s\right). \end{aligned} \tag{16}$$

Here

$$p(x) = \begin{cases} f_i^{\gamma}(x) & \text{(for (1))}, \\ 1 + \gamma\Psi_i(x) & \text{(for (3) and (4))}. \end{cases}$$

$$n \in \overline{1,k}; \quad i \in \overline{1,r_n}.$$

Now we consider *multiplicative algorithms of the* 1-*st order* which use the information on the first derivatives of functions $f(x)$.

The idea of these algorithms is based on using linear approximations on each iteration.

Let us represent (1) as

$$\varphi_i(x) = x_i(f_i(x) - 1) = 0, \quad i \in \overline{1,m}. \tag{17}$$

Let x^s be an approximation of the solution obtained at the s-th step of the iterative process. We linearize the functions $\varphi(x)$ with respect to x^s and consider a system of linear equations

$$\varphi_i(x^s) + \sum_{k=1}^{m} \frac{\partial \varphi_i(x^s)}{\partial x_k}(x_k - x_k^s) = 0, \quad i \in \overline{1,m}. \tag{18}$$

Denote

$$F(x^s) = \left[\frac{\partial \varphi_i(x^s)}{\partial x_k}\right] = \left[\begin{array}{ll} f_i(x^s) + x_i^s \dfrac{\partial f_i(x^s)}{\partial x_i}, & i = k \\ x_i^s \dfrac{\partial f_i(x^s)}{\partial x_k}, & i \neq k \end{array}\right]. \tag{19}$$

Then (17) and (18) give

$$x^{s+1} = x^s - F^{-1}(x^s)\left[x^s \otimes (f(x^s) - 1)\right]. \tag{20}$$

It is easy to see that this iterative process is the Newton method applied to (17)

Denote by $a_{ik}(x^s)$ the elements of inverse matrix $F^{-1}(x^s)$. Then (20) can be rewritten in the following multiplicative form

$$x_i^{s+1} = \sum_{k=1}^{m} x_k^s \left[\tilde{a}_{ik}(x^s) - a_{ik}(x^s) f_k(x^s)\right], \tag{21}$$

where $\tilde{a}_{ik}(x^s) = \begin{cases} a_{ik}(x^s) & (i \neq k), \\ a_{ii}(x^s) + 1 & (i = k). \end{cases}$

This iterative process uses the information on functions $f_i(x)$ and their derivatives $\dfrac{\partial f_i}{\partial x_k}$ (20).

4.2 Theorems on convergence of multiplicative algorithms

Let us consider the zero-order multiplicative algorithm (1.12). Let the starting point be $x^0 \in \mathfrak{G} \subset \mathcal{R} \subset R^m$, where $\mathfrak{G}$ is a compact set.

The algorithm (1.12) is called *$\mathfrak{G}$-convergent* if there exist a set $\mathfrak{G} \subset \mathfrak{P}$ and positive scalar numbers $a(\mathfrak{G})$ and γ such that for all $x^0 \in \mathfrak{G}$ and $0 < \gamma < a(\mathfrak{G})$ the algorithm converges to the solution x^* of (1.1), with the convergence being linear.

Note that the algorithm (1.12) generates a sequence of approximations which do not change the sign of the initial approximation x^0. If, for example, x^* is negative and $\mathfrak{G} = \{x^0 : x_i^0 > 0, \ i \in \overline{1,m}\}$, then $\mathfrak{G}$-convergence will not hold.

Auxiliary estimates.

Lemma 1. *Let an $m \times m$ matrix A satisfy a Hurwitz conditions (i.e. all its eigenvalues have negative real parts). Then there exist a matrix U, the norm $\|\bullet\|_U = \langle U\bullet, \bullet\rangle$ in the space of complex vectors and real numbers $\lambda_A > 0$ and $\gamma_A > 0$ such that*

$$\|(E+\gamma A)\, z\,\|_U \leq (1-\gamma\lambda_A)\ \|z\|_U \tag{1}$$

for all $\gamma \in (0, \gamma_A)$ and $z \in C^m$.

Proof.

Consider a matrix U which, together with A, satisfies the Lyapunov equation

$$UA + A'U = -E. \tag{2}$$

Since the matrix A is Hurwitz, this equation has a positive solution

$$U > 0 \text{ and } U' = U.$$

For $\|\bullet\|_U$ the following inequality holds

$$\mu_{\min}\|x\|^2 \leq \|x\|_U \leq \mu_{\max}\ \|x\|^2, \tag{3}$$

where $\mu_{\min}$ and $\mu_{\max}$ are minimum and maximum eigenvalues of U, respectively, and $\|\bullet\|$ is the Euclidean norm. Consider

$$\begin{aligned}\|(E+\gamma A)\, z\|_U &= \langle U\,(E+\gamma A)\, z,\ (E+\gamma A)\, z\rangle = \\ &= \|z\|_U + \gamma\langle (UA + A'U)\, z, z\rangle + \gamma^2\, \|Az\|_U.\end{aligned}$$

According to (2), we obtain

$$\|(E+\gamma A)\, z\|_U = \|z\|_U - \gamma\, \|z\|^2 + \gamma^2\, \|Az\|_U.$$

Let $\varepsilon_{\max}$ is the maximal eigenvalue of the matrix A.

Then we obtain from (3)

$$\|(E+\gamma A)\, z\|_U \leq \left(1 - \frac{\gamma}{\mu_{\max}} + \gamma^2 \frac{\mu_{\max}\varepsilon_{\max}}{\mu_{\min}}\right)\ \|z\|_U.$$

If

$$\gamma < \gamma_A = \frac{1}{2} \frac{\mu_{\min}}{\mu_{\max}^2 \varepsilon_{\max}}$$

then

$$\| (E + \gamma A) z \|_U \leq \left(1 - \frac{\gamma}{2\mu_{\max}}\right) \|z\|_U = (1 - \gamma \lambda_A) \|z\|_U,$$

where $\lambda_A = 1/2\mu_{\max}$.

Consider the matrix A depending on a scalar parameter γ.

Corollary 1. *Let there exist $b > 0$ such that for all $\gamma \in (0, b]$ the matrix $A(\gamma)$ is Hurwitz. Then there exist a matrix U, the norm $\| \bullet \|_U = \langle U \bullet, \bullet \rangle$ in the space C^m of complex vectors, and constants $\gamma_A > 0$ and $\lambda_A > 0$ such that*

$$\| (E + \gamma A(\gamma)) z \|_U \leq (1 - \gamma \widetilde{\lambda}_B) \|z\|_U$$

for all $\gamma \in (0, \gamma_0)$ and vectors $z \in C^m$, where $\gamma_0 = \min(\widetilde{\gamma}_A, b)$.

The proof follows obviously from Lemma 1. The difference is that the length of the interval for γ depends on both the maximum eigenvalue of $A(\gamma)$ and on the value $\gamma = b$ for which the matrix $A(\gamma)$ remains Hurwitz one.

Lemma 2. *Let A be a $m \times m$ symmetrical Hurwitz matrix and $B =$ diag(b_j), where $b_j > 0$, $j \in \overline{1, m}$. Then the matrix $C = BA$ is also Hurwitz.*

Proof. Let us consider the characteristic equation

$$\det (BA - \lambda E) = 0. \qquad (*)$$

It is equivalent to

$$\det (A - \lambda B^{-1}) = 0.$$

Let us determine the matrix $G = \text{diag}\sqrt{b_j}$ and multiply the latter equation by det G det G'. Then we obtain

$$\det G \det (A - \lambda B^{-1}) \det G' = \det (GAG' - \lambda E) = 0, \qquad (**)$$

since

$$GB^{-1}G' = E.$$

Hence the equations (*) and (**) are equivalent. The matrix GAG' is symmetrical. By the law of inertia of quadratic forms, the signatures of A and

GAG' coincide, i.e. GAG' is also Hurwitz. Since the characteristic equations (*) and (**) have the same roots, the matrix BA is Hurwitz.

Lemma 3. *For a matrix*

$$S = \begin{bmatrix} A & B \\ -B' & 0 \end{bmatrix}, \tag{4}$$

let A be a real $(m \times m)$ symmetrical Hurwitz matrix and
let B be a real $(m \times r)$ matrix with the rank r.
Then the matrix S is Hurwitz one.

Proof. Consider the eigenvector $\tilde{z} = \begin{bmatrix} \tilde{x} \\ \tilde{y} \end{bmatrix}$ of S corresponding to the eigenvalue λ. In general, the components of $\tilde{z}$ and λ are complex numbers.

Consider the form $\tilde{z}^* S\tilde{z}$ (* means transposition and transfer to complex-conjugate numbers). We obtain from (4)

$$\text{Re } \tilde{z}^* S\tilde{z} = \text{Re } \tilde{x}^* A \tilde{x} \leq 0, \tag{5}$$

(equality holds only for $\tilde{x} \equiv 0$), since

$$\text{Re } (\tilde{x}^* B\tilde{y} - \tilde{y}^* B' \tilde{x}) = 0.$$

On the other hand,

$$\text{Re } \tilde{z}^* S\tilde{z} = \text{Re } \lambda \, \|\tilde{z}\|^2. \tag{6}$$

Thus we obtain from (5) and (6) Re $\lambda < 0$ if $\tilde{x} \not\equiv 0$, and Re $\lambda = 0$ if $\tilde{x} \equiv 0$.

Let us prove that the component $\tilde{x}$ of $\tilde{z}$ is not zero. Assume that $\tilde{x} = 0$, while $\tilde{y} \not\equiv 0$. Then

$$S \begin{bmatrix} 0 \\ \tilde{y} \end{bmatrix} = B\tilde{y} = 0.$$

Since $\tilde{y} \not\equiv 0$, it follows from this equation that *rank* $B < r$, which contradicts the assumptions of the lemma. Therefore $\begin{pmatrix} 0 \\ \tilde{y} \end{pmatrix}$ is not an eigenvector of S and Re $\lambda < 0$.

Lemma 4. *Let the assumptions of Lemma* 3 *hold for the matrix* S (4). *Then the matrix* $Q = SC$, *where* $C = \text{diag}\,(c_i)$, $c_i > 0$, *is Hurwitz.*

Proof. Consider the characteristic equation

$$\det{(SC - \lambda E)} = 0.$$

It is equivalent to

$$\det(S - \lambda C^{-1}) = 0. \tag{7}$$

Consider the matrix $G = \operatorname{diag}(\sqrt{c_i})$. It is obvious that $GC^{-1}G' = E$. Let us multiply the right-hand sides of (7) by ıdet G and det G', respectively:

$$\det G \det(S - \lambda C^{-1}) \det G' = \det(GSG' - \lambda E) = 0, \tag{8}$$

where

$$GSG' = \begin{bmatrix} G_1AG_1' & G_1BG_2' \\ -G_2B'G_1' & 0 \end{bmatrix}; \quad G = \begin{bmatrix} G_1 & 0 \\ 0 & G_2 \end{bmatrix}.$$

It follows from (8) that

$$GSG' = SC. \tag{9}$$

By the law of inertia of quadratic forms, the signatures of A and G_1AG_1' coincide, hence the matrix G_1AG_1' is symmetrical and Hurwitz one. Since G_1 and G_2 are nondegenerate, $(G_1BG_2')' = G_2B'G_1'$ and their rank equals r. Thus the structure of GSG' is the same as of S (4) in Lemma 3, and, by this Lemma, GSG' is Hurwitz matrix. By referring to (9), we obtain the statement of Lemma 4.

Consider the algorithm

$$x^{s+1} = x^s + \gamma F(x^s), \quad x \in R^m, \quad F : R^m \to R^m. \tag{10}$$

Theorem 1. *Let*

(a) F *be a twice continuously differentiable function, let*

(b) $J(x^*) = \left[\dfrac{\partial F_i}{\partial x_j}\right]_{x^*}$ *be Hurwitz matrix, and let*

(c) *for any* ε *there exist* $\gamma(\varepsilon) > 0$ *such that* $\|x^s - x^*\| \leq \varepsilon$ *for all* $\gamma \in (0, \gamma(\varepsilon^0)]$ *and* $x \in \mathfrak{G}$, *beginning with some* $s > s_0$.

Then the algorithm (10) *is* $\mathfrak{G}$*-convergent.*

The proof of this theorem is similar to that of Theorem 3, which will be considered below in detail.

Consider the system of differential equations

$$\dot{x} = F(x), \tag{11}$$

obtained from (10) as $\gamma \to 0$.

Theorem 2. *Let*

(a) F *be a twice continuously differentiable function, let*

(b) $J(x^*) = \left[\dfrac{\partial F_i}{\partial x_j}\right]_{x^*}$ *be a Hurwitz matrix, let*

(c) $\mathfrak{G}$ *be a compact set and let*

(d) *there exist* $\varepsilon > 0$ *such that* $\|x(t) - x^*\| \le \varepsilon$ *for all* $x^0 \in \mathfrak{G}$, *beginning with same* $t > 0$, *where* $x(t)$ *is the solution of* (11) *with the initial state* $x(0) = x^0$.

Then the process (10) $\mathfrak{G}$*-converges.*

Proof. Let us show that the conditions of this theorem are sufficient for the conditions of Theorem 1 to be satisfied. Note that (a) and (b) are equivalent.

Choose a countable and everywhere dense subset of points $\{w_n\}$ in $\mathfrak{G}$. Let t_1 be the first moment when the trajectory of (11) comes to the $\frac{\varepsilon}{2}$ – neighborhood of x^* starting from w_1 (t_1 exists due to (d)). Choose a neighborhood $\mathfrak{G}_1$ of the point w_1 such that the points of this neighborhood come to the $\frac{3}{4}\varepsilon$ – neighborhood of x^* after t_1 (the condition (a) assures the continuous dependence of the trajectory on initial states).

Choose w_2 from the points outside of $\mathfrak{G}_1$. For w_2, we can determine t_2 and the neighborhood $\mathfrak{G}_2$ in the same way. Repeating the process, we obtain the covering of $\mathfrak{G}$ by the open sets $\mathfrak{G}_1, \mathfrak{G}_2, \ldots$. By (c), a finite sub-covering $\mathfrak{G}_{i_1}, \ldots, \mathfrak{G}_{i_m}$ can be chosen from this. Let $T = \min\{t_{i_1}, \ldots, t_{i_m}\}$. Then, by virtue of (a), we can choose $\gamma > 0$ so small that the trajectories of (11) can be approximated by the Euler piecewise-linear functions on the segment $[0, T]$ (i.e. by piecewise-linear functions which join the subsequent iterations neglecting $\frac{\varepsilon}{4}$). Then, by virtue of (d), the condition (c) of Theorem 1 will be satisfied.

Nonlinear equations.

Consider the system

$$f(x) = 1, \tag{12}$$

where

$$f:\ R^m \to R^m;\quad x \in R^m. \tag{12'}$$

Assume that (12) has the unique solution

$$x_i^* \neq 0,\quad i \in \overline{1,m}. \tag{13}$$

We will seek this solution by using the *zero-order multiplicative algorithm with the parallel structure:*

$$x_i^{s+1} = x_i^s f_i^\gamma (x^s); \quad \gamma > 0, \quad i \in \overline{1,m}. \tag{14}$$

Theorem 3. *Let*

(a) $f(x)$ *be twice continuously differentiable on* R^m, *let*

(b) *the matrix* $J(x^*) = \left[x_i \frac{\partial f_i}{\partial x_k} \right]_{x^*}$ *be Hurwitz one, and let*

(c) *there exist* $\varepsilon > 0$ *and* $\gamma(\varepsilon) > 0$ *such that* $\|x^s - x^*\| \leq \varepsilon$ *for all* $\gamma \in (0, \gamma(\varepsilon)]$ *and* $x \in \mathfrak{G}$, *beginning with some s.*

Then the algorithm (14) $\mathfrak{G}$*-converges.*

Proof.

I. Consider the neighborhood

$$\Omega_x = \{x : \|x - x^*\| \leq \delta\} \subset \mathfrak{G}.$$

According to (a), there exists $\delta > 0$ such that for all $u = x - x^* \in \Omega_u = \Omega_x \setminus x^*$ the algorithm (14) can be represented as

$$u^{s+1} = A(x^*) u^s + \gamma \widetilde{\Omega}(u^s), \tag{15}$$

where

$$A(x^*) = E + \gamma J(x^*); \tag{16}$$

$\widetilde{\Omega}(u^s)$ is a vector function characterizing the remainder of the Taylor series.

II. To continue the proof, we use Lemma 1 on the existence of matrix V, the norm $\| \bullet \|_V$ and real numbers $\gamma_B > 0$ and $\mu_B > 0$ such that

$$\|(E + \gamma J(x^*)) u\|_V \leq (1 - \mu_B \gamma) \|u\|_V$$

for all $\gamma \in (0, \gamma_B]$ and $u \in R^m$.

We have the following estimation from (15)

$$\|u^{s+1}\|_V \leq \|A(x^*) u^s\|_V + \gamma \|\widetilde{\Omega}(u^s)\|_V,$$

and by virtue of (3)

$$\|\widetilde{\Omega}(u)\|_V \leq \mu_{\max} \|\widetilde{\Omega}(u)\|^2,$$

where $\mu_{\max}$ is the maximal eigenvalue of matrix V.

For all $u \in \mathfrak{Q}_x$ there exists $c > 0$ such that

$$\|\widetilde{\Omega}(u)\| \leq c\|u\|^2 \leq \frac{c}{\mu_{\min}}\|u\|_V,$$

where $\mu_{\min}$ is the minimal eigenvalue of matrix V. The following estimate holds

$$\|u^{s+1}\|_V \leq \|u^s\|_V\left[(1-\gamma\mu_B)+\gamma c\,\kappa\|u^s\|_V\right],$$

where

$$\kappa = \frac{c\mu_{\max}}{\mu_{\min}^2}$$

Consider another neighborhood

$$\widetilde{\mathfrak{O}}_u = \{u: \ \|u\| \leq \varepsilon\}, \tag{17}$$

where

$$\varepsilon = \frac{\mu_B}{2c\kappa} < \delta. \tag{18}$$

Then the estimate

$$\|u^{s+1}\|_V \leq \left(1-\frac{\mu_B\gamma}{2}\right)\|u^s\|_V, \quad \gamma \in (0,\gamma_B). \tag{19}$$

holds in $\widetilde{\mathfrak{O}}_u$.

III. According to the condition (c) of Theorem 1, for ε (18) there exists $\gamma(\varepsilon) > 0$ such that the process (15) comes to the neighborhood $\widetilde{\mathfrak{O}}_u$ (17) for all $\gamma \in (0,\gamma(\varepsilon)]$. Denote $a(\mathfrak{G}) = \min(\gamma(\varepsilon),\ \gamma_B)$. Then for $\gamma \in (0, a(\mathfrak{G})]$ the algorithm (14) converges to x^* linearly. ■

Let us consider the *zero-order multiplicative algorithm with the coordinate-wise scheme* of iterations

$$x_i^{s+1} = x_i^s f_i^\gamma(x_1^s,\ldots,x_l^s,x_{l+1}^s,\ldots,x_{l+r}^s), \quad i \in \overline{1,l}; \tag{20}$$

$$x_{l+j}^{s+1} = x_{l+j}^s f_{l+j}^\gamma(x_1^{s+1},\ldots,x_l^{s+1},x_{l+1}^s,\ldots,x_{l+r}^s), \quad j \in \overline{1,r}. \tag{21}$$

This algorithm differs from (1.16) because the variables here are divided into two groups, namely $r_1 = l$ and $r_2 = m - l = r$.

Theorem 4. *Let the conditions of Theorem 3 be satisfied for the algorithm* (20) – (21). *Then it $\mathfrak{G}$-converges.*

To prove the theorem, let us reduce (20) – (21) to (15). By condition (a) of Theorem 3, there exists a neighborhood $\mathfrak{D}_x$ of x^*, $\mathfrak{D}_x \subset R^m$, where (20) – (21) can be represented as

$$x_i^{s+1} = x_i^* + \sum_{k=1}^{l} a_{ik}(x^*)(x_k^s - x_k^*) + \sum_{k=1}^{r} a_{i,l+k}(x^*)(x_{l+k}^s - x_{l+k}^*) + \\ + \gamma\Omega_1(x^s - x^*), \tag{22}$$

$$x_{l+j}^{s+1} = x_{l+j}^* + \sum_{k=1}^{l} a_{i+j,k}(x^*)(x_k^{s+1} - x_k^*) + \\ + \sum_{k=1}^{r} a_{l+j,l+k}(x^*)(x_{l+k}^s - x_{l+k}^*) + \\ + \gamma\Omega_2(x^{s+1} - x^*,\ x^s - x^*), \tag{23}$$

where

$$\begin{aligned} a_{ik}(x^*) &= 1 + \gamma b_{ik}(x^*), \quad i, k \in \overline{1,l}; \\ a_{i,l+k}(x^*) &= \gamma b_{i,l+k}(x^*), \quad i \in \overline{1,l}; \quad k \in \overline{1,r}; \\ a_{l+j,k}(x^*) &= \gamma b_{l+j,k}(x^*), \quad j \in \overline{1,r}; \quad k \in \overline{1,l}; \\ a_{l+j,l+k}(x^*) &= 1 + \gamma b_{l+j,l+k}(x^*), \quad j \in \overline{1,r}; \quad k \in \overline{1,r}; \end{aligned} \tag{24}$$

$$b_{\alpha\beta}(x^*) = x_\alpha^* \frac{\partial f_\alpha(x^*)}{\partial x_\beta}; \tag{25}$$
$$\beta, \alpha \in \overline{1,m}; \quad m = l + r.$$

Denote

$$\begin{aligned} B_1(x^*) &= \left[b_{ik}(x^*),\ i, k \in \overline{1,l}\right]; \\ B_2(x^*) = B_3'(x^*) &= \left[b_{i,l+k}(x^*),\ i \in \overline{1,l},\ k \in \overline{1,r}\right]; \\ B_4(x^*) &= \left[b_{l+j,l+k}(x^*),\ j, k \in \overline{1,r}\right]; \\ u_i = x_i - x_i^*; \quad i \in \overline{1,l}; &\quad v_j = x_{l+j} - x_{l+j}^*; \quad j \in \overline{1,r}. \end{aligned} \tag{26}$$

Then (20) – (21) can be written as follows:

$$u^{s+1} = (E + \gamma B_1(x^*))\, u^s + \gamma B_2(x^*)\, v^s + \gamma\, \Omega_1(u^s, v^s); \tag{27}$$

$$v^{s+1} = \gamma B_3(x^*)\, u^{s+1} + (E + \gamma B_4(x^*))\, v^s + \gamma\, \Omega_2(u^{s+1}, u^s, v^s). \tag{28}$$

Denote

$$w = \begin{bmatrix} u \\ v \end{bmatrix}; \quad T(x^*, \gamma) = \begin{bmatrix} E & 0 \\ \gamma B_3(x^*) & E \end{bmatrix}; \tag{29}$$

$$\Omega = \begin{bmatrix} \Omega_1 \\ \Omega_2 \end{bmatrix}. \tag{30}$$

Therefore (27, 28) can be represented as follows

$$w^{s+1} = (E + \gamma B(x^*, \gamma))\, w^s + \gamma T(x^*, \gamma)\, \Omega\,(w^{s+1}, w^s), \tag{31}$$

where

$$B(x^*, \gamma) = B(x^*) + \gamma D(x^*); \tag{32}$$

$$B(x^*) = \begin{pmatrix} B_1(x^*) & B_2(x^*) \\ B_3(x^*) & B_4(x^*) \end{pmatrix}; \tag{33}$$

$$D(x^*) = \begin{pmatrix} 0 & 0 \\ B_3(x^*)\, B_1(x^*) & B_3(x^*)\, B_2(x^*) \end{pmatrix}. \tag{34}$$

Let us consider the neighborhood

$$\mathfrak{D}_w = \{w : \; \|w\| \le \delta_1\}. \tag{35}$$

By condition (a) of Theorem 1, for all δ_1 and $w \in \mathfrak{D}_w$ there exists $c > 0$ such that

$$\|\,\Omega\,(w)\,\| \le c\,\|w\|^2 \le \frac{c}{\mu_{\min}} \|w\|_V. \tag{36}$$

Note that the norm of $T^{-1}(x^*, \gamma)$ (30) is equal to 1 (the matrix has the only eigenvalue 1 of multiplicity m).

Then according to Corollary 1, the estimate

$$\|w^{s+1}\|_V \le \left[\left(1 - \gamma \widetilde{\lambda}_B\right) + c\,\gamma\,\kappa\,\|w^s\|_V\right]\, \|w^s\|_V \tag{37}$$

where

$$\kappa = \frac{c\mu_{\max}}{\mu^2_{\min}}$$

holds for $w \in \mathfrak{D}_w$, for

$$\gamma \in (0, \gamma_0), \quad \gamma_0 = \min\,(\widetilde{\gamma}_B, b); \tag{38}$$

where $\gamma_0, \widetilde{\gamma}_B$ and $\widetilde{\lambda}_B$ are determined by Corollary 1.

Consider another neighborhood

$$\widetilde{\mathfrak{D}}_w = \{w : \ \|w\| \leq \delta\} \subseteq \mathfrak{D}_w, \tag{39}$$

where

$$\delta = \frac{\widetilde{\lambda}_B}{2c\kappa} < \delta_1. \tag{40}$$

By substituting (40) into (37), we obtain

$$\|w^{s+1}\|_V \leq \left(1 - \gamma \frac{\widetilde{\lambda}_B}{2}\right) \|w^s\|_V, \quad w \in \widetilde{\mathfrak{D}}_w. \tag{41}$$

It follows from the condition (b) of Theorem 1 that there exists $\gamma(\delta) > 0$ such that the process (31) comes to $\widetilde{\mathfrak{D}}_w$ for all $\gamma \in (0, \gamma(\delta))$.

Denote $a(\mathfrak{G}) = \min(\gamma_0, \gamma(\delta))$. Then the algorithm (20) – (21) $\mathfrak{G}$-converges for $\gamma \in (0, a(\mathfrak{G}))$.

Convex programming.

Consider the convex programming problem of the form

$$\min f(x); \quad g_i(x) \geq 0, \quad i \in \overline{1,r}, \quad x \in R^m_+, \quad m < r, \tag{42}$$

where $f(x)$ and $g_i(x)$, $i \in \overline{1,r}$, are twice continuously differentiable functions on R^m; $f(x)$ is strictly convex (its Hessian is positive-definite at x^*) and $g_i(x)$ $(i \in \overline{1,r})$ are concave functions.

Without loss of generality, let

$$\begin{aligned} &g_1(x^*) = \ldots = g_l(x^*) = 0; \quad x_1^* = \ldots = x_k^* = 0; \\ &g_{l+1}(x^*) > 0, \ldots, g_r(x^*) > 0; \quad x_{k+1}^* > 0, \ldots, x_m^* > 0. \end{aligned} \tag{43}$$

Let the gradients of active constraints be linearly independent at the point x^* (the regularity condition) (see Polyak, 1983, Chapter 9).

For problem (42), let us introduce the Lagrange function

$$L(x, \lambda) = P(x, \lambda) - \sum_{j=1}^{m} \lambda_{r+j} x_j, \tag{44}$$

$$P(x, \lambda) = f(x) - \sum_{i=1}^{r} \lambda_i g_i(x), \tag{45}$$

$$x \in R^m_+, \quad \lambda \in R^{m+r}.$$

Assume that the strict complementarity hold for (42). Then there exists a pair (x^*, λ^*) for which

$$L_{x_j}(x^*, \lambda^*) = \left.\frac{\partial L(x, \lambda)}{\partial x_j}\right|_{x^*,\lambda^*} = 0, \quad j \in \overline{1,m}; \tag{46}$$

$$\lambda_1^* > 0, \ldots, \lambda_l^* > 0; \quad \lambda_{r+1}^* > 0, \ldots, \lambda_{r+k}^* > 0; \quad l \leq r; \tag{47}$$

$$\lambda_{l+1}^* = \ldots = \lambda_r^* = \lambda_{r+k+1}^* = \lambda_{r+m}^* = 0; \quad k \leq m. \tag{48}$$

Note that the dimension of this system is $2m + r$.

Since we apply the multiplicative algorithm to problem (42), we can use the reduced Lagrange function (45). The optimality conditions in terms of $P(x, \lambda)$ are of the form

$$P_{x_j}(x^*, \lambda^*) \geq 0, \quad x_j^* P_{x_j}(x^*, \lambda^*) = 0, \quad x_j^* \geq 0, \quad j \in \overline{1,m}; \tag{49}$$

$$P_{\lambda_j}(x^*, \lambda^*) \leq 0, \quad \lambda_i^* P_{\lambda_i}(x^*, \lambda^*) = 0, \quad \lambda_i^* \geq 0, \quad i \in \overline{1,r}. \tag{50}$$

The dimension of this system is $m + r$.

Denote

$$Q_j(x, \lambda) = 1 - \gamma \frac{\partial P(x, \lambda)}{\partial x_j}, \quad j \in \overline{1,m}; \tag{51}$$

$$K_i(x, \lambda) = 1 + \gamma \frac{\partial P(x, \lambda)}{\partial \lambda_i}, \quad i \in \overline{1,r}; \quad \gamma > 0. \tag{52}$$

To solve (42), we consider two classes of multiplicative algorithms, namely (a) *with parallel organization of iterations*

$$x_j^{s+1} = x_j^s Q_j(x^s, \lambda^s), \quad j \in \overline{1,m}; \tag{53}$$

$$\lambda_i^{s+1} = \lambda_i^s K_i(x^s, \lambda^s), \quad i \in \overline{1,r}; \tag{54}$$

and

(b) *with coordinate-wise organization of iterations*

$$x_j^{s+1} = x_j^s Q_j(x^s, \lambda^s), \quad j \in \overline{1,m}; \tag{55}$$

$$\lambda_i^{s+1} = \lambda_i^s K_i(x^{s+1}, \lambda^s), \quad i \in \overline{1,r}. \tag{56}$$

According to the classification of convex programming algorithms suggested by Polyak (Polyak B. T., 1983, Chapter 9, Section 3, pp.251 – 254), the multiplicative algorithms (53) and (54) belong to the class of dual methods. The most popular representatives of this class are the gradient algorithm of Arrow-Hurwitz-Udzawa (AHU) and methods based on the augmented Lagrange function.

The methods of the first group use the operation of projection onto the nonnegative orthant with respect to dual variables corresponding to the constraints and nonnegativity conditions in (42).

Besides, the strong convexity of $f(x)$ in (42) is necessary for the AHU methods to converge ($\nabla^2 f(x) \geq a\, E$, E is the unit matrix and $a > 0$).

Methods based on the augmented Lagrange function have an auxiliary control parameter, the "penalty". The convergence can be proved for strongly convex functions $f(x)$ in (42), under an additional condition on the Hessians of $f(x)$ and $g_i(x)$ (a Lipschitz condition).

Comparing the algorithms mentioned above with that of (53, 54), we see that they do not have projection operations and are provided with only one control parameter, namely the step coefficient.

Below we show that the convergence of (53) – (54) can be proved under weaker assumptions on functions $f(x)$ and $g_1(x), \ldots, g_r(x)$; in particular, the strong convexity of $f(x)$ and a Lipschitz condition for Hessian of $f(x)$ and $g_i(x)$, $i \in \overline{1,r}$ are not necessary.

Below $\mathfrak{G}$-convergence of these algorithms is investigated.

The investigation is based on study of behaviour of the solution to the system of differential equations

$$\dot{x}_j = -x_j P_{x_j}(x, \lambda), \quad j \in \overline{1,m}; \tag{57}$$

$$\dot{\lambda}_i = \lambda_i P_{\lambda_i}(x, \lambda), \quad i \in \overline{1,r} \tag{58}$$

on

$$R_+^{r+m} = \left\{(x, \lambda) \,|\, x_j \geq 0; \quad \lambda_i \geq 0; \quad i \in \overline{1,r}; \quad j \in \overline{1,m}\right\}. \tag{59}$$

(see Aliev, Dubov, Izmailov, Popkov, 1985).

Lemma 5. *The poin (x^*, λ^*) is the singular point of the differential equations* (57, 58) *and it is asymptotically stable for any $(x^0, \lambda^0) \in R_+^{m+r}$.*

Proof. Let us define the function

$$V(x,\lambda) = \sum_{j=1}^{m}(x_j - x_j^*) - x_j^*(\ln x_j - \ln x_j^*) +$$
$$+\sum_{i=1}^{r}(\lambda_i - \lambda_i^*) - \lambda_i^*(\ln\lambda_i - \ln\lambda_i^*) \tag{60}$$

on R_+^{m+r}. The function $V(x,\lambda)$ can take infinite values. Besides, if $y=0$ then we put $y\ln y = 0$. The function V is strictly convex on R_+^{m+r}, attains its minimum at (x^*,λ^*) and the Hessian of V is non-degenerate at (x^*,λ^*). Hence the sets $\mathfrak{V}_a = \{(x,\lambda):\ V(x,\lambda) < a\}$ are bounded, and there exists a constant $c>0$ such that $\mathfrak{V}_c = \{(x,\lambda):\ V(x,\lambda) \le c\} \supset \mathfrak{V}_a$.

Let us determine the derivative of $V(x,\lambda)$ along the trajectories of (57, 58)

$$\frac{dV}{dt} = -\sum_{j=1}^{m}(x_j - x_j^*)\left(\frac{\partial f(x)}{\partial x_j} - \sum_{i=1}^{r}\lambda_i\frac{\partial g_i(x)}{\partial x_j}\right) -$$
$$-\sum_{i=1}^{r}(\lambda_i - \lambda_i^*)g_i(x) =$$
$$= \left[f(x) - f(x^*) - \sum_{j=1}^{m}(x_j - x_j^*)\frac{\partial f(x)}{\partial x_j}\right] +$$
$$+\left[\sum_{i=1}^{r}\lambda_i\left(\sum_{j=1}^{m}(x_j - x_j^*)\frac{\partial g_i(x)}{\partial x_j} - (g_i(x) - g_i(x^*))\right)\right] +$$
$$+\left[\sum_{i=1}^{r}\lambda_i^* g_i(x) - \sum_{i=1}^{r}\lambda_i g_i(x^*) - (f(x) - f(x^*))\right]. \tag{61}$$

Consider the sign of (61) on R_+^{m+r}. The term in the second bracket is non-positive due to the concavity of $g_i(x)$ $(i \in \overline{1,r})$. The term in the first bracket is negative for $x \ne x^*$ (due to the convexity of $f(x)$ and nondegeneracy of its Hessian at x^*) and is zero at $x = x^*$. Let us consider the term in the third bracket (61) and use the definition of a saddle point:

$$L(x^*,\lambda) \le L(x^*,\lambda^*) \le L(x,\lambda^*) \tag{62}$$

for all $\lambda \ge 0$ and $x \ge 0$. In particular, (62) should hold for $\lambda_{r+k+1} = \ldots = \lambda_{r+m} = 0$:

$$f(x^*) - \sum_{i=1}^{r} \lambda_i g_i(x^*) \le f(x^*) \le f(x) - \sum_{i=1}^{r} \lambda_i^* g_i(x) - \sum_{j=1}^{k} \lambda_{r+j}^* x_j.$$

Since $\sum_{j=1}^{k} \lambda_{r+j}^* x_j \ge 0$, the right inequality can be strengthened as follows

$$f(x^*) - \sum_{i=1}^{r} \lambda_i g_i(x^*) \le f(x^*) \le f(x) - \sum_{i=1}^{r} \lambda_i^* g_i(x).$$

Thus

$$f(x^*) - f(x) - \sum_{i=1}^{r} \lambda_i g_i(x^*) + \sum_{i=1}^{r} \lambda_i^* g_i(x) \le 0,$$

i.e. the third bracket in (61) is nonpositive.

Thus (61) is nonpositive on R_+^{m+r}. Moreover, if $x \ne x^*$ then (61) is negative; if $x = x^*$ then (in this case the first and second brackets in (61) are zero) the third bracket in (61), by virtue of (47, 48), is

$$-\sum_{i=l+1}^{r} \lambda_i g_i(x^*),$$

i.e. it is either zero (if $\lambda_{l+1} = \lambda_{l+1}^* = \ldots = \lambda_r = \lambda_r^* = 0$) or negative.

Thus

$$\frac{dV}{dt} \begin{cases} < 0 & ((x, \lambda) \in R_+^{m+r} \setminus \mathfrak{L}), \\ = 0 & ((x, \lambda \in \mathfrak{L}), \end{cases} \tag{63}$$

where

$$\mathfrak{L} = \{(x, \lambda) \in R_+^{m+r} \mid x = x^*, \quad \lambda_{l+1} = \ldots = \lambda_r = 0\}. \tag{64}$$

Note that $(x^*, \lambda^*) \in \mathcal{L}$.

If $l = 0$ then $\mathfrak{L} = (x^*, \lambda^*)$ (64), and V is the Lyapunov function for (57, 58), i.e. the derivative of V along the solution of (57, 58) is negative and strictly separated from zero.

Now let $l \ge 1$. Note that in this case $l+k \le m$, since $\nabla g_1(x^*), \ldots, \nabla g_l(x^*)$, $e_1, \ldots, e_k$ are linearly independent. In this case, we consider $P(x, \lambda)$ (45).

Let us calculate the derivative of $P(x, \lambda)$ along the trajectories of (57, 58):

$$\frac{dP}{dt} = -\sum_{j=1}^{m} x_j \left(\frac{\partial f(x)}{\partial x_j} - \sum_{i=1}^{r} \lambda_i \frac{\partial g_i(x)}{\partial x_j} \right)^2 + \sum_{i=1}^{r} \lambda_i g_i^2(x).$$

By virtue of (46 – 48), we obtain for the set $\mathfrak{L}$ (64)

$$\frac{dP}{dt} = -\sum_{j=k+1}^{m} x_j^* \left(\frac{\partial f(x^*)}{\partial x_j} - \sum_{i=1}^{l} \lambda_i \frac{\partial g_i(x^*)}{\partial x_j} \right)^2 =$$

$$= -\sum_{j=k+1}^{m} x_j^* \left(\frac{\partial f(x^*)}{\partial x_j} - \sum_{i=1}^{l} \lambda_i^* \frac{\partial g_i(x^*)}{\partial x_j} + \sum_{i=1}^{l} (\lambda_i^* - \lambda_i) \frac{\partial g_i(x^*)}{\partial x_j} \right)^2 .$$

By virtue of (47, 48)

$$P_{x_{k+1}}(x^*, \lambda^*) = L_{x_{k+1}}(x^*, \lambda^*) = \ldots = P_{x_m}(x^*, \lambda^*) = L_{x_m}(x^*, \lambda^*) = 0.$$

We obtain from these

$$\frac{dP}{dt} = -\sum_{j=k+1}^{m} x_j^* \left(\sum_{i=1}^{r} (\lambda_i^* - \lambda_i) \frac{\partial g_i(x^*)}{\partial x_j} \right)^2 . \tag{65}$$

It follows from the linear independence of $\nabla g_1(x^*), \ldots, \nabla g_l(x^*), e_1, \ldots, e_k$ that the last $m-k$ coordinates of the vectors $\nabla g_1(x^*), \ldots, \nabla g_l(x^*), e_1, \ldots, e_k$ are linearly independent (note that $m - k \geq l \geq 1$). Thus, (65) is a continuous quadratic function of $\lambda^* - \lambda$. We obtain

$$\frac{dP}{dt} \begin{cases} < 0 & ((x, \lambda) \in \mathfrak{L} \setminus (x^*, \lambda^*)), \\ = 0 & ((x, \lambda) = (x^*, \lambda^*)). \end{cases} \tag{66}$$

Denote the δ-neighborhood of $\mathfrak{L}$ (64) by $\mathfrak{L}_\delta$, and the ε-neighborhood of the point (x^*, λ^*) by $\mathfrak{O}_\varepsilon$.

Consider the function

$$V_\eta(x, \lambda) = V(x, \lambda) + \eta P(x, \lambda), \tag{67}$$

$$(x, \lambda) \in R_+^{m+r}; \quad \eta > 0,$$

and the sets

$$\mathfrak{V}_c = \{(x, \lambda) \mid V(x, \lambda) \leq c\},$$

$$\mathcal{M}_c(\eta) = \{(x, \lambda) \mid V_\eta(x, \lambda) \leq c\}.$$

Since the functions V and P are convex, there exists $\varepsilon > 0$ such that

$$\mathfrak{V}_c \subset \mathcal{M}_{c+\varepsilon}(\eta) \subset \mathfrak{V}_{c+2\varepsilon}. \tag{68}$$

By the continuity of (65), there exists $\delta > 0$ such that

$$\frac{dP}{dt} < 0 \quad \text{for} \quad (x, \lambda) \in (\mathfrak{V}_c \cap \mathfrak{L}_\delta) \setminus \mathfrak{D}_\varepsilon .$$

According to (68), there exists $\eta^* > 0$ such that

$$\eta \left| \frac{dP}{dt} \right| \leq \frac{1}{2} \left| \frac{dV}{dt} \right|$$

for all $\eta \in (0, \eta^*)$, where $(x, \lambda) \in \mathcal{W} = (\mathcal{M}_{c+\varepsilon}(\eta) \cap \mathfrak{L}_\delta) \setminus \mathfrak{D}_\varepsilon$.

Hence the function $V_\eta(x, \lambda)$ for $(x, \lambda) \in \mathcal{W}$ is a Lyapunov function for the system (57, 58), because its time derivative along the trajectories of (57, 58) is negative and strictly separated from zero; it becomes zero only at the point (x^*, λ^*).

Thus, for (57, 58) there exists a constant $\eta > 0$ and the function

$$v(x, \lambda) = \begin{cases} V(x, \lambda) & \text{for} \quad (x, \lambda) \in R_+^{m+r} \\ V(x, \lambda) + \eta P(x, \lambda) & \text{for} \quad (x, \lambda) \in \mathfrak{L} \end{cases}$$

whose time derivative is negative for all $(x, \lambda) \in R_+^{m+r}$ and is zero at (x^*, λ^*), and this finishes the proof of the lemma. ■

Lemma 6. *The Jacobian of* (57, 58) *is Hurwitz matrix at* (x^*, λ^*).

Proof. The matrix J is

$$J = \begin{bmatrix} \dfrac{\partial}{\partial x_p}(-x_q P_{x_q}) & \dfrac{\partial}{\partial x_p}(\lambda_s P_{\lambda_s}) \\ \dfrac{\partial}{\partial \lambda_t}(-x_q P_{x_q}) & \dfrac{\partial}{\partial \lambda_t}(\lambda_s P_{\lambda_s}) \end{bmatrix}_{(x^*, \lambda^*)}, \tag{69}$$

where $p, q \in \overline{1, m}$; $s, t \in \overline{1, r}$, p and t correspond to the rows, and q and s correspond to the columns of J. Note that, by virtue of (43 – 48), the relations

$$P_{x_j}(x^*, \lambda^*) = L_{x_j}(x^*, \lambda^*) + \lambda^*_{r+j} \geq 0, \quad j \in \overline{1, k};$$

$$P_{x_j}(x^*, \lambda^*) = L_{x_j}(x^*, \lambda^*), \quad j \in \overline{k+1, m}.$$

hold. Let us rewrite (69), taking into account these equations. Then

$$J = \left[\begin{array}{c|c|c|c}
\begin{matrix}-\lambda^*_{r+1} & & \\ & \ddots & \\ & & -\lambda^*_{r+k}\end{matrix} & \begin{matrix}\cdots\cdots\\ \cdots\cdots \\ \cdots\cdots\end{matrix} & & 0 \\ \hline
 & H & U & \\ \cline{2-3}
0 & V & 0 & \\ \cline{2-4}
 & \begin{matrix}\cdots\\ \cdots\\ \cdots\end{matrix} & \begin{matrix} \\ 0 \\ \\ \end{matrix} & \begin{matrix}-g_{l+1} & & \\ & \ddots & \\ & & -g_r\end{matrix}
\end{array}\right]_{(x^*,\lambda^*)}, \qquad (70)$$

where the elements of the matrices H, U and V are equal to

$$h_{pq} = -x_q^* P_{x_p x_q}; \quad u_{ps} = -\lambda_s^* \frac{\partial g_s}{\partial x_p}; \quad v_{tq} = x_q^* \frac{\partial g_t}{\partial x_q}; \qquad (70')$$

and

$$p, q \in \overline{k+1, m}; \quad s, t \in \overline{l+1, r},$$

respectively.

The dots in (70) can be replaced by zeros, with J remaining stable. Indeed, the Raus-Hurwitz criterion of matrix stability links the matrix stability with the properties of coefficients in the characteristic polynom. These coefficients are the sums of main minors. It follows from (70) that none of these minors will be changed in replacing the dots by zeros. Thus, instead of proving the stability of J, we can prove the stability of J_1, with J_1 differing from J by zeros and an orthogonal transformation of basis (this does not change the eigenvalues and hence does not influence the stability of J):

$$J = \left[\begin{array}{c|cc}
\begin{matrix}-\lambda^*_{r+1} & & & & & \\ & \ddots & & 0 & & \\ & & -\lambda^*_{r+k} & & & \\ & & & -g_{l+1}(x^*) & & \\ & 0 & & & \ddots & \\ & & & & & -g_m(x^*)\end{matrix} & & 0 \\ \hline
 & H & U \\ \cline{2-3}
0 & V & 0
\end{array}\right]. \qquad (71)$$

The negative-definiteness and nondegeneracy of the matrix $-P_{x_p x_p}(x^*, \lambda^*)$ follows from the convexity of f, concavity of g_i $(i \in \overline{1,r})$ and non-degeneracy of the Hessian of f at x^* . Besides, the matrix $\nabla g(x^*)$ is of full rank (a corollary of the regularity conditions) and the numbers x_q^*, λ_s^* $(q \in \overline{k+1,m};\ s \in \overline{1,l};\ \lambda_{r+1}^*, \ldots, \lambda_{r+k}^*;\ g_{l+1}(x^*), \ldots, g_r(x^*)$ are positive.

The spectrum of J_1 consists of $-\lambda_{r+1}^*, \ldots, -\lambda_{r+k}^*, -g_{l+1}(x^*), \ldots, -g_m(x^*) < 0$ and eigenvalues of the matrix

$$\widetilde{J}_1 = \begin{bmatrix} H & U \\ V & 0 \end{bmatrix},$$

whose elements are given by the equalities (70′).

Theorem 6. *For any bounded set* $\mathfrak{G} \subset \operatorname{int} R_+^{m+r}$ *the algorithm* (55, 54) $\mathfrak{G}$*-converges to the point* (x^*, λ^*) *solving the problem* (42).

The proof follows from Lemmas 3–5 and Theorem 2.

Now let us study the $\mathfrak{G}$*-convergence of algorithm* (55, 56). Consider a neighborhood $\mathfrak{O}$ of (x^*, λ^*) in $\mathfrak{G} \subset R_+^{m+r}$ such that for all $(x, \lambda) \in \mathfrak{O}$ the algorithm (55, 56) can be represented as follows

$$x^{s+1} = x^* + [E + \gamma M_x(x^*, \lambda^*)]\,(x^s - x^*) + \gamma M_\lambda(x^*, \lambda^*)(\lambda^s - \lambda^*) + \\ + \gamma \Omega_1(x^s - x^*,\ \lambda^s - \lambda^*), \tag{72}$$

$$\lambda^{s+1} = \lambda^* + [E + + \gamma U_\lambda(x^*, \lambda^*)]\,(\lambda^s - \lambda^*) + \\ + \gamma U_x(x^*, \lambda^*)(x^{s+1} - x^*) + \gamma \Omega_2(x^{s+1} - x^*, \lambda^* - \lambda^*),$$

where the elements of $(m \times m)$ matrix $M_x(x^*, \lambda)$ are

$$m_{jp}^x = -x_j^* \left.\frac{\partial^2 P}{\partial x_p\, \partial x_j}\right|_*, \quad p, j \in \overline{1,m}; \tag{73}$$

the elements of $(m \times r)$ matrix $M_\lambda(x^*, \lambda^*)$ are

$$m_{jn}^\lambda = -x_j^* \left.\frac{\partial^2 P}{\partial \lambda_n\, \partial x_j}\right|_*, \quad n \in \overline{1,r}; \quad j \in \overline{1,m}; \tag{74}$$

the elements of $(r \times m)$ matrix $U_x(x^*, \lambda^*)$ are

$$u_{ip}^x = \lambda_i^* \left.\frac{\partial^2 P}{\partial x_p\, \partial \lambda_i}\right|_* \quad p \in \overline{1,m}; \quad i \in \overline{1,r}; \tag{75}$$

the elements of $(r \times r)$ matrix $U_\lambda(x^*, \lambda^*)$ are

$$u_{in}^{\lambda} = \lambda_i^* \frac{\partial^2 P}{\partial \lambda_n \partial \lambda_i}, \quad n, i \in \overline{1, r}; \tag{76}$$

$\Omega_1(x, \lambda)$ and $\Omega_2(x, \lambda)$ are vector functions characterizing the corresponding remainders of Taylor series.

Let us introduce the notations

$$u = x - x^*; \quad v = \lambda - \lambda^*; \quad w = \begin{bmatrix} u \\ v \end{bmatrix}. \tag{77}$$

Then (72) can be represented as

$$w^{s+1} = (E + \gamma A(\gamma))\, w^s + \gamma T \Omega\,(w^s, w^{s+1}), \tag{78}$$

where

$$A(\gamma) = A + \gamma D; \tag{79}$$

$$A = \begin{bmatrix} M_x & M_\lambda \\ U_x & U_\lambda \end{bmatrix}; \quad D = \begin{bmatrix} 0 & 0 \\ U_x M_x & U_x M_\lambda \end{bmatrix}; \quad T = \begin{bmatrix} E & 0 \\ \gamma U_x & E \end{bmatrix}. \tag{80}$$

Lemma 7. *The matrix A is Hurwitz one.*

To prove the theorem, it is sufficient to write down the matrix A according to (73) – (76) and to confirm that its structure coincides with that of matrix J (70) (see also Lemma 6).

Lemma 8. *The matrix $U_x M_\lambda$* (74, 75) *is Hurwitz one.*

Proof. From (74) and (75) we obtain

$$U_x M_\lambda = -\mathrm{diag}\,(\lambda_i^*)\, S, \tag{81}$$

where

$$S = \left[\sum_{j=1}^{m} x_j^* \left(\frac{\partial g_n}{\partial x_j} \frac{\partial g_i}{\partial x_j} \right)_* \right]. \tag{81'}$$

The matrix S is symmetrical and, by the regularity conditions (linear independence of constraint gradients at x^*), is strictly positive-definite (i.e. its eigenvalues have strictly positive real parts).

Then it follows from (81) that the matrix $U_x M_\lambda$ has eigenvalues with non-positive real parts.

Lemma 9. *The matrix $A(\gamma)$ is Hurwitz one for all $\gamma > 0$.*

Proof. It follows from (80) that the spectrum of the matrix D consists of zero eigenvalue with multiplicity m and non-positive eigenvalues of the matrix $U_x M_\lambda$ (81). Since A is Hurwitz (Lemma 7), $A(\gamma)$ is also Hurwitz, according to (79).

Theorem 7. *Let the condition (c) of Theorem* 1 *be satisfied for* (55, 56). *Then the algorithm* (55, 56) *$\mathfrak{G}$-converges.*

The proof follows obviously from Theorem 1 and Lemma 9.

4.3 Analysis of multiplicative algorithms for macrosystems with linear consumption of resources

Macrosystems with complete resource consumption. For MSS of this class, the set $\mathfrak{D}$ of feasible states is characterized by the linear consumption function in (2.4.3). To study the properties of computational methods, the set $\mathfrak{D}$ should be described after norming the parameters of the linear consumption function.

Since $q_k > 0$ ($k \in \overline{1,r}$), the feasible set $\mathfrak{D}$ can be represented as

$$\mathfrak{D} = \left\{ N : \sum_{n=1}^{m} \tilde{t}_{kn} N_n = 1, \quad k \in \overline{1,r} \right\}, \tag{1}$$

where

$$\tilde{t}_{kn} = t_{kn}/q_k. \tag{1'}$$

Assume that $T = [\tilde{t}_{kn}]$ has the rank r. This representation of $\mathfrak{D}$ causes changes in expressions determining the solutions of the problems of constrained entropy maximization for $F-$, $E-$ and $B-$MSS. The corresponding expressions can be easily obtained by using the techniques of Chapter 2.

The stationary state for F-model is

$$N_n^*(\lambda) = \frac{G_n}{1 + b_n \exp\left(\sum_{j=1}^{r} \lambda_j \tilde{t}_{jn}\right)}, \quad n \in \overline{1,m}; \tag{2}$$

$$b_n = 1/\widetilde{a}_n, \quad \widetilde{a}_n = a_n/(1-\alpha_m).$$

The Lagrange multipliers $\lambda_1, \ldots, \lambda_r$ included into this equality are determined from the following system of equations:

$$\Phi_k(\lambda) = \sum_{n=1}^{m} \frac{\widetilde{t}_{kn} G_n}{1 + b_n \exp\left(\sum_{j=1}^{r} \lambda_j \widetilde{t}_{jn}\right)} = 1, \quad k \in \overline{1,r}. \tag{3}$$

The stationary state for B-model is

$$N_n^*(\lambda) = \frac{\widetilde{a}_n}{\exp\left(\sum_{j=1}^{r} \lambda_j \widetilde{t}_{jn}\right)}, \quad \widetilde{a}_n = a_n G_n, \quad n \in \overline{1,m}. \tag{4}$$

The Lagrange multipliers $\lambda_1 \ldots, \lambda_r$ are determined from the following system of equations

$$M_k(\lambda) = \sum_{n=1}^{m} \frac{\widetilde{a}_n \widetilde{t}_{kn}}{\exp\left(\sum_{j=1}^{r} \lambda_j \widetilde{t}_{jn}\right)} = 1, \quad k \in \overline{1,r}. \tag{5}$$

The stationary state for E-model is

$$N_n^*(\lambda) = \frac{G_n}{c_n \exp\left(\sum_{j=1}^{r} \lambda_j \widetilde{t}_{jn}\right) - 1}, \quad c_n = 1/a_n; \quad n \in \overline{1,m}. \tag{6}$$

The Lagrange multipliers $\lambda, \ldots, \lambda$ are determined from the following system of equations:

$$\Psi_k(\lambda) = \sum_{n=1}^{m} \frac{\widetilde{t}_{kn} G_n}{c_n \exp\left(\sum_{j=1}^{r} \lambda_j \widetilde{t}_{jn}\right) - 1} = 1, \quad k \in \overline{1,r}. \tag{7}$$

In (3), (5) and (7) we use non-negative variables

$$z_j = \exp(\lambda_j), \quad j \in \overline{1,r}.$$

We obtain

(a) for F-MSS:

$$\varphi_k(z) = \sum_{n=1}^{m} \frac{\widetilde{t}_{kn} G_n}{1 + b_n \prod_{j=1}^{r} z_j^{\widetilde{t}_{jn}}} = 1, \quad k \in \overline{1,r}; \tag{8}$$

(b) for B-MSS:

$$m_k(z) = \sum_{n=1}^{m} \frac{\widetilde{\alpha}_n \widetilde{t}_{kn}}{\prod_{j=1}^{r} z_j^{\widetilde{t}_{jn}}} = 1, \quad k \in \overline{1,r}; \tag{9}$$

(c) for E-MSS:

$$\psi_k(z) = \sum_{n=1}^{m} \frac{\widetilde{t}_{kn} G_n}{c_n \prod_{j=1}^{r} z_j^{\widetilde{t}_{jn}} - 1} = 1, \quad k \in \overline{1,r}. \tag{10}$$

To solve the equations (8), (9) and (10), we use the multiplicative algorithm (2.25):

$$z_k^{s+1} = z_k^s f_k^\gamma(z^s), \tag{11}$$

where

$$f_k(z) = \begin{cases} \varphi_k(z) & (F\text{-MSS (8)}); \\ m_k(z) & (B\text{-MSS (9)}); \\ \psi_k(z) & (E\text{-MSS (10)}); \end{cases} \tag{12}$$

$z^0 = \{z_1^0, \dots, z_r^0\} \in \widetilde{\mathfrak{G}} \subset \operatorname{int} R_+^r$ and $\widetilde{\mathfrak{G}}$ is a compact set.

The additive form of the algorithm (11) will be used along with the algorithm itself, namely

$$\lambda_k^{s+1} = \lambda_k^s + \gamma \ln F_k(\lambda^s), \tag{13}$$

where

$$F_k(\lambda) = \begin{cases} \Phi_k(\lambda) & (F\text{-MSS (3)}); \\ M_k(\lambda) & (B\text{-MSS (5)}); \\ \Psi_k(\lambda) & (E\text{-MSS (7)}); \end{cases} \tag{14}$$

$$\lambda^0 = \{\lambda_1^0, \dots, \lambda_r^0\} \in \mathfrak{G}$$

and $\mathfrak{G}$ is the image of the set $\widetilde{\mathfrak{G}}$ in R^r.

The dual function

$$\widetilde{L}(\lambda) = H(N^*(\lambda)) + \sum_{k=1}^{r} \lambda_k \left(1 - \sum_{n=1}^{m} \widetilde{t}_{kn} N_n^*(\lambda)\right), \tag{15}$$

is important in studying the convergence of (11). Here

$$H\left(N^*(\lambda)\right) = \begin{cases} H_F(N^*(\lambda)) & (F\text{-MSS}); \\ H_B(N^*(\lambda)) & (B\text{-MSS}); \\ H_E(N^*(\lambda)) & (E\text{-MSS}); \end{cases} \tag{16}$$

$\mathfrak{G}$-convergence of the algorithm (11)–(12) can be proved by using Theorem 2.2. Consider the system of differential equations obtained from (13) as $\gamma \to 0$:

$$\dot{\lambda} = \ln F(\lambda), \tag{17}$$

where $\ln F(\lambda) = \{\ln F_1(\lambda), \ldots, \ln F_r(\lambda)\}$.

The Jacobian of this system is

$$J(\lambda^*) = \left[\frac{\partial F_k(\lambda)}{\partial \lambda_j}\right]_{\lambda^*}, \tag{17'}$$

since $F_k(\lambda^*) = 1$, $k \in \overline{1,r}$.

Let us consider the F-MSS. From (3) and (14) we obtain

$$J_{kj}(\lambda) = \frac{\partial \Phi_k}{\partial \lambda_j} = -\sum_{n=1}^{m} \tilde{t}_{kn} \tilde{t}_{jn} \alpha_n < 0; \quad k, j \in \overline{1,r}, \tag{18}$$

where

$$\alpha_n = \frac{G_n b_n \exp\left(\sum_{j=1}^{r} \lambda_j \tilde{t}_{jn}\right)}{\left[1 + b_n \exp\left(\sum_{j=1}^{r} \lambda_j \tilde{t}_{jn}\right)\right]^2} > 0, \quad n \in \overline{1,m}. \tag{18'}$$

Consider the symmetrical matrix $J = \left[\frac{\partial \Phi_k}{\partial \lambda_j}\right]$ and quadratic form $K = \langle J\xi, \xi \rangle$ for any vectors $\xi \in R^r$:

$$K = -\sum_{k,j=1}^{r} \sum_{n=1}^{m} \alpha_n \tilde{t}_{kn} \tilde{t}_{jn} \xi_k \xi_j = -\sum_{n=1}^{m} \alpha_n \left(\sum_{k=1}^{r} \tilde{t}_{kn} \xi_k\right)^2 \leq 0. \tag{19}$$

Since the matrix $T = [\tilde{t}_{kn}]$ is of full rank, $K = 0$ only for $\xi \equiv 0$ i.e. the quadratic form K is negative-definite.

The matrix J is symmetrical, therefore its eigenvalues $\mu_j (j \in \overline{1,r})$ are real. Reducing K to the main axis, we obtain

$$K = \sum_{j=1}^{r} \mu_j u_j^2 \begin{cases} < 0 & (u \in R^r), \\ = 0 & (u \equiv 0). \end{cases} \tag{20}$$

where μ_j are the eigenvalues of J. We obtain from (20) that $\mu_j < 0$, i.e. the matrix J is Hurwitz one, and condition (b) of Theorem 2.2 is satisfied for algorithm the (13), (14, F-MSS).

Lemma 1. *The solution of* (17) *is asymptotically stable for any* $\lambda \in R^r$.

Proof. Consider the dual function (15). Its gradient components are

$$\frac{\partial \widetilde{L}}{\partial \lambda_k} = \sum_{n=1}^{m} \left. \frac{\partial H_F}{\partial N_n^*} \right|_{N^*} \frac{\partial N_n^*}{\partial \lambda_k} - \sum_{k=1}^{r} \lambda_k \widetilde{t}_{kn} \frac{\partial N_n^*}{\partial \lambda_k} + \\ + \left(1 - \sum_{n=1}^{m} \widetilde{t}_{kn} N_n^*(\lambda) \right). \tag{21}$$

By the optimality conditions,

$$\frac{\partial H_F}{\partial N_n^*} - \sum_{k=1}^{r} \lambda_k \widetilde{t}_{kn} = 0, \quad n \in \overline{1,m}.$$

Therefore, taking (2) and (3) into account, we have

$$\frac{\partial \widetilde{L}}{\partial \lambda_k} = 1 - \Phi_k(\lambda), \quad k \in \overline{1,r}. \tag{22}$$

It follows from this that the function $\widetilde{L}(\lambda)$ (15) has the Hessian

$$\widetilde{\Gamma} = \left[\frac{\partial^2 \widetilde{L}}{\partial \lambda_j \, \partial \lambda_k} \right] = -J = \left[\frac{\partial \Phi_k}{\partial \lambda_j} \right].$$

According to (19), the matrix $\widetilde{\Gamma}$ is strictly positive-definite, and therefore the function $\widetilde{L}(\lambda)$ (15) is strictly convex. Hence it has the unique minimum at a finite point λ^*.

Let us consider the time derivative of $\widetilde{L}(\lambda)$ along the trajectories of the system (17)

$$\frac{d\widetilde{L}}{dt} = \sum_{k=1}^{r} \frac{\partial \widetilde{L}}{\partial \lambda_k} \dot{\lambda}_k = \sum_{k=1}^{r} [1 - \Phi_k(\lambda)] \ln \Phi_k(\lambda). \tag{23}$$

We obtain

$$\frac{d\widetilde{L}}{dt}\begin{cases} < 0, & \Phi_k(\lambda) \neq 1 \quad \text{or} \quad \dfrac{\partial \widetilde{L}}{\partial \lambda_k} \neq 0 \\ = 0, & \Phi_k(\lambda) = 1 \quad \text{or} \quad \dfrac{\partial \widetilde{L}}{\partial \lambda_k} = 0, \quad k \in \overline{1,r}. \end{cases} \tag{23'}$$

By the uniqueness of the minimum point of $\widetilde{L}(\lambda)$, its gradient is zero at this unique point. Hence

$$\frac{d\widetilde{L}}{dt}\begin{cases} < 0, & \text{for all } \lambda \in R^r, \text{ (excluding infinity)} \\ & \text{and } \lambda \neq \lambda^*; \\ = 0, & \text{for } \lambda = \lambda^*. \end{cases} \tag{24}$$

Thus, $\widetilde{L}(\lambda)$ (15) is a Lyapunov function for the system (17)

Theorem 1. *The algorithm* (13) (*and hence* (11)) $\mathfrak{G}$-*converges for F-MSS* (12)–(14).

The *proof* follows from the conditions (a), (b) and (c) of Theorem 2.2 being satisfied. The condition (d) of Theorem 2.2 is satisfied by Lemma 1.

Let us consider the B-MSS. From (5) we obtain

$$J_{kj}(\lambda) = \frac{\partial M_k}{\partial \lambda_j} = -\sum_{n=1}^{m} \beta_n \tilde{t}_{kn} \tilde{t}_{jn} < 0, \quad k, j \in \overline{1,r}, \tag{25}$$

where

$$\beta_n = \frac{\widetilde{a}_n}{\exp\left(\sum_{j=1}^{r} \lambda_j \tilde{t}_{jn}\right)} > 0, \quad n \in \overline{1,m}. \tag{25'}$$

Comparing the equations (25), (25′) and (18), (18′), we see that they have the same structure. Therefore the Jacobian $J(\lambda^*)$ for B-MSS is Hurwitz one. It is easy to see that Lemma 1 holds in this case as well.

Hence the following theorem holds.

Theorem 2. *The algorithm* (13) $\widetilde{\mathfrak{G}}$-*converges* (*and hence the algorithm* (11) $\mathfrak{G}$-*converges*) for *B-MSS* (14).

Let us consider the E-MSS. The determination of the macrosystem stationary state for the E-model is reduced to solving the equations (7). But, unlike F- and B-models (3) and (5), the functions $\psi_k(\lambda)$ in (7) are discontinuous on the set

$$\Lambda_0 = \bigcup_{n=1}^{m} \{\lambda : \ \delta_n(\lambda) = 0\}. \tag{26}$$

where

$$\delta_n(\lambda) = c_n \exp\left(\sum_{j=1}^{r} \lambda_j \tilde{t}_{jn}\right) - 1. \tag{26'}$$

Consider the system of equalities (1) describing the feasible set $\mathcal{D}$. Denote

$$t_{\min} = \min \tilde{t}_{kn}; \quad N_{\max} = \frac{1}{t_{\min}}.$$

Then it follows from (6) that

$$0 \le N_n^*(\lambda) \le N_{\max}, \quad n \in \overline{1,m},$$

This condition is satisfied for the set

$$\Lambda = \bigcup_{n=1}^{m} \left\{\lambda : \ \delta_n(\lambda) \ge \frac{G_n}{N_{\max}}\right\}, \tag{27}$$

where $\delta_n(\lambda)$ is given by the equality (26′). The comparison of (26) and (27) shows that $\Lambda_0 \cup \Lambda = \varnothing$, i.e. the functions $\Psi_k(\lambda)$ (7) are continuously differentiable on the set Λ (27).

Now we consider the Jacobian for E-MSS. From (7) and (14) we obtain

$$J_{kj}(\lambda) = \left[\frac{\partial \Psi_k}{\partial \lambda_j}\right] = -\sum_{n=1}^{m} \tilde{t}_{kn} \tilde{t}_{jn} \kappa_n, \tag{28}$$

where

$$\kappa_n = \frac{c_n G_n \exp\left(\sum_{j=1}^{r} \lambda_j \tilde{t}_{jn}\right)}{\delta_n^2(\lambda)}, \quad \lambda \in \Lambda. \tag{28'}$$

It is easy to see that $J_{kj}(\lambda) < 0$ for $\lambda \in \Lambda$, $\kappa_n > 0$, and it can also be shown, as before for F-MSS (see (19), (20)), that the matrix $J(\lambda) = [J_{kj}(\lambda)]$ is Hurwitz one; here $J_{kj}(\lambda)$ are given by the equalities (28) and (28′).

Lemma 2. *The solutions of system* (17) *are asymptotically stable for all* $\lambda^0 \in \Lambda$ (*the set* Λ *is given by* (27)).

Proof. Consider the function $\tilde{L}(\lambda)$ (15) on the set Λ (27). Since it is continuously differentiable on this set, the components of its gradient are

$$\frac{\partial \tilde{L}}{\partial \lambda_k} = 1 - \Psi_k(\lambda), \quad k \in \overline{1,r},$$

where $\Psi_k(\lambda)$ are given by (7). The Hessian of $\widetilde{L}(\lambda)$ is

$$\widetilde{\Gamma} = \left[\frac{\partial^2 \widetilde{L}}{\partial \lambda_j \, \partial \lambda_k}\right] = -J$$

and, according to (28) and (28′), it is positive-definite on the set Λ (27). Therefore the function $\widetilde{L}(\lambda)$ (15) is strictly convex for $\lambda \in \Lambda$. The function $\widetilde{L}(\lambda)$ attains its unique minimum on the set Λ (27) at the point λ^*, i.e. the solution λ^* of the system

$$\frac{\partial \widetilde{L}}{\partial \lambda_k} = 1 - \widetilde{\Psi}_k(\lambda) = 0, \quad k \in \overline{1,r}, \tag{29}$$

satisfies the condition

$$\lambda^* \in \Lambda. \tag{30}$$

Consider the time derivative of $\widetilde{L}(\lambda)$ (15) along the trajectories of system (17). By virtue of (29) and (30), we have

$$\frac{d\widetilde{L}}{dt} \begin{cases} < 0 & \text{for all } \lambda \in \Lambda \text{ (excluding the infinity) and } \lambda \neq \lambda^*, \\ = 0 & \text{for } \lambda = \lambda^*. \end{cases}$$

Hence $\widetilde{L}(\lambda)$ is a Lyapunov function for the system (17), (14) for $\lambda^0 \in \Lambda$.

Thus the theorem holds.

Theorem 3. *The algorithm* (13) (*and hence the algorithm* (11)) $\mathfrak{G}$-*converges for E-MSS* (12), (14), *and* $\mathfrak{G} \subset \Lambda$ (27), (26′).

The proof follows from Theorem 2.2 and Lemma 2.

Consider the B-MSS with transport constraints. In applied analysis of network transportation flows, the following formulation of B-MSS is used:

$$H_B(N) = -\sum_{i,j=1}^{l} N_{ij} \ln \frac{N_{ij}}{e a_{ij} G_{ij}} \rightarrow \max,$$

$$\sum_{i=1}^{l} N_{ij} = P_j, \quad j \in \overline{1,l}; \tag{31}$$

$$\sum_{j=1}^{l} N_{ij} = Q_i, \quad i \in \overline{1,l}.$$

Here i and j are the numbers of communication network nodes. A node is characterized by the capacities P_j and Q_j of occupied and free states, respectively. The matrix $[N_{ij}]$ is formed from the occupation numbers for communications (i, j) with the capacities G_{ij}. It is assumed that the occupation is random with the apriori probabilities a_{ij} (see Shmulyan, Imelbayev 1978).

Consider the Lagrange function for this problem:

$$L(N, \lambda) = H_B(N) + \sum_{j=1}^{l} \lambda_j \left(P_j - \sum_{i=1}^{l} N_{ij} \right) + \sum_{i=1}^{l} \lambda_{l+i} \left(Q_i - \sum_{j=1}^{l} N_{ij} \right).$$

Its stationarity conditions are:

$$\begin{gathered} N_{ij}^* = \widetilde{a}_{ij}(-\lambda_j - \lambda_{l+i}); \\ \sum_{i=1}^{l} N_{ij}^* = P_j; \quad \sum_{j=1}^{l} N_{ij}^* = Q_i; \\ \widetilde{a}_{ij} = a_{ij} G_{ij}; \quad j \in \overline{1,l}; \quad i \in \overline{1,l}. \end{gathered} \tag{32}$$

Substituting the expressions for N_{ij}^* into the last group of equalities, we obtain the following equations

$$\frac{P_j}{z_j \sum_{i=1}^{l} \widetilde{a}_{ij}\, z_{l+i}} = 1, \quad j \in \overline{1,l};$$

$$\frac{Q_i}{z_{l+i} \sum_{j=1}^{l} \widetilde{a}_{ij}\, z_j} = 1, \quad i \in \overline{1,l}$$

for the variables

$$z_j = \exp(-\lambda_j); \quad z_{l+i} = \exp(-\lambda_{l+i}).$$

By multiplying the equations of the first group by $z_1, \ldots, z_l$, and the equations of the second group by $z_{l+1}, \ldots, z_{2l}$, respectively, we obtain

$$z_j = \frac{P_j}{\sum_{i=1}^{l} \widetilde{a}_{ij}\, z_{l+i}}, \quad j \in \overline{1,l};$$

$$z_{l+i} = \frac{Q_i}{\sum_{j=1}^{l} \widetilde{a}_{ij}\, z_j}, \quad i \in \overline{1,l}.$$

To solve this, we use the coordinate-wise scheme of iterations. We have

$$z_j^{s+1} = \frac{P_j}{\sum_{i=1}^{l} \widetilde{a}_{ij}\, z_{l+i}^{s}}, \quad j \in \overline{1,l}; \tag{33}$$

$$z_{l+i}^{s+1} = \frac{Q_i}{\sum_{j=1}^{l} \widetilde{a}_{ij}\, z_j^{s+1}}, \quad i \in \overline{1,l}. \tag{33'}$$

This is the well-known "balancing algorithm" proposed by Sheleikhovsky in 1934 and then studied by Bregman in 1967 (see Sheleikhovsky, 1946; Bregman, 1967).

As shown above, the "balancing algorithm" is a modification of the multiplicative algorithm with the coordinate-wise scheme of iterations and the coefficient $\gamma = 1$.

Macrosystems with incomplete consumption of resources. This class is often used in applied problems.

Consider an MSS with linear and incomplete consumption of resources (1.3, 1.2)

$$H(N) \to \max,$$

$$\sum_{n=1}^{m} \widetilde{t}_{kn} N_n \le 1, \quad k \in \overline{1,r}; \tag{34}$$

$$\widetilde{t}_{kn} = \frac{1}{q_k} t_{kn}, \quad n \in \overline{1,m},$$

where $H(N)$ is the generalized information entropy for macrosystems with Fermi-, Einstein- and Boltzmann-states, respectively.

Assume that the feasible set

$$\mathfrak{D}\left\{N : \sum_{n=1}^{m} \widetilde{t}_{kn} N_n \le 1, \quad k \in \overline{1,r}\right\} \subset \mathfrak{N}$$

is nonempty. Recall that $\mathfrak{N} = R_+^m$ for B- and E-MSS and

$$\mathfrak{N} = \{N : \ 0 \le N_n \le G_n, \quad n \in \overline{1,m}\} \quad \text{for} \quad F\text{-MSS}.$$

We consider the Lagrange function

$$L(N,\lambda) = H(N) + \sum_{k=1}^{r} \lambda_k \left(1 - \sum_{n=1}^{m} \widetilde{t}_{kn} N_n\right) \tag{35}$$

and use the optimality conditions formulated in terms of the Lagrange function saddle-point (2.45):

$$\begin{gathered} \nabla_{N_n} L(N^*,\lambda^*) = 0, \quad n \in \overline{1,m}; \\ \nabla_{\lambda_k} L(N^*,\lambda^*) \geq 0, \\ \lambda_k^* \nabla_{\lambda_k} L(N^*,\lambda^*) = 0, \\ \lambda_k^* \geq 0, \quad k \in \overline{1,r}, \end{gathered} \tag{36}$$

where

$$\begin{gathered} \nabla_{N_n} L(N^*,\lambda^*) = \frac{\partial H}{\partial N_n} - \sum_{k=1}^{r} \lambda_k^* \widetilde{t}_{kn}, \\ \nabla_{\lambda_k} L(N^*,\lambda^*) = 1 - \sum_{n=1}^{m} \widetilde{t}_{kn} N_n^*. \end{gathered} \tag{36'}$$

The conditions of existence of a saddle-point in (34) are

$$\begin{gathered} \frac{\partial H(N^*)}{\partial N_n} - \sum_{k=1}^{r} \lambda_k^* \widetilde{t}_{kn} = 0, \quad n \in \overline{1,m}; \\ \lambda_k^* \left(1 - \sum_{n=1}^{m} \widetilde{t}_{kn} N_n^*\right) = 0, \\ 1 - \sum_{n=1}^{m} \widetilde{t}_{kn} N_n^* \geq 0, \\ \lambda_k^* \geq 0, \quad k \in \overline{1,r}. \end{gathered} \tag{37}$$

The first group of equations in (37) can be solved explicitly for N^* (see (2), (4) and (6)). Therefore the conditions (37) can be simpler:

$$\lambda_k^* \left(1 - \sum_{n=1}^{r} \widetilde{t}_{kn} N_n^*(\lambda)\right) = 0, \tag{38}$$

$$1 - \sum_{n=1}^{m} \widetilde{t}_{kn} N_n^*(\lambda) \geq 0, \quad \lambda_k \geq 0, \quad k \in \overline{1,r},$$

where $N^*(\lambda)$ is given by (2), (4) and (6) for F-, B- and E-MSS, respectively.

Denote

$$\widetilde{L}(\lambda) = L(N^*(\lambda), \lambda). \tag{39}$$

Then

$$\frac{\partial \widetilde{L}}{\partial \lambda} = 1 - \sum_{n=1}^{m} \widetilde{t}_{kn} N_n^*(\lambda),$$

and (38) can be rewritten as

$$\begin{aligned} &\lambda_k^* \nabla_{\lambda_k} \widetilde{L}(\lambda^*) = 0, \\ &\lambda_k^* \geq 0, \quad \nabla_{\lambda_k} \widetilde{L}(\lambda^*) \geq 0, \quad k \in \overline{1,r}. \end{aligned} \tag{40}$$

To determine nonnegative solutions of (40), we use the multiplicative algorithm

$$\lambda_k^{s+1} = \lambda_k^s (1 - \gamma \nabla_{\lambda_k} \widetilde{L}(\lambda^s)), \quad k \in \overline{1,r}. \tag{41}$$

Theorem 4. *The algorithm* (41) *$\mathfrak{G}$-converges to the solution of* (34).

The proof uses an auxiliary system of differential equations

$$\dot{\lambda}_k = -\lambda_k \nabla_{\lambda_k} \widetilde{L}(\lambda), \quad k \in \overline{1,r} \tag{42}$$

and Theorem 2.2.

Consider the function $\widetilde{L}(\lambda)$ (39). It shown before that $\widetilde{L}(\lambda)$ is strictly convex and bounded from below for all the three classes of MSS.

Determine the time derivative of this function along the trajectories of (42)

$$\frac{d\widetilde{L}}{dt} = -\sum_{k=1}^{r} \left(\nabla_{\lambda_k} \widetilde{L}(\lambda)\right)^2 \lambda_k \leq 0 \quad \text{for} \quad \lambda \in R_+^r.$$

It follows from the optimality conditions that $\dfrac{d\widetilde{L}}{dt} = 0$ only for $\lambda = \lambda^*$. Hence $\widetilde{L}(\lambda)$ is a Lyapunov function for system (42). It follows from this that the condition (d) of Theorem 2.2 is satisfied.

Consider the Jacobian of (41):

$$J(\lambda) = \frac{\partial}{\partial \lambda_i} \left[-\lambda_k \nabla_{\lambda_k} \widetilde{L}(\lambda)\right] =$$

$$= \begin{cases} -\nabla_{\lambda_k} \widetilde{L}(\lambda) - \lambda_k \nabla^2_{\lambda_k} \widetilde{L}(\lambda) & (k = i); \\ -\lambda_k \nabla^2_{\lambda_k \lambda_i} \widetilde{L}(\lambda) & (k \neq i). \end{cases} \tag{43}$$

Here $\nabla_\lambda \widetilde{L}(\lambda^*) \geq 0$ and the Hessian $\nabla_\lambda^2 \widetilde{L}(\lambda^*)$ is positive-definite at the point λ^* of constrained minimum of $\widetilde{L}(\lambda)$. It follows from this that $J(\lambda)$ (43) is negative-definite. Hence its eigenvalues are real and negative, therefore the condition (c) of Theorem 2.2 is also satisfied, and consequently the algorithm (41) $\mathfrak{G}$-converges.

4.4 Multiplicative algorithms for MSS of macrosystems with nonlinear consumption of resources

Consider an MSS with mixed consumption of resources (Chapter 2, Section 4):

$$H(N) \to \max;$$
$$\widetilde{\varphi}_k(N) = 1, \quad k \in \overline{1,l}; \tag{1}$$
$$\widetilde{\psi}_k(N) \leq 1, \quad k \in \overline{1,p}; \quad p + l = r < m,$$

where

$$\widetilde{\varphi}_k(N) = \frac{1}{q_k}\varphi_k(N), \quad k \in \overline{1,l};$$
$$\widetilde{\psi}_k(N) = \frac{1}{q_k}\psi_k(N), \quad k \in \overline{1,p}.$$

The functions $H(N)$ are the generalized information entropies of Fermi-Dirac, Boze-Einstein or Boltzmann, respectively. The feasible set is

$$\mathfrak{D} = \{N : \widetilde{\varphi}_k(N) = 1, \quad k \in \overline{1,l}; \quad \widetilde{\psi}_k(N) \leq 1, \quad k \in \overline{1,p}\} \subset \mathfrak{N}, \tag{2}$$

where $\mathfrak{N} = R_+^m$ for B- and E-MSS and

$$\mathfrak{N} = \{N : 0 \leq N_n \leq G_n, \quad n \in \overline{1,m}\} \quad \text{for} \quad F\text{-MSS}.$$

The set $\mathfrak{D}$ is characterized by the equality constraints and can have rather complex configurations for arbitrary consumption functions. Even if consumption functions have the typical properties (monotonicity and convexity (concavity)), the set $\mathfrak{D}$ (2) becomes nonconvex (nonconcave) because of the equality constraints in (1).

Linearization methods. It follows from Section 3 that the MSS with linear consumption (both complete and incomplete) are quite simple, and multiplicative algorithms are efficient computational tools for solving the corresponding optimization problems.

Therefore it looks logical to use decomposition of the general problem (1) into a sequence of problems with linear consumption functions.

An obvious approach to this is the linearization of the feasible set $\mathfrak{D}$ (2).

Assume that (1) has the solution N^*, there exists a subset $\mathfrak{G}$ of $\mathfrak{D}$ (2) containing N^* and there are no other local maximums of $H(N)$ in $\mathfrak{G}$. We will seek N^* by an iterative process. Let the process state be characterized by the point N^s at the step s. In a neighborhood of this point, we linearize the functions $\widetilde{\varphi}_k(N)$ $(k \in \overline{1,l})$ and $\widetilde{\psi}_k(N)$ $(k \in \overline{1,p})$:

$$\widetilde{\varphi}_k(N^s) + \langle \nabla \widetilde{\varphi}_k(N^s), \quad N - N^s \rangle = 1, \quad k \in \overline{1,l}; \tag{3}$$

$$\widetilde{\psi}_k(N^s) + \langle \nabla \widetilde{\psi}_k(N^s), \quad N - N^s \rangle \leq 1, \quad k \in \overline{1,p}; \tag{4}$$

Assume that all the consumption functions in (1) are monotone increasing, i.e.

$$\begin{aligned} \frac{\partial \widetilde{\varphi}_k}{\partial N_n} &\geq 0, \quad k \in \overline{1,l}; \\ \frac{\partial \widetilde{\psi}_k}{\partial N_n} &\geq 0, \quad k \in \overline{1,p}. \end{aligned} \tag{5}$$

The equalities (3) and (4) describe a polyhedral set approximating the feasible set $\mathfrak{D}$ near the point N^s.

Consider the problem:

$$H(N) \to \max, \tag{6}$$

$$\langle \nabla \widetilde{\varphi}_k(N^s),\ N - N^s \rangle = q_k(N^s) = 1 - \widetilde{\varphi}_k(N^s), \quad k \in \overline{1,l}; \tag{7}$$

$$\langle \nabla \widetilde{\psi}_k(N^s),\ N - N^s \rangle \leq g_k(N^s) = 1 - \widetilde{\psi}_k(N^s), \quad k \in \overline{1,p}; \tag{8}$$

and denote its solution by N_*^s Note that N_*^s always exists and is bounded, under certain conditions. Therefore the next approximation of the solution of (1) can be determined as

$$N^{s+1} = N^s + \alpha N_*^s, \tag{9}$$

where α is nonnegative constant.

Then we repeat the linearization of feasible set near the point N^{s+1} and solve new problem (6)–(8).

Note that this linearization method differs from the traditional ones (see Polyak 1983, Pshenichny 1983), where the initial problem is reduced to a quadratic programming problem (with quadratic objective function and linear constraints).

In this case at each step of the iterative process (9) we solve the entropy maximization problem on the polyhedral set

$$H(N) \rightarrow \max; \tag{10}$$

$$\widehat{\mathfrak{D}} = \left\{ N : \sum_{n=1}^{m} t_{kn}^{(1)} N_n = h_k, \quad k \in \overline{1,l}\,; \right.$$
$$\left. \sum_{n=1}^{m} t_{kn}^{(2)} N_n \leq w_k, \quad k \in \overline{1,p} \right\} \subset \mathfrak{N}; \tag{11}$$

$$\widehat{\mathfrak{D}} \neq \varnothing;$$

$$\mathfrak{N} = \begin{cases} N: \; N_n \geq 0, & n \in \overline{1,m} \quad \text{(for } B\text{- and } E\text{-MSS);} \\ N: \; 0 \leq N_n \leq G_n, & n \in \overline{1,m} \quad \text{(for } F\text{-MSS),} \end{cases}$$

where

$$t_{kn}^{(1)} \geq 0, \quad k \in \overline{1,l}; \quad t_{kn}^{(2)} \geq 0, \quad k \in \overline{1,p}; \quad n \in \overline{1,m},$$

$$T = \begin{bmatrix} t_{kn}^{(1)} \\ t_{kn}^{(2)} \end{bmatrix}; \quad \text{rank } T = r = l + p; \tag{12}$$

Below we call (10), (11) and (12) the *linear entropy module* (*LEM*). By comparing (6) – (8) and LEM, we can see that

$$t_{kn}^{(1)} = \left. \frac{\partial \widetilde{\varphi}_k}{\partial N_n} \right|_{N^s}; \quad h_k = 1 - \widetilde{\varphi}_k(N^s), \quad k \in \overline{1,l};$$
$$t_{kn}^{(2)} = \left. \frac{\partial \widetilde{\psi}_k}{\partial N_n} \right|_{N^s}; \quad w_k = 1 - \widetilde{\psi}_k(N^s), \quad k \in \overline{1,p}. \tag{13}$$

The LEM is characterized by the operator

$$P_D^H(N^s) = N_*^s = \arg\max(H(N) \mid N \in \widehat{\mathfrak{D}}(N^s)), \tag{14}$$

where $\widehat{\mathfrak{D}}(N^s)$ is the feasible set (11) with the parameters determined with respect to the point N^s.

Then the iterative process (9) can be represented as

$$N^{s+1} = N^s + \alpha \arg\max(H(N) \mid N \in \widehat{\mathfrak{D}}(N^s)). \tag{15}$$

Such iterative processes are widely used in optimization theory. In particular, they are considered in (Antipin, 1989). The results obtained can be used in proving the convergence of such processes.

From the computational point of view, the computation of values of the operator P_D^H (14), i.e. the determination of LEM's solution, is the main part of the algorithm (15). Let us consider this problem in detail.

The Lagrange function for the LEM is

$$\begin{aligned} L(N, \lambda, \mu) &= H(N) + \sum_{k=1}^{l} \lambda_k \left(h_k - \sum_{n=1}^{m} t_{kn}^{(1)} N_n \right) + \\ &+ \sum_{k=1}^{p} \mu_k \left(w_k - \sum_{n=1}^{m} t_{kn}^{(2)} N_n \right). \end{aligned} \tag{16}$$

The optimality conditions for the LEM are

$$\begin{aligned} \nabla_{N_n} L(N^*, \lambda^*, \mu^*) &= \left. \frac{\partial H}{\partial N_n} \right|_{N^*} - \sum_{k=1}^{l} \lambda_k^* t_{kn}^{(1)} \\ &- \sum_{k=1}^{p} \mu_k^* t_{kn}^{(2)} = 0, \quad n \in \overline{1, m}; \end{aligned} \tag{17}$$

$$\nabla_{\lambda_k} L(N^*, \lambda^*, \mu^*) = h_k - \sum_{n=1}^{m} t_{kn}^{(1)} N_n^* = 0, \quad k \in \overline{1, l}; \tag{18}$$

$$\nabla_{\mu_k} L(N^*, \lambda^*, \mu^*) = w_k - \sum_{n=1}^{m} t_{kn}^{(2)} N_n^* \geq 0, \quad k \in \overline{1, p}; \tag{19}$$

$$\mu_k^* \nabla_{\mu_k} L(N^*, \lambda^*, \mu^*) = \mu_k^* \left(w_k - \sum_{n=1}^{m} t_{kn}^{(2)} N_n^* \right) = 0, \ \mu_k^* \geq 0, \ k \in \overline{1, p}. \tag{20}$$

The structure of the entropy functions in (1) is such that the equations (17) can be solved for the variables N^*. In terms of the Lagrange function (16) these are

(a) for B – MSS

$$N_n^*(\lambda^*, \mu^*) = \frac{\widetilde{a}_n}{\exp\left(\sum_{k=1}^{l} \lambda_k^* t_{kn}^{(1)} + \sum_{k=1}^{p} \mu_k^* t_{kn}^{(2)}\right)}; \tag{21}$$

$$\widetilde{a}_n = a_n G_n, \quad n \in \overline{1, m},$$

(b) for F – MSS

$$N_n^*(\lambda^*, \mu^*) = \frac{G_n}{1 + b_n \exp\left(\sum_{k=1}^{l} \lambda_k^* t_{kn}^{(1)} + \sum_{k=1}^{p} \mu_k^* t_{kn}^{(2)}\right)}; \tag{22}$$

$$b_n = 1/\widetilde{a}_n, \quad \widetilde{a}_n = a_n/(1 - a_n); \quad n \in \overline{1, m},$$

(c) for E – MSS

$$N_n^*(\lambda^*, \mu^*) = \frac{G_n}{C_n \exp\left(\sum_{k=1}^{l} \lambda_k^* t_{kn}^{(1)} + \sum_{k=1}^{p} \mu_k^* t_{kn}^{(2)}\right) - 1}; \tag{23}$$

$$C_n = 1/a_n, \quad n \in \overline{1, m}.$$

Thus the conditions (18)–(20) form the following system of equalities and inequalities:

$$\Phi_k(\lambda^*, \mu^*) = \frac{1}{h_k} \sum_{n=1}^{m} t_{kn}^{(1)} N_n^*(\lambda^*, \mu^*) = 1, \quad k \in \overline{1, l}; \tag{24}$$

$$\left.\begin{aligned} &\mu_k^* \nabla_{\mu_k} \widetilde{L}(\lambda^*, \mu^*) = 0; \\ &\nabla_{\mu_k} \widetilde{L}(\lambda^*, \mu^*) = \nabla_{\mu_k} L(N^*(\lambda^*, \mu^*), \lambda^*, \mu^*) = \\ &\qquad = w_k - \sum_{n=1}^{p} t_{kn}^{(2)} N_n^*(\lambda^*, \mu^*) \geq 0, \\ &\mu_k^* \geq 0, \quad k \in \overline{1, p}. \end{aligned}\right\} \tag{25}$$

Equations similar to (24) were considered in the MSS with linear and complete consumption of resources; those similar to (25) were considered in MSS with linear and incomplete consumption of resources.

If we substitute

$$z_k = e^{\lambda_k}, \quad k \in \overline{1,l},$$

to (24), the system (24)–(25) can be solved by the following multiplicative algorithm

$$z_k^{v+1} = z_k^v \, \widetilde{\Phi}_k^\gamma \left(z^v, \mu^v\right); \tag{26}$$

$$\mu_k^{v+1} = \mu_k^v \left(1 - \rho \nabla_{\mu_k} \widehat{L}\left(z^v, \mu^v\right)\right), \tag{27}$$

where

$$\widetilde{\Phi}_k(z, \mu) = \Phi_k(\ln z, \mu);$$

$$\nabla_{\mu_k} \widehat{L}\left(z, \mu\right) = \nabla_{\mu_k} \widetilde{L}\left(\ln z, \mu\right).$$

The proof of convergence is based on transforming (26) to the additive form

$$\lambda_k^{v+1} = \lambda_k^v + \gamma \ln \Phi_k(\lambda^v, \mu^v) \tag{28}$$

(to do this, we need $h_k \geq 0$, i.e. $\varphi_k(N^s) < 1$ (13)). Then the following system of differential equations should be considered:

$$\begin{aligned} \dot{\lambda}_k &= \gamma \ln \Phi_k(\lambda, \mu), & k \in \overline{1,l}; \\ \dot{\mu}_k &= -\rho \nabla_{\mu_k} \widehat{L}\left(\lambda, \mu\right), & k \in \overline{1,p}. \end{aligned} \tag{29}$$

Combined method. Let us consider an MSS with incomplete consumption of resources (without equality constraints in (1)) and represent it as follows:

$$H\left(N\right) \to \max; \tag{30}$$

$$\mathfrak{D} = \left\{N : \ \widetilde{\varphi}_k(N) \leq 1, \ k \in \overline{1,r}\right\} \subset \mathfrak{N}; \quad \mathfrak{D} \neq \varnothing, \tag{31}$$

where

$$\mathfrak{N} = \begin{cases} \{N : \ N_n \geq 0, & n \in \overline{1,m}\} \ \ (B\text{- and } E\text{- models}); \\ \{N : \ 0 \leq N_n \leq G_n, & n \in \overline{1,m}\} \ \ (F\text{- models}); \end{cases} \tag{32}$$

$$\widetilde{\varphi}_k(N) = \frac{1}{q_k} \varphi_k(N), \quad q_k \geq \varepsilon_k > 0, \quad k \in \overline{1,r}.$$

Algorithms for solving (30)–(32) are based on the optimality conditions formulated in terms of the saddle-point of the Lagrange function

$$L(N, \lambda) = H(N) + \sum_{k=1}^{r} \lambda_k (1 - \widetilde{\varphi}_k(N)). \tag{33}$$

Assume that the optimality conditions for (30)–(32) include the conditions of saddle-point existence, and these are strengthened by the strict complementarity conditions:

$$\nabla_{N_n} L(N^*, \lambda^*) = 0, \quad n \in \overline{1, m}; \tag{34}$$

$$\left.\begin{array}{l} \nabla_{\lambda_k} L(N^*, \lambda^*) \geq 0, \\ \lambda_k^* \nabla_{\lambda_k} L(N^*, \lambda^*) = 0, \\ \lambda_1^* > 0, \ldots, \lambda_l^* > 0; \quad \lambda_{l+1}^* = \ldots = \lambda_r^* = 0, \end{array}\right\} \tag{35}$$

where l is the number of active constraints at (N^*, λ^*);

$$\nabla_{N_n} L(N, \lambda) = \frac{\partial H}{\partial N_n} - \sum_{k=1}^{r} \lambda_k \frac{\partial \widetilde{\varphi}_k}{\partial N_n}$$

and (36)

$$\nabla_{\lambda_k} L(N, \lambda) = 1 - \widetilde{\varphi}_k(N).$$

By comparing (34) and (35) with the general conditions of existence of a saddle-point for the reduced Lagrange function (2.55), we can see that, instead of $\nabla_{N_n} L(N^*, \lambda^*) \leq 0$ and $N_n^* \nabla_{N_n} L(N^*, \lambda^*) = 0$, we have the only condition of gradient $L(N, \lambda)$ being zero with respect to the primal variables.

To solve (34) and (35), we use a combined algorithm (the gradient search with respect to the primal variables and multiplicative algorithm with respect to the dual variables):

$$N_n^{s+1} = N_n^s + \gamma \nabla_{N_n} L(N^s, \lambda^s), \quad n \in \overline{1, m}; \tag{37}$$

$$\lambda_k^{s+1} = \lambda_k^s \left(1 - \gamma \nabla_{\lambda_k} L(N^s, \lambda^s)\right), \quad k \in \overline{1, r} \tag{38}$$

with the initial conditions

$$N^0, \lambda^0 \in \mathfrak{G}. \tag{39}$$

To study the convergence of (37)–(38), we consider an auxiliary system of differential equations

$$\dot{N}_n = \nabla_{N_n} L(N,\lambda), \quad n \in \overline{1,m}; \tag{40}$$

$$\dot{\lambda}_k = -\lambda_k \nabla_{\lambda_k} L(N,\lambda), \quad k \in \overline{1,r} \tag{41}$$

with the initial conditions (39).

Lemma 1. *In (30)–(31), let the functions $\widetilde{\varphi}_k(N)$ $(k \in \overline{1,r})$ be strictly convex and let the matrix $\left[\dfrac{\partial \widetilde{\varphi}_k}{\partial N_n}\right]$ be non-degenerate. Then the solution of (40)–(41) is asymptotically stable for $(N^0, \lambda^0) \in \operatorname{int}(\mathfrak{N} \times R^r_+)$.*

Proof. We define the function

$$V(N,\lambda) = \frac{1}{2}\sum_{n=1}^{m}(N_n - N_n^*)^2 + \sum_{k=1}^{r}(\lambda_k - \lambda_k^*) - \lambda_k^*(\ln\lambda_k - \ln\lambda_k^*)$$

on $R^m \times R^r_+$. It is strictly convex and attains its minimum (equal to zero) at the point N^*, λ^*.

Consider its time derivative along the trajectories of (40)–(41)

$$\frac{dV}{dt} = \sum_{n=1}^{m}(N_n - N_n^*)\nabla_{N_n} L(N,\lambda) + \sum_{k=1}^{r}(\lambda_k - \lambda_k^*)\nabla_{\lambda_k} L(N,\lambda).$$

Taking (36) into account, we obtain

$$\frac{dV}{dt} = \sum_{n=1}^{m}(N_n - N_n^*)\left(\frac{\partial H}{\partial N_n} - \sum_{k=1}^{r}\lambda_k \frac{\partial \widetilde{\varphi}_k}{\partial N_n}\right) -$$

$$-\sum_{k=1}^{r}(\lambda_k - \lambda_k^*)(1 - \widetilde{\varphi}_k(N)) =$$

$$= \left[H(N^*) - H(N) + \sum_{n=1}^{m}(N_n - N_n^*)\frac{\partial H}{\partial N_n}\right] +$$

$$+ \left[\sum_{k=1}^{r}\lambda_k\left(-\sum_{n=1}^{m}(N_n - N_n^*)\frac{\partial \widetilde{\varphi}_k}{\partial N_n} - (1-\widetilde{\varphi}_k(N)) + (1-\widetilde{\varphi}_k(N^*))\right)\right] +$$

$$+ \left[\sum_{k=1}^{r}\lambda_k^*(1 - \widetilde{\varphi}_k(N)) - \sum_{k=1}^{r}\lambda_k(1 - \widetilde{\varphi}_k(N^*)) - H(N^*) + H(N))\right].$$

Since the entropy in the MSS is a strictly concave function, we have

$$H(N^*) - H(N) + \sum_{n=1}^{m}(N_n - N_n^*)\frac{\partial H(N)}{\partial N_n} \leq 0,$$

and equality is attained only for $N = N^*$.

It follows from the strict convexity of the consumption functions $\widetilde{\varphi}_k(N)$ that

$$\left(1-\widetilde{\varphi}_k(N^*)\right)-(1-\widetilde{\varphi}_k(N))-\sum_{n=1}^{m}(N_n-N_n^*)\frac{\partial\,\widetilde{\varphi}_k(N)}{\partial\,N_n}\leq 0,$$

and equality is attained only for $N=N^*$.

Let us consider the third term in the expression for $\dfrac{dV}{dt}$ and use the definition of the saddle-point:

$$H(N)+\sum_{k=1}^{r}\lambda_k^*(1-\widetilde{\varphi}_k(N))\leq H(N^*)\leq H(N^*)+\sum_{k=1}^{r}\lambda_k(1-\widetilde{\varphi}_k(N^*)).$$

Omitting the inner part of this inequality, we obtain

$$H(N)-H(N^*)+\sum_{k=1}^{r}\lambda_k^*(1-\widetilde{\varphi}_k(N))-\sum_{k=1}^{r}\lambda_k(1-\widetilde{\varphi}_k(N^*))\leq 0.$$

Note that the left-hand side of this expression is zero on the subset

$$\mathcal{L}=\{(N,\lambda):\ N_n=N_n^*,\ n\in\overline{1,m};\quad \lambda_{l+1}=\lambda_{l+1}^*=0,\ldots,\lambda_r=\lambda_r^*=0;$$

$$\lambda_1>0,\ldots,\lambda_l>0\},$$

where l is the number of active inequalities.

Therefore

$$\frac{dV}{dt}\begin{cases}<0 & \text{for all } \{N,\lambda\}\in\operatorname{int}((\mathfrak{N}\times R_+^r)\setminus\mathcal{L});\\ =0 & \text{for all } \{N,\lambda\}\in\mathcal{L}.\end{cases}$$

If $l=0$ then $V(N,\lambda)$ is a Lyapunov function for (40)–(41).

Consider the case $l>0$. Recalling the proof of Lemma 2.5, we introduce

$$\widetilde{V}(N,\lambda)=V(N,\lambda)+\delta\,\widetilde{L}(N,\lambda),$$

where

$$\widetilde{L}(N,\lambda)=-L(N,\lambda)\quad\text{for}\quad \lambda\in R_+^r,\quad N\in\mathfrak{N};$$

and $L(N,\lambda)$ is given by (33), $\delta>0$.

Determine the derivative of $\widetilde{L}(N,\lambda)$ along the trajectories of (40)–(41):

$$\frac{d\widetilde{L}}{dt}=-\sum_{n=1}^{m}(\nabla_{N_n}L(N,\lambda))^2+\sum_{k=1}^{r}\lambda_k(1-\widetilde{\varphi}_k(N))^2.$$

On the set $\mathcal{L}$ we have:

$$\frac{d\widetilde{L}}{dt} = -\sum_{n-1}^{m} \left(\frac{\partial H(N^*)}{\partial N_n} - \sum_{k=1}^{l} \lambda_k \frac{\partial \widetilde{\varphi}_k(N^*)}{\partial N_n} \right)^2 =$$

$$= -\sum_{n=1}^{m} \left[\frac{\partial H(N^*)}{\partial N_n} - \sum_{k=1}^{l} \lambda_k^* \frac{\partial \widetilde{\varphi}_k(N^*)}{\partial N_n} + \sum_{k=1}^{l} (\lambda_k^* - \lambda_k) \frac{\partial \widetilde{\varphi}_k(N^*)}{\partial N_n} \right]^2 =$$

$$= -\sum_{n=1}^{m} \left(\sum_{k=1}^{r} (\lambda_k^* - \lambda_k) \frac{\partial \widetilde{\varphi}_k(N^*)}{\partial N_n} \right)^2 .$$

Since $\left[\frac{\partial \widetilde{\varphi}_k(N^*)}{\partial N} \right]$ is assumed to be non-degenerate,

$$\frac{d\widetilde{L}}{dt} = \begin{cases} < 0 & \text{for} \quad \lambda \in \mathcal{L}; \\ = 0 & \text{for} \quad \lambda = \lambda^*. \end{cases}$$

Using the scheme of the proof of Lemma 2.5, we can show that there exists $\delta > 0$ for which $\widetilde{V}(N, \lambda)$ is a Lyapunov function for (40)–(41). The statement of Lemma 1 follows immediately from this.

Let us consider the system (40)–(41) and represent the Jacobian of its right-hand sides as

$$J(N, \lambda) = \begin{bmatrix} A & B \\ C & D \end{bmatrix}, \tag{42}$$

where

A is an $(m \times m)$ matrix with the elements

$$a_{np} = \frac{\partial^2 L}{\partial N_p \partial N_n} = \frac{\partial^2 H}{\partial N_p \partial N_n} - \sum_{k=1}^{r} \lambda_k \frac{\partial^2 \widetilde{\varphi}_k}{\partial N_p \partial N_n}, \qquad n, p \in \overline{1, m};$$

B is an $(m \times r)$ matrix with the elements

$$b_{ns} = -\frac{\partial^2 L}{\partial \lambda_s \partial N_n} = -\frac{\partial \widetilde{\varphi}_s}{\partial N_n}, \quad n \in \overline{1, m}, \quad s \in \overline{1, r};$$

C is an $(r \times m)$ matrix with the elements

$$c_{kp} = \frac{\partial}{\partial N_p} \left(-\lambda_k \frac{\partial L}{\partial \lambda_k} \right) = \lambda_k \frac{\partial \widetilde{\varphi}_k}{\partial N_p}, \quad k \in \overline{1, r}, \quad p \in \overline{1, m};$$

and D is an $(r \times r)$ matrix with the elements

$$d_{ss} = \frac{\partial}{\partial \lambda_s} \left(-\lambda_k \frac{\partial L}{\partial \lambda_k} \right) = -(1 - \widetilde{\varphi}_s), \quad s \in \overline{1, r}.$$

Taking into account the strict complementarity conditions (35), we obtain the matrix $J(N^*, \lambda^*)$ in the form

$$J(N^*, \lambda^*) = \left[\begin{array}{c|c|c} \widetilde{A} & U & W \\ \hline V & 0 & 0 \\ \hline 0 & 0 & \widetilde{D} \end{array}\right], \tag{43}$$

where

$\widetilde{A}$ is an $(m \times m)$ matrix with the elements

$$a_{np} = \frac{\partial^2 H(N^*)}{\partial N_p \partial N_n} - \sum_{k=1}^{l} \lambda_k^* \frac{\partial^2 \widetilde{\varphi}_k(N^*)}{\partial N_p \partial N_n} < 0, \quad p, n \in \overline{1, m},$$

since $H(N)$ is strictly concave and $\widetilde{\varphi}_k(N)$ is strictly convex;
U is an $(m \times l)$ matrix with the elements

$$u_{ns} = -\frac{\partial \widetilde{\varphi}_s(N^*)}{\partial N_n}, \quad s \in \overline{1, l}, \quad n \in \overline{1, m};$$

W is an $(m \times (r - l))$ matrix with the elements

$$w_{ns} = -\frac{\partial \widetilde{\phi}_s(N^*)}{\partial N_n}, \quad s \in \overline{l+1, r}, \quad n \in \overline{1, m};$$

V is an $(l \times m)$ matrix with the elements

$$v_{kp} = \lambda_k^* \frac{\partial \widetilde{\varphi}_s(N^*)}{\partial N_n}, \quad k \in \overline{1, l}, \quad p \in \overline{1, m};$$

$\widetilde{D}$ is an $(r - l) \times (r - l)$ matrix with the elements

$$d_{ss} = -(1 - \widetilde{\varphi}_s(N^*)), \quad s \in \overline{l+1, r}.$$

We can see from the structure of (15) that its characteristic polynom would not change if we replaced W by the zero matrix and considered the characteristic polynom of the matrix

$$J_1(N^*, \lambda^*) = \left[\begin{array}{c|c|c} \widetilde{A} & U & 0 \\ \hline V & 0 & 0 \\ \hline 0 & 0 & \widetilde{D} \end{array}\right], \tag{44}$$

The matrix spectrum consists of the eigenvalues equal to $-(1-\widetilde{\varphi}_s(N^*)) < 0$, $s \in \overline{l+1, r}$ and eigenvalues of

$$F = \begin{bmatrix} \widetilde{A} & U \\ V & 0 \end{bmatrix}. \tag{45}$$

Represent F as

$$F = RS_1,$$

where

$$R = \left[\begin{array}{ccc|ccc} 1 & & 0 & & & \\ & \ddots & & & 0 & \\ 0 & & 1 & & & \\ \hline & & & \lambda_1^* & & 0 \\ & 0 & & & \ddots & \\ & & & 0 & & \lambda_l^* \end{array}\right]; \quad S_1 = \begin{bmatrix} \widetilde{A} & U \\ -U' & 0 \end{bmatrix}.$$

The matrix S_1 has the same structure and properties as S (2.4), $\widetilde{A}$ is a negative-definite symmetrical matrix and U is of full rank l. According to Lemma 2.3, the matrix F (45) is Hurwitz one, since S_1 is Hurwitz and $\lambda_1^* > 0, \ldots, \lambda_l^* > 0$.

Therefore the following lemma holds.

Lemma 2. *If the functions $\varphi_k(N)$ are strictly convex and the matrix $\left[\dfrac{\partial \varphi_k}{\partial N_n}\right]$ is non-degenerate then*

$$J(N, \lambda) = \begin{bmatrix} \dfrac{\partial}{\partial N_p}(\nabla_{N_n} L) & \dfrac{\partial}{\partial \lambda_j}(\nabla_{N_n} L) \\ \dfrac{\partial}{\partial N_p}(-\lambda_k \nabla_{\lambda_k} L) & \dfrac{\partial}{\partial \lambda_j}(-\lambda_k \nabla_{\lambda_k} L) \end{bmatrix}$$

is Hurwitz matrix.

Thus all the conditions of Theorem 2.2 are satisfied for the system (40)–(41), and consequently the algorithm (37)–(38) $\mathfrak{G}$-converges.

4.5 Experimental study of multiplicative algorithms for MSS

Theoretical analysis of various modifications of multiplicative algorithms used to determine the stationary states of macrosystems allows us to determine the classes of MSS and feasible areas of initial points and parameters for which these algorithms converge.

The convergence theorems proved above present certain guidelines in solving practical problems.

First of all, they answer the question of whether the algorithm converges and whether this fact is typical for the MSS under consideration. But trying to use the algorithm directly, we encounter the problems outstanding the statements formulated and proved in the previous sections.

For example, all the theorems state that there exists a finite interval $(0, \gamma_A)$ such that the process converges in some sense for all $\gamma \in (0, \gamma_A)$. But what is this interval in real problems? Do the situations actually exist when this interval is extremely small? Can the algorithm converge outside the interval, since all the statements are sufficient?

The convergence theorems include certain conditions which should be checked to establish the convergence. These conditions often involve the solution to be sought. Thus it is difficult or even impossible to check them apriori.

It should be noted that the convergence does not ensure that the algorithm can be implemented, because the algorithm converging theoretically may be unacceptable in the computation costs.

And, finally, an ideal model was used in proving the theorems and all computation errors influencing the iterative process were absent. All these points are typical for any computational method (see Polyak 1983, pp. 39–43) and justify the need of experimental studies of the above multiplicative algorithms.

Consider the widely-applied MSS with linear consumption of resources:

$$H(N) \to \max$$

Complete consumption of the resources	Incomplete consumption of the resources
$\sum_{n=1}^{m} \tilde{t}_{kn} N_n = 1;$	$\sum_{n=1}^{m} \tilde{t}_{kn} N_n \leq 1;$

$$\tilde{t}_{kn} = t_{kn}/q_k, \quad n \in \overline{1,m}; \quad k \in \overline{1,r}. \tag{1}$$

The entropy function $H(N)$ can be of the form:

$$H(N) = \begin{cases} -\sum_{n=1}^{m} N_n \ln \frac{N_n}{a_n G} & (B\text{–MSS}); \\ -\sum_{n=1}^{m} N_n \ln \frac{N_n(1-a_n)}{a_n} + \\ +(G - N_n)\ln(G - N_n) & (F\text{–MSS}); \\ -\sum_{n=1}^{m} N_n \ln \frac{N_n}{a_n} - \\ -(G + N_n)\ln(G - N_n) & (E\text{–MSS}). \end{cases} \tag{2}$$

Various families of test problems were developed for experimental purposes.

Test 1. External control parameters for the test problem are

m, the number of variables;

r, the number of constraints;

G, the capacity of the subset of close states and

v and d, the parameters of the feasible set for MSS.

The internal parameters for the test problem are formed as follows

$$a_n = \frac{\left|\sin \frac{2\pi}{\sqrt{m}} n\right|}{\sum_{n=1}^{m} \left|\sin \frac{2\pi}{\sqrt{m}} n\right|} \tag{3}$$

$$\tilde{t}_{k1} = \frac{d\, b_{k1}^{(1)}}{G C_k}$$

$$\tilde{t}_{kn} = \begin{cases} \frac{b_{kn}^{(1)}}{C_k} & (2 \leq n \leq r); \\ \frac{b_{kn}^{(2)}}{C_k} & (r < n \leq m), \end{cases} \tag{4}$$

where

$$b_{nn}^{(1)} = 1 + a_n; \quad b_{kn}^{(1)} = (1 + a_n)\, e^{(-v\,|k-n|)}; \quad k \neq n; \quad n \in \overline{1,r};$$

$$b_{kn}^{(2)} = \sqrt{k}\ln(1 + a_n); \quad k \in \overline{1,r}; \quad n \in \overline{r+1,m}; \tag{5}$$

and

$$C_k = \sum_{n=1}^{m} b_{kn}.$$

The rules for obtaining the matrix T (4, 5) provide the fullness of its rank.

Test 2. The external control parameters for the test problem are

l, the parameter of problem dimension ($m = l^2$, $r = 2l$);

d, α and β, the parameters of the feasible set.

The internal parameters for the test problem are formed as follows

$$a_n = \left|\sin\frac{\pi}{\sqrt{7}}n\right|; \quad n \in \overline{1,m};$$

$$t_{k,i+j} = 1; \quad i = (k-1)\,l; \quad k, j \in \overline{1,l};$$

$$t_{j+1,i+j} = 1; \quad i = (k-1)\,l; \quad j \in \overline{1,l-1};$$

$$t_{kn} = 0; \quad k \neq \overline{1,l}; \quad n \neq i+j;$$

$$k \neq j + l; \quad n = i + j;$$

$$t_{rn} = a_n; \quad n \in \overline{1,m};$$

$$q_k = \sum_{n=1}^{m} t_{kn}(dn^2 - bn + c);$$

$$b = 2\,\alpha\, d\, m/3; \quad c + \beta\frac{b^2}{4d}; \quad k \in \overline{1,r}.$$

In this family of test problems, the feasible set is given by the equalities or inequalities, with their right-hand sides being equal to real numbers $q_k (k \in \overline{1,r})$, and the matrix T consists of (0, 1).

The experiments were carried out for F-, B- and E-MSS with complete and incomplete consumption of resources. The multiplicative algorithms with a parallel scheme of iterations (3.11) were considered for the MSS with ***complete resource consumption***, namely

$$z_k^{s+1} = z_k^s\, f_k^{\gamma}(z^s); \quad k \in \overline{1,r}, \tag{6}$$

where $f_k(z)$ are given by (3.12) for F-, B-, and E-MSS, respectively, and a special coordinate-wise scheme was used. The meaning of this scheme is as follows.

Consider the system of equations

$$f_k(z) = 1; \quad k \in \overline{1,r}; \tag{7}$$

as in (4.3), (4.5) and (4.7), which represent the optimality conditions for the MSS. Denote by z^s the vector of dual variables at the iteration step s. Among the equations (7), choose the equation with the maximum deviation

$$k1 = \arg\max_k |f_k(z^s) - 1|. \tag{8}$$

The next, $(s+1)$th, iteration is of the form

$$\begin{aligned} z_{k1}^{s+1} &= z_{k1}^s f_{k1}^\gamma(z_1^s, \ldots, z_r^s); \\ z_k^{s+1} &= z_k^s f_k^\gamma(z_1^s, \ldots, z_{k1-1}^s, z_{k1}^{s+1}, \ldots, z_r^s), \quad k \neq k1, \quad k \in \overline{1,r}. \end{aligned} \tag{9}$$

For the MSS *with incomplete consumption of resources*, we consider the multiplicative algorithm with the parallel scheme as in (4.41):

$$\lambda_k^{s+1} = \lambda_k^s\{1 - \gamma[1 - F_k(\lambda^s)]\}, \tag{10}$$

where

$$F_k(\lambda) = \begin{cases} \Phi_k(\lambda) & (F\text{-MSS (4.3)}); \\ M_k(\lambda) & (B\text{-MSS (4.5)}); \\ \Psi_k(\lambda) & (E\text{-MSS (4.7)}). \end{cases} \tag{11}$$

The coordinate-wise scheme of iterations is realized in the same way (8, 9):

$$\begin{gathered} k1 = \arg\max_k |\lambda_k^s\ [1 - F_k(\lambda^s)]|; \\ \lambda_{k1}^{s+1} = \lambda_{k1}^s\{1 - \gamma\ [1 - F_{k1}(\lambda^s)]\}; \\ \lambda_k^{s+1} = \lambda_k^s\{1 - \gamma\ [1 - f_k(\lambda_1^s, \ldots, \lambda_{k1-1}^s, \lambda_{k1}^{s+1}, \ldots, \lambda_r^s)]\}, \\ k \in \overline{1,r}; \quad k \neq k1. \end{gathered} \tag{12}$$

The computer experiments were carried out by using the program and its modification given in Appendix. The program is written on Turbo Pascal 5.5.

The program allows one to study numerically the MSS (1) of three classes, namely B, E, and F, with complete (C) and incomplete (U) consumption of resources by applying the algorithms (P) - (6, 10) or (K) - (10, 12).

Initial approximations can be determined by the user (mode InitCond = 0) or automatically (mode InitCond = 1). In the latter case, $(r+1)$ variants of starting points are generated and their coordinates are determined by two external parameters, namely *iniz* and p.

The iterations are controlled by using the expression

$$\text{norm} = \begin{cases} \sqrt{\sum_{k=1}^{m}(1 - f_k(z^s))^2} & \text{(MSS of type } C\text{)}. \\ \sqrt{\sum_{k=1}^{r}\left[\lambda_k^s(1 - F_k(\lambda^s))\right]^2} & \text{(MSS of type } U\text{)}. \end{cases}$$

The stopping rule is

$$\text{norm} \leq \text{delt}.$$

The input variables of the program are

r, the number of constraints;

m, the number of variables;

d, the parameter of test problem;

gamma, the parameter of algorithm;

delt, the maximum error;

g, the capacity of subsets of close states;

ClassMode, the class of models (B, E, F);

TypeMode, the type of models (C, U);

ClastAlg, the class of algorithms (P, K);

RegWrite, the write mode (S, L);

InitCond, the mode for generating the starting points (0 - manually, 1 - automatically);

iniz, the parameter of starting points;

p, the parameter of starting points;

v, the parameter of test problem and

$z[1], \ldots, z[r]$, the array of starting points for input by user.

To obtain the output data, the program provides two modes:

$RegWrite = S$ for $r \leq 3, \quad m \leq 8$ and

$RegWrite = L$ for r, m bounded by the dimensions of real arrays (*real*).

The regime S outputs the computed values of the variables $z_k = z[k]$ or $\lambda_k = L[k]$ after each iteration, the number of iterations before attaining the solution and primal $z[k]$, $L[k]$ and dual $N_n = nn[n]$ solutions. The error (*norm*) of the solution obtained and the coordinates of the unconstrained entropy maximum (*absmax*) are also output.

The output files are

A). $RegWrite = S$;

fot $'NameModel'$: the table of input data and parameters of the test problem; the table of iterations and computed solutions for primal and dual variables;

rst $'NameModel'$: the table of input data and parameters of the test problem; the table of final values with respect to primal and dual variables;

B). $RegWrite = L$;

rst $'NameModel'$: the table of input data; the table of final values with respect to primal and dual variables;

C). $RegWrite = L$ or S;

out $'NameModel'$: the table of the results (stationary macrostate) with respect to primal variables.

$NameMode\,1 := 'M' + ClassMode + TypeMode + ClastAlg.$

The first objective of the experiment was to compare the MSS.
The MSS TEST 2 with the parameters

$$l = 5 \ (m = 25, \quad r = 10);$$

$$G = 50;$$

$$d = 1; \quad \alpha = 1,4; \quad \beta = 1,1$$

was chosen as basic.

Stationary macrostate has been determined by the multiplicative algorithm (8, 9). The results are represented in fig. 4.1 for the MSS with complete consumption of resources and in fig. 4.2 for the MSS with incomplete consumption of resources.

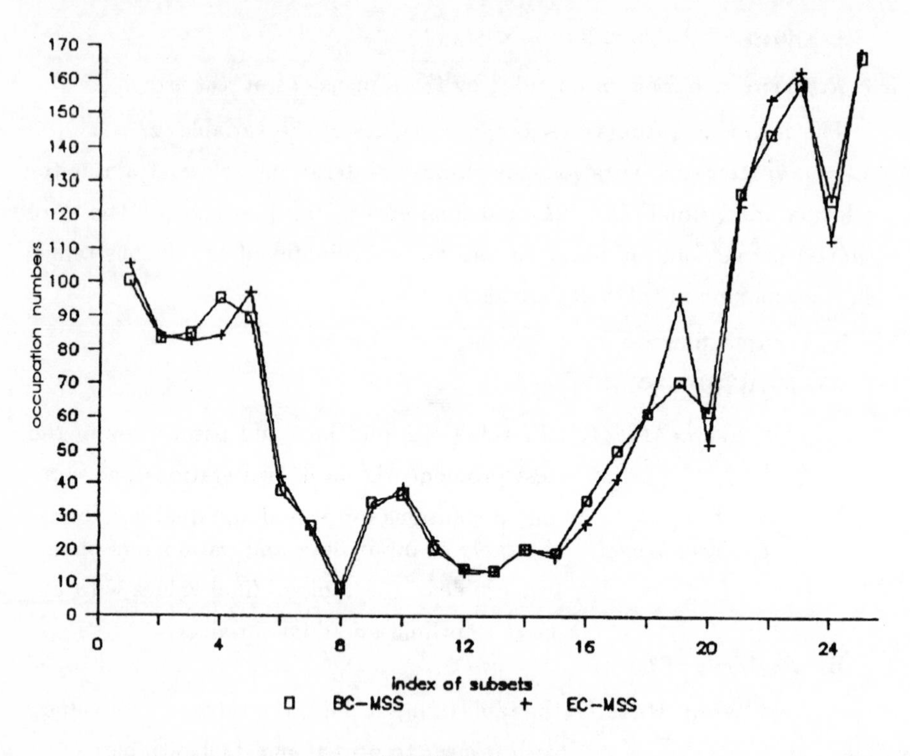

FIGURE 4.1.

Analysis shows that various MSS with complete consumptions (C) have quite close stationary states. Significant difference is observed for MSS with incomplete consumption (see Fig. 4.2).

It should be noted that in these experiments the feasible sets and apriori probabilities of state occupation were the same for all models.

The second objective of the experiment was to investigate the computational properties of multiplicative algorithms. The MSS TEST 1 with the parameters

$$m = 7; \quad r = 5 \ (r = 3); \quad d = 30 \ (d = 2.0); \quad v = 0.66;$$

$$G = 50.0;$$

was basic in this case.

During the experiments, we studied how the number of iterations depends on starting points and estimated the computational speed of the algorithms of types P and K. Six initial points were generated:

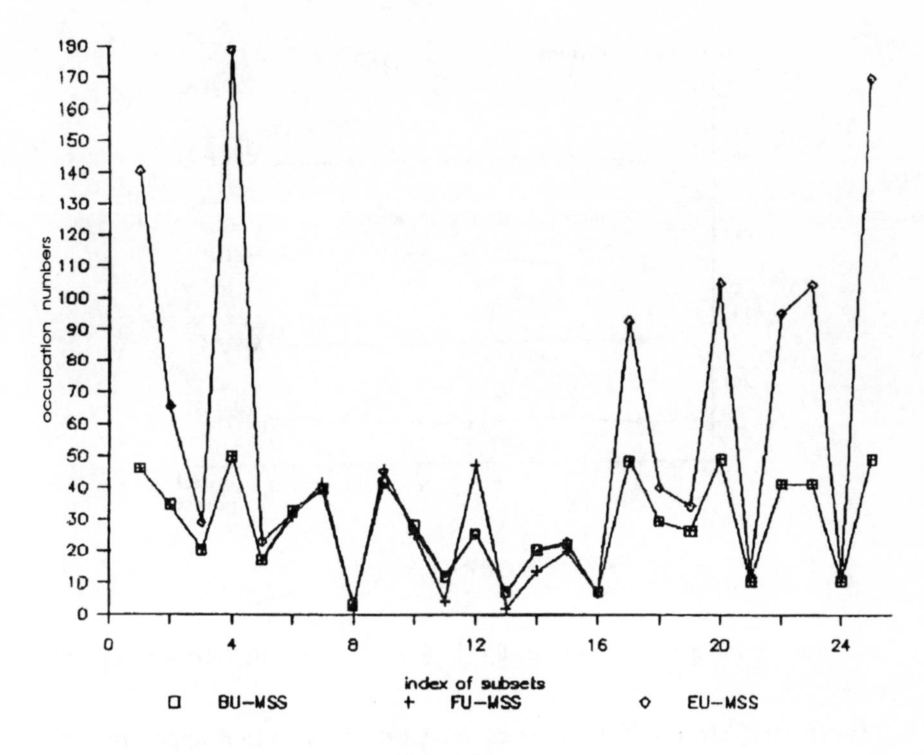

FIGURE 4.2.

$1: \ z[k] = a, \quad k \in \overline{1,5};$

$2: \ z[1] = b, \quad z[k] = a, \quad k \in \overline{2,5};$

$3: \ z[1] = z[2] = b, \quad z[k] = a; \quad k \in \overline{3,5};$

$4: \ z[1] = z[2] = z[3] = b, \quad z[4] = z[5] = a;$

$5: \ z[1] = \ldots = z[4] = b, \quad z[5] = a;$

$6: \ z[k] = b, \quad k \in \overline{1,5},$

where

$a = 10, \quad b = 100$ for MSS with complete consumption and

$a = 1, \quad b = 50$ for MSS with incomplete consumption.

The analysis of the computations shows that

(a) the multiplicative algorithms of type K (8, 9, 12) are significantly more efficient than those of type P (6, 10, 11). For some problems (MECP, MECK) they are tenfold faster. It seems that these problems are characterized by "ravenous" properties. If the problem is sufficiently symmetrical, the algorithms of type K are also faster (MEUP, MEUK) (see fig. 4.3);

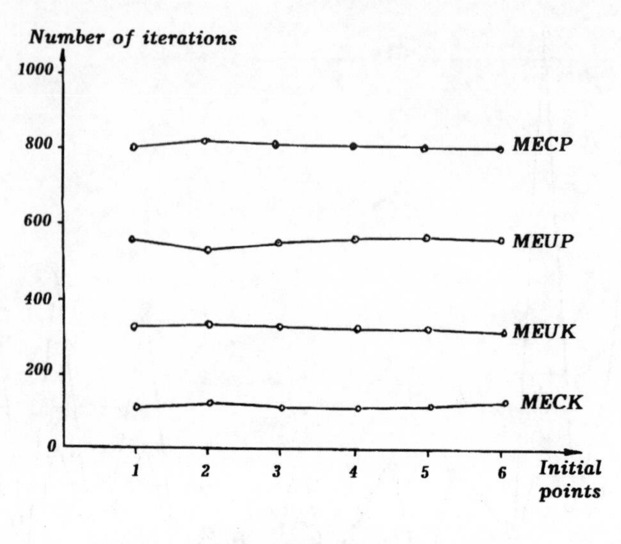

FIGURE 4.3.

(b) the starting points practically do not influence the number of iterations;

(c) the trajectories of the process are constructed so that approximately one third of all time is necessary to attain the 10% – neighborhood (by the norm of deviations) from any starting point.

CHAPTER
FIVE

PROBABILISTIC HIERARCHICAL STRUCTURES

5.1 Definition of probabilistic hierarchy

Hierarchical ordering occurs in many problems. The order relations between a pair of elements or a pair of groups consisting of homogeneous elements are those of "better", "more important", "more significant", and so on. Applying these relations to an arbitrary set, one can obtain a series of elements (or groups of element), with each element being "better", "more important", "more significant", etc., than the following one. Usually, the places in such a series are interpreted as subordination levels, and an element (or a group of elements) on the k-th level is subjected to the elements of the $(k-1)$-th level.

A hierarchical order can be considered as the means used by the nature or a man to order one or another phenomenon, object or problem and to make them purposeful. There is no doubt about the significance of hierarchical order for the progress of human being.

It is worth to mention C. Arrow here, who said that "among all the creations of human being, the use of a purposeful organization is one of its greatest and earliest inventions" (Arrow, 1964).

Usually, a hierarchical order is needed in studying and designing complex systems for structuring such systems (Mesarovich, Mako, Takahara, 1973). Hierarchical, or multilevel structure of a complex problem arises in decomposing the system, grouping the parts into homogeneous classes, and setting a sequence in their functioning. Such an approach is typical in building large systems of distribution of material resources (power supply systems, water economy systems), systems of organization management, many technological systems (ethylene and sulfuric acid production, conveyer systems, etc.).

However, a hierarchical order in not a characteristic feature of only "complex" systems or problems. It can arise in quite "simple" situations. To illustrate this, a system of recurrent equations can be considered, each being very simple (e.g. a linear algebraic equation). But only on solving the first (solvable) of them, one can proceed to the following equation and so on, i.e. there exists a strict hierarchy of the equations.

Many papers are devoted to study of hierarchical order, hierarchical organization (orderliness), and hierarchical structure. These are dealt with both the general questions of construction and study of hierarchy (Bernussou, Titli, 1982; Singh, Titli, 1978) and with the problems arising in particular fields (Segall, 1979; Milner, 1975; Ovsievich, 1979). In these papers, the strict hierarchical order is studied in which the distribution of elements (or groups of elements) among the subordination levels is fixed.

In many real situations, however, such a distribution is subject to changes (variations) which have a stochastic character. In order to describe these phenomena, we introduce the notion of probabilistic hierarchy.

Let us consider a universal finite set $\mathfrak{D}$ with the elements $D_\mu (\mu \in \overline{1,m})$. We call the sequence of elements D_μ ordered by means of a strict dominance a "*hierarchical structure* $\mathfrak{S}$" (or a structure with a strict hierarchy), namely

$$\mathfrak{S} = (D_{\mu_{11}}, \ldots, D_{\mu_{1s_1}}) \succ \ldots \succ (D_{\mu_{k1}}, \ldots, D_{\mu_{ks_k}}), \tag{1}$$

where μ_{ij} is a number of element on the i-th level; s_i is the number of elements on the i-th level; k is the number of levels ($k \leq m$) and $\mu_{ij} \in \overline{1,m}$; $i \in \overline{1,k}$; $j \in \overline{1,s_i}$.

If only one element is at each level then

$$\mathfrak{S} = \{D_{\mu_1} \succ D_{\mu_2} \succ \ldots \succ D_{\mu_m}\}, \tag{1'}$$

where $\mu_i \in \overline{1,m}$; and i is the level number.

The definitions (1) and (1′) mean that the subordination between the elements or groups of elements is fixed in the structures having the strict hierarchy.

If the subordination among the elements is not fixed and is random then the element D_j enters the i-th level with the probability w_{ij}. Thus the hierarchical structure $\mathfrak{S}$ (1) happens to be a complex stochastic event consisting in the fact that "the elements $D_{\mu_{11}}, \ldots, D_{\mu_{1s_1}}$ are on the first level; $D_{\mu_{21}}, \ldots, D_{\mu_{2s_2}}$ are on the 2-nd, ..., and $D_{\mu_{k1}}, \ldots, D_{\mu_{ks_k}}$ are on the k-th one".

The contingency of an element D_j entering a level i gives rise to an ensemble $\mathfrak{T}$ of hierarchical structures.

If only one element can be on each level, and the number of these levels is equal to m, then the number of hierarchical structures in $\mathfrak{T}$ is equal $q = m!$.

By reasoning in the way similar to that stated above, we can obtain the realization probability for every structure in the ensemble $\mathfrak{T}$ and thus define the probability distribution function $P(\mathfrak{S}_\nu)$ on the ensemble $\mathfrak{T}$, where $\mathfrak{S}_\nu \in \mathfrak{T}$; $\nu = \{(\mu_{11}, \ldots, \mu_{1s_1}), \ldots, (\mu_{k1}, \ldots, \mu_{ks_k})\}$ for the structure of type (1) and $\nu = \{\mu_1, \ldots, \mu_k\}$ for the structure of type (1′); $\nu \in \overline{1,q}$.

We call the ensemble of hierarchical structures with the probability distribution function $P(\mathfrak{S}_\nu)$, $\mathfrak{S}_\nu \in \mathfrak{T}$, defined on it the "*probabilistic hierarchical structure* $\mathfrak{H}$", namely

$$\mathfrak{H} = \{\mathfrak{T}, P(\mathfrak{S}_\nu) \,|\, \mathfrak{S}_\nu \in \mathfrak{T}, \quad \nu \in \overline{1,q}\}. \tag{2}$$

Note that the term "structure" used in this section should not be understood literally. Here it is used in a wider sense. This could be an order or a sequence of ranks between some objects and materialized links between the objects could be absent. Along with such an interpretation, the term "structure" will also be used in the common sense, where the real communication channels exist and substance, energy or information are transferred between the objects.

5.2 Classification and characteristics of probabilistic hierarchical structures

It follows from the definition of probabilistic hierarchy (1.2) that all its properties are reflected in the distribution function $P(\mathfrak{S})$ ($\mathfrak{S}_\nu \in \mathfrak{T}$, $\nu \in \overline{1,q}$).

The ensemble $\mathfrak{T}$ of hierarchical structures is a random (nonordered) set which makes the investigation difficult. Many problems of analyzing would be simpler if the ensemble could be ordered somehow. For these purposes, the distribution function $P(\mathfrak{S}_\nu)$ can be used.

A probabilistic hierarchical structure is said to be "*stochastically ordered*" if the hierarchical structures $\mathfrak{S}_\nu$ from $\mathfrak{T}$ can be renumbered so that their probabilities will decrease strictly monotonously, i.e.

$$P(\mathfrak{S}_{\nu_1}) > \ldots > P(\mathfrak{S}_{\nu_q}). \tag{1}$$

The stochastic order means, in particular, that each element $\mathfrak{S}_\nu$ in the ensemble $\mathfrak{T}$ can be realized with the probability different from those of the other elements.

A probabilistic hierarchical structure is said to be "*stochastically half-ordered*" if

$$P(\mathfrak{S}_{\nu_1}) \geq P(\mathfrak{S}_{\nu_2}) \geq \ldots \geq P(\mathfrak{S}_{\nu_q}), \tag{2}$$

where at least one inequality is strict.

In such probabilistic hierarchical structures, some (but not all) elements $\mathfrak{S}_\nu$ from $\mathfrak{T}$ can have the same probabilities of realization.

And finally if

$$P(\mathfrak{S}_\nu) = \widetilde{P} = 1/q \tag{3}$$

then such a probabilistic hierarchical structure is said to be "*stochastically disordered*".

It is seen from the classification that the elements of ensembles $\mathfrak{T}$ of the first two classes differ by the probabilities of their realization; because of this, these numerical characteristics can be used for comparison and choice of an appropriate hierarchical structure.

The elements of ensemble $\mathfrak{T}$ of the third class (3) are equivalent in this sense.

For the ensembles $\mathfrak{T}$ of the first and second classes, the conditions (1) and (2) define merely the monotonicity or strict monotonicity of the probability distribution. The decrease rate of $P(\mathfrak{S}_\nu)$, however, can turn out to be so small that even slight errors in calculating $P(\mathfrak{S}_\nu)$ can break these conditions.

So the definitions (1) or (2) are not sufficient to define the class to which a probabilistic hierarchical structure belongs. It is necessary that the difference between $P(\mathfrak{S}_\nu)$ and the homogeneous distribution (3) is not lower than a certain threshold.

We shall use the information metrics (Stratonovich, 1975) to estimate the distance between $P(\mathfrak{S}_\nu) = P(\nu)$ and the homogeneous distribution $\widetilde{P} = 1/q$

$$I[P(\nu)] = \sum_{\nu=1}^{q} P(\nu)[\ln P(\nu) - \ln \widetilde{P}]. \tag{4}$$

The right-hand side of the inequality is nonnegative for any $P(\nu)$. Indeed, let us denote $\widetilde{P}/P(\nu) = \xi$ and consider

$$\mathcal{M}(\xi) = \sum_{\nu=1}^{q} P(\nu)(\widetilde{P}/P(\nu)) = 1,$$

and

$$\mathcal{M}(\ln \xi) = \sum_{\nu=1}^{q} P(\nu)\ln(\widetilde{P}/P(\nu)),$$

where $\mathcal{M}$ is the operator of mathematical expectation.

According to Jensen's inequality,

$$\mathcal{M}(\ln \xi) \leq \ln(\mathcal{M}\xi).$$

Therefore

$$\sum_{\nu=1}^{q} P(\nu)\ln(\widetilde{P}/P(\nu)) \leq 0,$$

and consequently

$$I[P(\nu)] \geq 0.$$

It follows from (4) that $I = 0$ if $P(\nu) = \widetilde{P}$, i.e. the minimum of the information metrics (4) is attained on the stochastically disordered probabilistic hierarchical structure.

Let us calculate the information metrics for a strict hierarchical structure. In this case, the probability distribution function is

$$P(\nu) = P_1(\nu) = \begin{cases} 1 & (\tilde{\nu} = \nu), \\ 0 & (\tilde{\nu} \neq \nu), \end{cases} \tag{5}$$

and

$$I[P_1(\nu)] = \ln q.$$

Therefore

$$I[P(\nu)] - I[P_1(\nu)] = \sum_{\nu=1}^{q} P(\nu) \ln P(\nu) \leq 0$$

for any distribution function $P(\nu)$, with equality attained only for $P(\nu) = P_1(\nu)$. Hence

$$I[P(\nu)] \leq I[P_1(\nu)], \tag{6}$$

i.e. the strict hierarchical structure has the maximum information metrics $I[P(\nu)]$ (4).

We shall define *the ordering level* ω *of a probabilistic hierarchical structure* by using the information metrics (4):

$$\omega = I / \ln q. \tag{7}$$

This numerical characteristic varies from 0 (for a stochastically disordered probabilistic hierarchical structure) to 1 (for a deterministic (strict (5)) hierarchical structure). This can be used to determine of whether a particular probabilistic hierarchical structure belongs to the class of stochastically ordered structures.

Even if this is the case, it is difficult, however, to use the whole ensemble of hierarchical structures. It is more convenient to choose a structure which would reflect the properties of the ensemble (in some sense).

Since a probabilistic hierarchical structure is characterized by the distribution function $P(\nu)$, such representative structure should reflect some features of $P(\nu)$. For example, it can correspond to the maximum value of $P(\nu)$. But such the most probable structure has the number $\nu = 1$ for stochastically ordered and half-ordered structures and does not depend on the way the probability distribution function varies for $\nu > 1$. Therefore the most probable structure will be the same for both $P(\nu)$ close to the homogeneous distribution and $P(\nu)$ very different from it; in this sense this structure will give little or no information about the properties of the ensemble as a whole.

To choose such a representative structure, numerical characteristics of the distribution like moments, particularly, the "mean" could be used. However, two complications are here, namely the mean of a discrete random value is a real number and therefore does not define any structure from the ensemble; the mean value exists for any discrete distribution including the homogeneous one, though it is clear that no representative element exists for a stochastically disordered probabilistic structure.

We see that it would be logical to associate the representativity of a structure with the information that can be contained in it.

Consider the following probabilistic hierarchical structure

$$\mathfrak{H} = \{\mathfrak{T},\ P(\mathfrak{S}_\nu) \mid \mathfrak{S}_\nu \in \mathfrak{T},\ \ \nu \in \overline{1,q}\}.$$

If we can find a representative structure $\mathfrak{S}_{\nu^*}$ having the number ν^*, the hierarchical structure can be characterized by the distribution function

$$P^*(\nu) = \sigma(\nu - \nu^*) = \begin{cases} 1 & (\nu^* = \nu), \\ 0 & (\nu^* \neq \nu), \end{cases} \tag{8}$$

$$\nu^* \in \overline{1,q}.$$

Let us consider the difference

$$\Phi(\nu, \nu^*) = |P(\nu) - P^*(\nu)| \tag{9}$$

and define the normed function

$$\varphi(\nu, \nu^*) = \Phi(\nu, \nu^*)/2[1 - P(\nu^*)], \tag{10}$$

where

$$2[1 - P(\nu^*)] = \sum_{\nu=1}^{q} \Phi(\nu, \nu^*).$$

The function $\varphi(\nu, \nu^*)$ is the normed distribution function of "absolute errors" (9), with the parameter ν^* fixed. Usually, numerical characteristics, for example the information contained in φ, are introduced to estimate its properties. Use of the information as a numerical characteristic of distribution of "absolute errors" is justified by the fact that the representativity of a structure is desired to be associated with the information on the distribution $P(\nu)$.

Thus let us define the information

$$\widetilde{I}[\varphi(\nu,\nu^*)] = -\sum_{\nu=1}^{q} \varphi(\nu,\nu^*) \ln \varphi(\nu,\nu^*). \tag{11}$$

By substituting the expression for $\varphi(\nu,\nu^*)$ (10) into this equality, we obtain

$$\begin{aligned} \widetilde{I}[P(\nu^*)] &= \frac{\widetilde{I}_0}{2[1-P(\nu^*)]} + \frac{1}{2} \ln 2[1-P(\nu^*)] + \\ &+ \frac{P(\nu^*)}{2[1-P(\nu^*)]} \ln P(\nu^*) - \frac{1}{2} \ln \frac{1}{2}, \end{aligned} \tag{12}$$

where $\widetilde{I}_0 = -\sum\limits_{\nu=1}^{q} P(\nu) \ln P(\nu)$.

By checking the first- and second-order conditions directly, we can see that the function $\widetilde{I}(P)$ (12) has a minimum at a certain point $P_0 \in \overline{0,1}$. If $P(\nu^*)$ is a strictly monotonous function (a stochastically ordered structure), the only unique value ν_0^* corresponds to P_0.

We call a hierarchical structure corresponding to $\nu^* = \nu_0^*$ the "*information structure*", i.e.

$$\nu_0^* = \arg\min \widetilde{I}[P(\nu^*)]. \tag{13}$$

Consider two marginal cases, viz the stochastically disordered structure ($P(\nu) = 1/q$) and the strict structure ($P(\nu) = P_1(\nu)$ (5)). In the first case, the information (12) is $\widetilde{I} = \ln 2 + \frac{1}{2}\ln(q-1)$. Hence it is seen that $\widetilde{I}$ is constant for all ν^*. Therefore there is no information structure for a stochastically disordered structure.

For the strict structure (5),

$$\widetilde{I} = \begin{cases} \ln 2 & (\nu^* \neq \widetilde{\nu}), \\ -\infty & (\nu^* \neq \widetilde{\nu}), \end{cases}$$

i.e. the functional $\widetilde{I}[P(\nu^*)]$ attains the minimum value $(-\infty)$ for $\nu_0^* = \widetilde{\nu}$ (the information structures coincides with the strict one).

The numerical characteristics of stochastic order (the ordering degree and the information structure) allow one to analyze constructively the properties of probabilistic hierarchical structure.

One of the purposes of such an analysis is to determine the links between the ordering degree, the information structure and the most probable structure.

From (4) and (7) we have

$$\widetilde{I} = (1-\omega)\ln q.$$

By substituting this expression into (12), we obtain

$$\widetilde{I}[P(\nu^*)] = \frac{(1-\omega)\ln q}{2[1-P(\nu^*)]} + \frac{1}{2}\ln[1-P(\nu^*)] + \\ + \frac{P(\nu^*)\ln P(\nu^*)}{2[1-P(\nu^*)]} - \frac{1}{2}\ln\frac{1}{2}.$$

This equality links the ordering degree ω and the information structure ν_0^* (13).

Example 1. Consider the probabilistic hierarchical structure with the exponential distribution $P(\nu) = ae^{-\beta\nu}$, where $a = \dfrac{e^{-\beta}-1}{e^{-\beta}(e^{-\beta q}-1)}$; and $\beta \geq 0$.

Let us determine the dependence of the values ω, ν_0^*, and $P_{\max}$ on the parameter β.

From (12) we have

$$\widetilde{I}(\nu^*) = -\frac{e^{-\beta}[(\alpha-\gamma) - \alpha e^{-\beta} + e^{\beta q}((q+1)\gamma - \alpha) - e^{-\beta(q+1)}(\gamma q - \alpha)]}{2(e^{-\beta}-1)^2(1-ae^{-\beta\nu^*})} + \\ + \frac{1}{2}\ln 2(1-ae^{-\beta\nu^*}) + \frac{ae^{-\beta\nu^*}(\ln a - \beta\nu^*)}{2(1-ae^{-\beta\nu^*})} + C,$$

where $C = -\dfrac{1}{2}\ln\dfrac{1}{2}$; $\alpha = a\ln a$ and $\gamma = a\beta$.

By replacing ν^* by the real variable y, we find that the minimum of $\widetilde{I}(y)$ is attained at the point

$$y_0 = \frac{1}{\beta}\times \\ \times\left(\ln a - \frac{e^{-\beta}[(\alpha-\gamma) - \alpha e^{-\beta} + e^{-\beta q}(\gamma(q+1) - \alpha) - e^{-\beta(q+1)}(\gamma q - \alpha)]}{(1-e^{-\beta})^2}\right).$$

β	0.1	0.2	0.4	0.6	0.8	1.0
y_0	3.1293	2.9382	2.4338	2.0480	1.7663	1.5655
$\min I$	1.4844	1.4641	1.3756	1.2495	1.1186	1.1903
ν_0^*	3	3	2	2	2	2
ω	0.0126	0.0314	0.1138	0.2204	0.3300	0.4290

TABLE 5.1.

Then the minimum of $\widetilde{I}(\nu^*)$ is attained at the point

$$[y_0] = \begin{cases} [y_0]^- = n & (\xi \leq \frac{1}{2}), \\ [y_0]^+ = n+1 & (\xi > \frac{1}{2}), \end{cases}$$

where n is the nearest integer less than y_0; $\xi = y_0 - n$.

The ordering degree is

$$\omega = 1 + \frac{e^{-\beta}\left[(\alpha-\gamma) - \alpha e^{-\beta} + e^{-\beta q}\left(\gamma(q+1) - \alpha\right) - e^{-\beta(q+1)}(\gamma q - \alpha)\right]}{(1-e^{-\beta})^2 \ln q}.$$

Given in Table 5.1 are the calculations of the ordering degree ω and the information structure ν_0^* for the probabilistic hierarchical structure containing $q = 6$ of three-level structures. Note that the most probable structure corresponds to $\nu = 1$.

It is seen from this table that the information structure ν_0^* differs significantly from the most probable one for small β, i.e. when the function $P(\nu)$ decreases slowly.

When β increases, the information structure approximates to the most probable one and the ordering degree approximates to 1 thus reflecting the qualitative characteristics of the given probabilistic hierarchical structure.

In fact, if the distribution function $P(\nu)$ of probabilistic hierarchical structure slightly differs from the homogeneous distribution, the most probable structure $\nu = 1$ cannot be regarded as a representative of the ensemble, since the other structures have the realization probabilities close to $P_{\max}$. The structure with $\nu > 1$ could be more representative. In this case, the more the function $P(\nu)$ slopes, the more biasing to the right ν_0^* is. This tendency is reflected in Table 5.1.

5.3 Models of ensembles

Statistical properties of probabilistic hierarchical structures are characterized by the distribution functions $P(\mathfrak{S}_\nu) = P(\nu)$. In modeling these properties, the aim is to reproduce the corresponding distribution functions in which the desired properties of a probabilistic hierarchical structure and, on the other hand, the natural mechanisms of their forming could be reflected.

Recall that a probabilistic hierarchical structure consists of the elements $D_1, \ldots, D_m$ that are distributed independently and stochastically with the probabilities $w_{11}, \ldots, w_{1k}; \ldots; w_{m1}, \ldots, w_{mk}$ among k levels thus forming various chains $\{D_{\mu 1}, \ldots, D_{\mu k}\}$. If the matrix of probabilities $W = [w_{ij}]$ is given, the probabilities of the chains $P(\nu)$ being realized are determined completely.

In reality, however, W is not known and should be found by using the information about forming the chains in probabilistic hierarchical structure and about the desired properties of the ensemble of structures being formed.

The latter is usually related to the ensemble characteristic, namely to the distribution function of the probabilities of realizing the chains $P(\mathfrak{S}_\nu) = P(\nu)$.

The desired properties of $P(\nu)$ can be described in terms of its approximation to a fixed distribution function, under a constraint on some of its characteristics.

We shall use information criteria to evaluate the quality of such an approximation. One of these was already used (Section 2) in estimating the information distance between the distribution functions of stochastically ordered or disordered probabilistic hierarchical structures.

We shall denote by $P_0(\nu)$ the desired distribution function. Then the information distance between $P_0(\nu)$ and $P(\nu)$ can be determined as follows

$$I[P(\nu)] = \sum_{\nu=1}^{q} P(\nu) \ln \frac{P(\nu)}{P_0(\nu)}, \tag{1}$$

where the distribution function $P(\nu)$ satisfied the norming condition

$$\sum_{\nu=1}^{q} P(\nu) = 1. \tag{2}$$

The resources are consumed in forming a probabilistic hierarchical structure with the distribution function "similar" to the desired function $P_0(\nu)$ in sense (1). Assume that there are r types of such resources. Resource consumption will be characterized by the appropriate consumption functions $\varphi_1, \ldots, \varphi_r$. In the simplest case the consumption functions depend only on the corresponding structure, i.e. $\varphi_s = \varphi_s(\nu)$, $s \in \overline{1,r}$. However, the situations occur when the consumption functions depend not only on the structure $\mathfrak{S}$, but also on the probability of its realization, i.e. $\varphi_s = \varphi_s(\nu, P(\nu))$.

Since the structures $\mathfrak{S}_\nu$ are stochastic, it is logical to use the mean consumption with respect to the ensemble

$$\Pi_s = \sum_{\nu=1}^{q} \varphi_s(\nu) P(\nu).$$

Denote by $q_1, \ldots, q_r$ the stocks of r resources, respectively. In forming a probabilistic hierarchical structure, the stocks of recourses can be consumed completely ($\Pi_s = q_s$, $s \in \overline{1,r}$) and incompletely ($\Pi_s \leq q_s$, $s \in \overline{1,r}$).

Thus a mathematical model of forming the desired distribution function over the ensemble of structures can be represented in the following form

$$I = \sum_{\nu=1}^{q} P(\nu) \ln \frac{P(\nu)}{P_0(\nu)} \to \min; \tag{3}$$

$$\sum_{\nu=1}^{q} P(\nu) = 1, \quad P(\nu) \in \mathfrak{D}; \tag{4}$$

$$\mathfrak{D} = \begin{cases} \left\{ P(\nu) : \sum_{\nu=1}^{q} \varphi_s(\nu) P(\nu) = q_s, \quad s \in \overline{1,r} \right\}, & (5) \\ \left\{ P(\nu) : \sum_{\nu=1}^{q} \varphi_s(\nu) P(\nu) \leq q_s, \quad s \in \overline{1,r} \right\}, & (6) \end{cases}$$

Note that the model (3–6) is formally similar to the MSSs considered in Chapter 2. (The entropy objective function in MSS have the sign opposite to that of the information distance in the given model).

Consider the simplest modification of (3–5) where the recourses are not constrained. Let us introduce the notation

$$p_\nu = P(\nu); \quad a_\nu = P_0(\nu); \quad H(p) = -I(p); \quad p = \{p_1, \ldots, p_q\}. \tag{7}$$

Then we obtain

$$\begin{gathered} H(p) = -\sum_{\nu=1}^{q} p_\nu \ln \frac{p_\nu}{a_\nu} \to \max, \\ \sum_{\nu=1}^{q} p_\nu = 1. \end{gathered} \tag{8}$$

The solution of this problem is defined by the stationary point of the Lagrange function $L = H + \lambda \left(1 - \sum_{\nu=1}^{q} p_\nu\right)$, namely

$$\frac{\partial L}{\partial p_k} = -\ln p_k - 1 + \ln a_k - \lambda = 0, \quad k \in \overline{1, q}.$$

Hence

$$p_k^0 = a_k e^{-1-\lambda}, \quad k \in \overline{1, q}.$$

By substituting this expression into the constraint equation in (8) and taking into account that $\sum_{k=1}^{q} a_k = 1$, $e^{-1-\lambda} = 1$, we obtain

$$p_k^0 = a_k, \quad k \in \overline{1, q}. \tag{9}$$

In particular, when the stochastically disordered structure is "desired", i.e. $a_k = P_0(k) = \frac{1}{q}$ then

$$p_k^0 = P_0(k) = \frac{1}{q}.$$

Thus if the consumed recourses are not constrained, the distribution function of probabilistic hierarchical structure coincides with the "desired" one.

Consider a ***probabilistic hierarchical structure with its formation completely consuming one resource*** ($r = 1$). In addition to (7), we introduce the following notation

$$\varphi_1(\nu) = c_\nu, \quad q_1 = C. \tag{10}$$

Then the model (3–5) is of the form

$$H(p) = -\sum_{\nu=1}^{q} p_\nu \ln \frac{p_\nu}{a_\nu} \to \max,$$

$$\sum_{\nu=1}^{q} p_\nu = 1. \tag{11}$$

$$\sum_{\nu=1}^{q} c_\nu p_\nu = C, \quad p_\nu \geq 0, \quad \nu \in \overline{1,q}.$$

To facilitate the analysis, we reduce this model to the standard form

$$H(p_1, \ldots, p_q) \to \max;$$

$$\sum_{\nu=1}^{q} p_\nu = 1; \tag{12}$$

$$\sum_{\nu=1}^{q} \beta_\nu p_\nu = 1,$$

where $\beta_\nu = c_\nu / C$. The Lagrange function for this problem is

$$L = H(p) + \lambda \left(1 - \sum_{\nu=1}^{q} p_\nu\right) + \varepsilon \left(1 - \sum_{\nu=1}^{q} \beta_\nu p_\nu\right).$$

The stationary value of Lagrange's function is attained at the point with the coordinates

$$p_k^0 = a_k e^{-1-\lambda-\varepsilon\beta_k}, \quad k \in \overline{1,q}.$$

By substituting these equalities into the constraints of problem (12), we obtain

$$p_k^0 = \frac{a_k e^{-\varepsilon\beta_k}}{\sum_{k=1}^{q} a_k e^{-\varepsilon\beta_k}}, \qquad k \in \overline{1,q};$$

$$\sum_{k=1}^{q} a_k(\beta_k - 1) e^{-\varepsilon\beta_k} = 0. \tag{13}$$

It is clear from the last equation that the necessary condition for a solution to exist is the following

$$\beta_{\min} < 1 < \beta_{\max},$$

where

$$\beta_{\min} = \min_k \beta_k \quad \text{and} \quad \beta_{\max} = \max_k \beta_k.$$

Let us transform the equation (13) to the form

$$\phi(\varepsilon) = \frac{\sum_{k=1}^{q} a_k \beta_k e^{-\varepsilon\beta_k}}{\sum_{k=1}^{q} a_k e^{-\varepsilon\beta_k}} = 1$$

and introduce a new variable

$$z = e^{-\varepsilon} \geq 0.$$

We obtain

$$\Phi(z) = \frac{\sum_{k=1}^{q} a_k \beta_k z^{\beta_k}}{\sum_{k=1}^{q} a_k z^{\beta_k}} = 1.$$

To define the solution of (13), we can use the multiplicative algorithm (Chapter 4):

$$z^{s+1} = z^s \Phi^\gamma(z^s). \tag{14}$$

Example 1. Consider a model of an urban system consisting of three subsystems, namely the "basic employment sector", the "service sector", and the "population" (see Popkov, 1979). This set of subsystems is usually used in modeling the functionally-spatial structure of a town.

The input of "service" is the spatial distribution of the consumers N_Π of the subsystem objects, while the output is the distribution of the working places E in "service". The model of this subsystem links N_Π and E in the following way

$$E = S(N_\Pi),$$

where S is the subsystem operator.

The state of the "basic sector" is defined by the spatial distribution of workers W, while its output are the distribution of working population among the residence places N_w and the distribution of capacities of labor places G. These are related by the equation

$$W = B(G, N_w),$$

where B is the subsystem operator.

Finally, the output of "population" is the spatial distribution of population P among the residence places, while its input is the spatial distribution of working places K, i.e.

$$P = N(K),$$

where N is the subsystem operator.

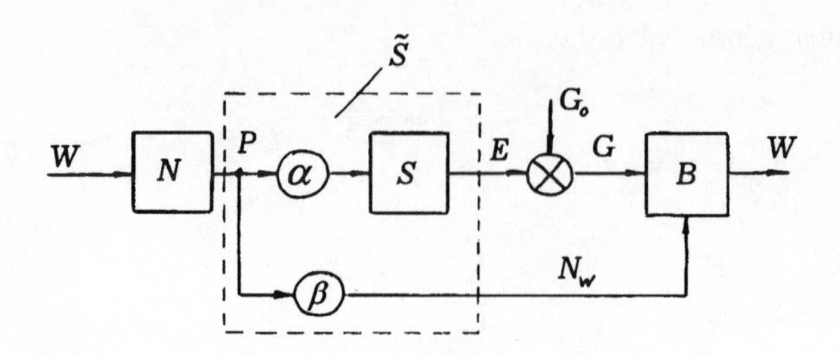

FIGURE 5.1a.

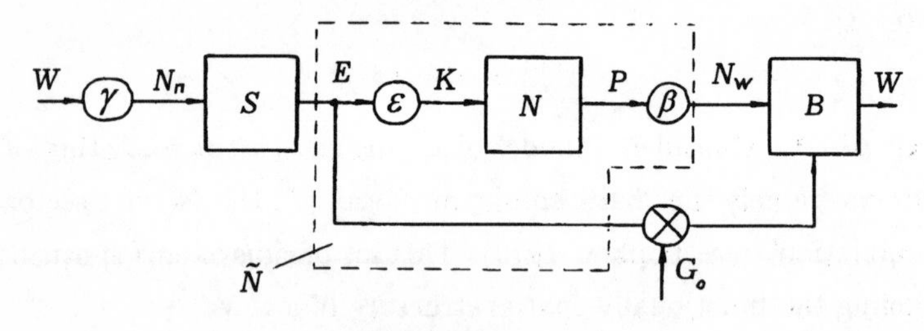

FIGURE 5.1b.

For definiteness sake, we assume that the system model should reproduce a change in the spatial distribution of workers, i.e.

$$W = A(W),$$

where A is the model operator.

It is easy to note that the operator A can be formed from the subsystem operators S, B, and N at least in two ways.

(a) The working places K are balanced with respect to the workers W ($W = K$); the consumers of the "service" are a part of the whole population ($N_{\Pi} = \alpha P$); the capacity of the labor places G consists of a predetermined value and working places E in the service ($G = G_0 + E$); the working population N_{w} is a part of the whole population ($N_{\mathrm{w}} = \beta P$). As a result, we have

$$A = B \circ \widetilde{S} \circ N,$$

where $\widetilde{S} = \{\alpha S, \beta\}$, α and β are scalar.

Such a representation of A corresponds to the structure given in fig. 5.1 (a).

(b) The consumers of "service" N_Π are determined by the workers $W(N_\Pi = \gamma W)$; as before, the capacity of labor places consists of a predetermined value and working places E in the "service" $(G = G_0 + E)$; the general number of working places K is determined by the working places E in service $(K = \varepsilon E)$; the working population is $N_w = \beta P$. In this case, the operator A is of the form (fig. 5.1 b)

$$A = B \circ \widetilde{N} \circ S,$$

where $\widetilde{N} = \{\varepsilon N \beta, 1\}$; ε and β are scalar.

Each of these schemes defines an order between the subsystems S, B, and N, and, in this sense, can be considered as a hierarchical structure. Such hierarchical structures describes a quite real process of using service objects by the inhabitants.

The first hierarchical structure characterizes the use of objects in the basic sector (working population) and service (consumers of the subsystem "service") by the residents ($N_\Pi = \alpha P$, $N_w = \beta P$ and P is the population distributed among the residence places).

The second hierarchical structure characterizes another process in which the objects of "service" are used only by the working population $(N_\Pi = \gamma W)$.

It is clear that both are behavioural hypotheses. Even if the urban population has only two behaviour standards mentioned above, these are realized simultaneously.

Thus we cannot prefer any of the structures (fig. 5.1 (a), (b)). Hence the system model should have a "variable" structure, in order to be more adequate to the behavioural standards considered.

Such a "variance" can be realized by using the models of probabilistic hierarchical structures.

Only two behavioural standards were considered as an example. However, six standards and therefore six hierarchical chains are theoretically possible for three subsystems. This means that we can make six models analogous to those shown in fig. 5.1 (a) and 5.1 (b), i.e. the output of each of the three subsystems can be the output of the system model. We shall assume that the output of the system model is the first level of the hierarchical chain; thereby the second and third levels are defined automatically.

Usually, there is no information about the form of the desired distribution function $P_0(\nu)$ on the ensemble of hierarchical structures; so let us assume that $P(\nu) = \text{const}$. Then, according to (12),

$$H(p) = -\sum_{\nu=1}^{6} p_\nu \ln p_\nu + \text{ const}, \tag{15}$$

$$\sum_{\nu=1}^{6} p_\nu = 1, \qquad \sum_{\nu=1}^{6} \beta_\nu p_\nu = 1.$$

The function of relative costs β_ν ($\nu \in \overline{1,6}$) is an exogenous variable in this model. This function evaluates the behavioural standards (being the basis of the corresponding hierarchical chains) in the units of relative costs.

In this case, the costs of forming the chains are defined by entering the μ-th subsystem in the level ρ, since the distribution of subsystems among the levels is the mapping of the corresponding behavioural hypothesis.

Denote by $C_{\mu,\rho}$ the relative costs related to entering the μ-th subsystem in the level ρ. Then

$$\beta_\nu = \beta(\mu_1^\nu, \mu_2^\nu, \mu_3^\nu) = \sum_{\rho=1}^{3} C_{\mu_\rho^\nu, \rho}, \tag{16}$$

where $\{\mu_1^\nu, \mu_2^\nu, \mu_3^\nu\} \to \nu$.

Consider typical kinds of relative cost functions.

(a) **Linear cost:**

$$\beta_\nu = A + B(\nu - 1), \tag{17}$$

where $A \geq 0$.

If $B > 0$ then the structure having the number $n = q! = 6$ will be the most "expensive"; if $B < 0$, the most "expensive" is the structure having the number 1. It is supposed that $A \geq Bq$. Substituting β_ν into (13), we obtain

$$p_\nu^0 = \frac{e^{-\varepsilon(A+B(\nu-1))}}{\sum_{\nu=1}^{6} e^{-\varepsilon(A+B(\nu-1))}}. \tag{18}$$

The equation for the Lagrange multiplier ε is of the form

$$\Psi(s) = \sum_{\nu=1}^{q-1} s^\nu \gamma_\nu + \gamma_0 = 0, \tag{19}$$

where $s = e^{-\varepsilon B} \geq 0$; $\gamma_\nu = \gamma_{\nu-1} + B$; $\gamma_0 = A - 1$.

Only a positive root has a sense among the roots of this equation. Recall that the number of positive roots of an algebraic equation is equal to the number of sign changes in the sequence of its coefficients $\gamma_{m-1}, \gamma_{m-2}, \ldots, \gamma_1$.

Consider the case $B > 0$. In this case, if $\gamma_0 > 0$ then all the coefficients γ_ν ($\nu \in \overline{1, q-1}$) are positive, and there is no positive root in equation (19).

If $\gamma_0 < 0$ then the signs can change only one time. For this it is necessary that the integer solution k^* of the auxiliary equation

$$A + Bk^* - 1 = 0$$

belongs to the interval $\overline{1, q}$, i.e.

$$1 \le k^* = E\left[\frac{1-A}{B}\right] \le q - 1, \tag{20}$$

where $E(\cdot)$ is an integral part of the number in brackets.

Now let us proceed to the case $B < 0$. In this case, if $\gamma_0 < 0$ then there are no sign changes in the sequence of coefficients, and the equation (19) has no positive roots. On the contrary, if $\gamma_0 > 0$ then one sign change is possible if the condition (20) is satisfied.

Thus (19) has the only positive root s_p if the condition (20) is satisfied and one of the following equalities hold

$$\begin{aligned} \gamma_0 &= A - 1 < 0, \quad \text{if} \quad B > 0, \\ \gamma_0 &= A - 1 > 0, \quad \text{if} \quad B < 0. \end{aligned} \tag{21}$$

The parameter ε is defined as

$$\varepsilon = -\frac{1}{B} \ln s_p.$$

From this expression it is seen that the value $s_p = 1$ is critical, since, for $B > 0$, $\varepsilon > 0$ if $s_p < 1$, and $\varepsilon < 0$ if $s_p > 1$; on the contrary, for $B < 0$, $\varepsilon > 0$ if $s_p > 1$, and $\varepsilon < 0$ if $s_p < 1$.

Therefore not only the existence of a positive root of the equation (19) is important, but also the estimate of its value.

For this purpose, let us use a simple method based on the uniqueness of the positive root of (19). If the values of $\Psi(s)$ are defined in (19) at $s = 1$ and $s = 0$, the position of a positive root can be determined by using the signs of $\Psi(1)$ and $\Psi(0)$.

Chains $\mu_1^\nu, \mu_2^\nu, \mu_3^\nu$	123	132	213	231	312	321
$\beta(\mu_1^\nu, \mu_2^\nu, \mu_3^\nu)$	0.97	1.59	0.66	2.21	1.90	1.28
ν (variant 1)	2	4	1	6	5	3
p_ν^0	0.2544	0.0785	0.4581	0.0242	0.0436	0.1412
ν (variant 2)	5	3	6	1	2	4
$\widetilde{p}_\nu^0$	0.2544	0.0785	0.4581	0.0242	0.0436	0.1412

TABLE 5.2.

According to (19), we obtain

$$\Psi(1) = \sum_{\nu=1}^{q-1}(\gamma_0 + \nu B) + \gamma_0 = q\left[\gamma_0 + \frac{B}{2}(q-1)\right].$$

It follows from this that $\Psi(1) > 0$ if $\gamma_0 > -\frac{B}{2}(q-1)$, and $\Psi(1) < 0$ if $\gamma_0 < -\frac{B}{2}(q-1)$. Let us substitute $s = 0$ into (19). We obtain $\Psi(0) = \gamma_0$, and therefore, according to (21), we have

$$\Psi(0) < 0 \quad \text{for} \quad B > 0,$$

$$\Psi(0) > 0 \quad \text{for} \quad B < 0.$$

By comparing the conditions obtained, we can see that if $B > 0$ then the equation (19) has a single positive root $s_p < 1$ under the following inequality holding

$$-\frac{B}{2}(q-1) < A - 1; \tag{22}$$

if $B < 0$ then the equation (19) has a single positive root $s_p > 1$ under the inequality (22) holding.

Thus if these conditions hold then the system with the linear cost function (17) has the optimal hierarchical distribution of structures (18) with the parameter $\varepsilon > 0$ regardless of the increase ($B > 0$) or decrease ($B < 0$) of the cost.

This is a very important fact confirming the property of the distribution function $P_0^{(2)}(\nu)$ that its maximum corresponds to the minimum of the cost function. In system terms, this means that the most probable system structure requires the minimum forming costs.

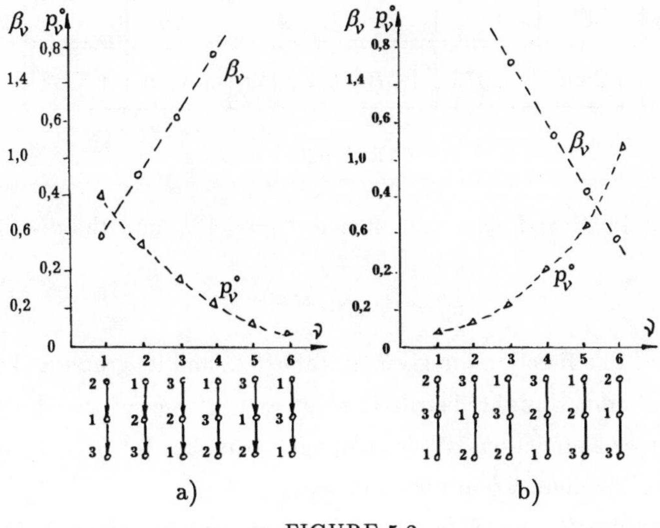

FIGURE 5.2.

Now let us proceed to the construction of an ensemble of models consisting of three subsystems of urban system. Let the relative cost function β_ν be linear (17), with the parameters $A = 0.66$ and $B = 0.31$; and let the minimal cost correspond to the structure having the number $\nu = 1$ (variant 1). The values of β_ν are given in Table 5.2.

The cost function (17) has $B > 0$. Therefore, according to the condition (22), the equation (19) is of the form

$$0.55s^5 + 0.41s^4 + 0.27s^3 + 0.13s^2 - 0.01s - 0.15 = 0,$$

where $s = 0.8694^\varepsilon$; $s_p = 0.55$. Hence $\varepsilon^* = 4.198$, and the distribution function will have the form

$$p_\nu^0 = 0.8246e^{-0.588\nu}.$$

The values of this function are given in Table 5.2, and its graph is shown in fig. 5.2 (a).

Consider another rule for the chain numeration where $\nu = 1$ will correspond to the minimum cost (see Table 5.2, variant 2). In this case the relative cost function remains linear (17), but with other parameters, namely $A = 2.21$ and $B = -0.31$. According to (22), the equation (19) is of the form

$$-0.15s^5 - 0.01s^4 + 0.13s^3 + 0.27s^2 + 0.041s + 0.55 = 0,$$

ν	1	2	3	4	5	6
$\widetilde{I}$	1.3564	1.2573	1.2763	1.3080	1.3356	1.3562

TABLE 5.3.

where $s = 1.1503^{\varepsilon}$ and $s_p = 1.8$. Hence $\varepsilon^* = 4.198$, and the distribution function is

$$\widetilde{p}_\nu^0 = 0.0134e^{0.588\nu}.$$

The values of this function are given in Table 5.2, and its graph is shown in fig. 5.2 (b). Comparing the distributions p_ν^0 and $\widetilde{p}_\nu^0$, we see that the values of distributions as functions of the chains (i.e. the sets $\{\mu_1^\nu, \mu_2^\nu, \mu_3^\nu\}$) do not change when the numeration rules change.

The found distribution function of a probabilistic hierarchical structure can be used in developing a model of the urban system having the strict structure, by using the notions of ordering degree and information structure.

According to (2.7),

$$\omega = \frac{1}{\ln 6} \sum_{\nu=1}^{6} P^0(\nu) \ln P^0(\nu) + 1 = 0.2214.$$

Let us define the information structure; for this purpose we calculate the values of functional (2.12). These are given in Table 5.3. It follows from the table that the minimum value of $\widetilde{I}$ corresponds to $\nu^* = 2$. Therefore the information structure does not coincide with the most probable one for the distribution found.

(b) **Logarithmic cost:**

$$\beta_\nu = A + B\nu + D \ln \left(\sum_{k=0}^{N} \gamma_k \nu^k \right), \quad \gamma_0 = 1. \tag{23}$$

Substituting this expression into (13), we obtain

$$p_\nu^0 = \frac{\left(\sum_{k=0}^{N} \gamma_k \nu^k \right)^{-\varepsilon D} e^{-\varepsilon(A+B\nu)}}{\sum_{\nu=1}^{m} \left(\sum_{k=0}^{N} \gamma_k \nu^k \right)^{-\varepsilon D} e^{-\varepsilon(A+B\nu)}}. \tag{24}$$

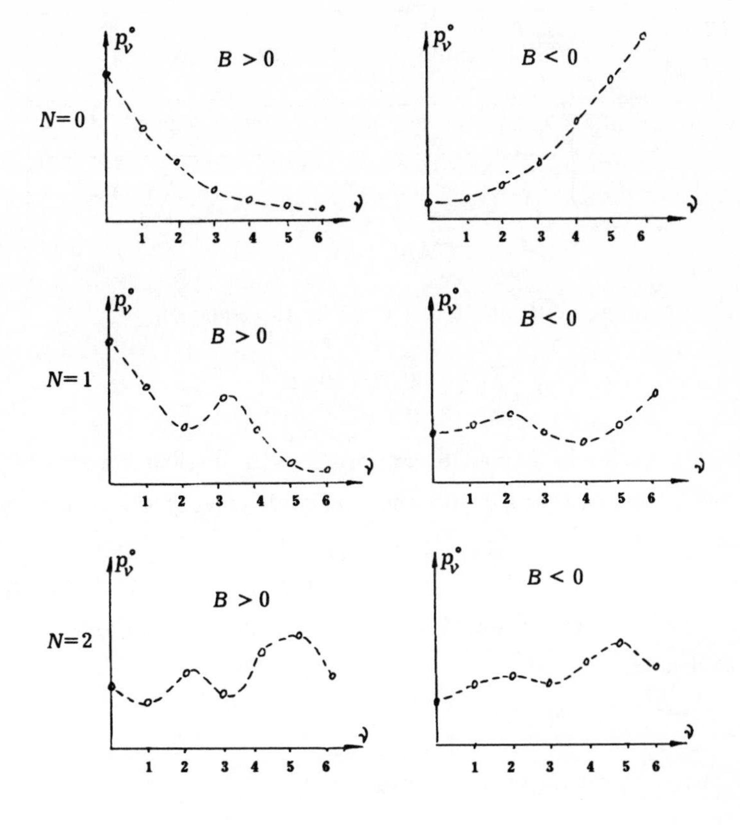

FIGURE 5.3.

This distribution function is a logical generalization of the distribution function (18). The qualitative graphs of the functions p_ν^0 are shown in fig. 5.3 for $N = 0, 1, 2$.

Let us return to the problem of forming the ensemble of models for the urban system with the relative cost function $\beta_\nu = \widetilde{\beta}_\nu C$, where β_ν is as in (23), with the parameters $A = 1.5$, $B = 0.2$, $N = 1$, $\gamma_1 = e^B$, $\gamma_0 = 0$ and $D = -1$. Its values at $C = 1$ are given in Table 5.4 (fig. 5.4).

According to (24), the optimal function of probability distribution is

$$p_\nu^0 = \frac{\nu^\varepsilon e^{-\varepsilon B\nu}}{\sum_{\nu=1}^{6} \nu^\varepsilon e^{-\varepsilon B\nu}}, \tag{25}$$

Chains $\mu_1^\nu, \mu_2^\nu, \mu_3^\nu$	123	132	213	231	312	321
$\beta(\mu_1^\nu, \mu_2^\nu, \mu_3^\nu)$	2.591	1.182	0.595	4.286	0.8303	1.718
ν	2	4	6	1	5	3

TABLE 5.4.

where the Lagrange multiplier ε is defined by the equation

$$\sum_{\nu=1}^{6}(\beta_\nu - 1)\nu^\varepsilon e^{-\varepsilon B\nu} = 0. \tag{26}$$

Since $\beta_\nu = \widetilde{\beta}_\nu C$, the Lagrange multiplier ε is an implicit function of the parameter C, which characterizes the change of scale of the relative cost function.

If we refer to the model (11), we can see that the parameter C characterizes a stock of the resource consumed to form the hierarchical chains.

Let us denote

$$\phi(\varepsilon, C) = \sum_{\nu=1}^{6}(\widetilde{\beta}_\nu C - 1)\nu^\varepsilon e^{-\varepsilon B\nu}.$$

Then it is convenient to rewrite (26) in the form

$$\phi(\varepsilon, C) = 0. \tag{27}$$

For $C_0 = 1$ the solution of this equation is $\beta_0 = 1$. In this case, according to (25), we have

$$p_\nu^0 = \frac{\nu e^{-0.2\nu}}{9.2492}. \tag{28}$$

Consider the way in which the solution of (27) behaves as the function of C in the neighborhood of the point $C_0 = 1$, $\varepsilon_0 = 1$.

Using the technique of studying the properties of an implicit function (see Chapter 3), we obtain that

$$\beta(C) = \beta_0 + \left.\frac{\partial\beta}{\partial C}\right|_{C_0}(C_0 - C), \tag{29}$$

in the neighborhood of the point $C_0 = 1$, $\beta = 1$, where

$$\left.\frac{\partial\beta}{\partial C}\right|_{C_0} = -\left.\frac{\partial\phi}{\partial C}\right|_{C_0} \Big/ \left.\frac{\partial\phi}{\partial\beta}\right|_{\beta_0, C_0}.$$

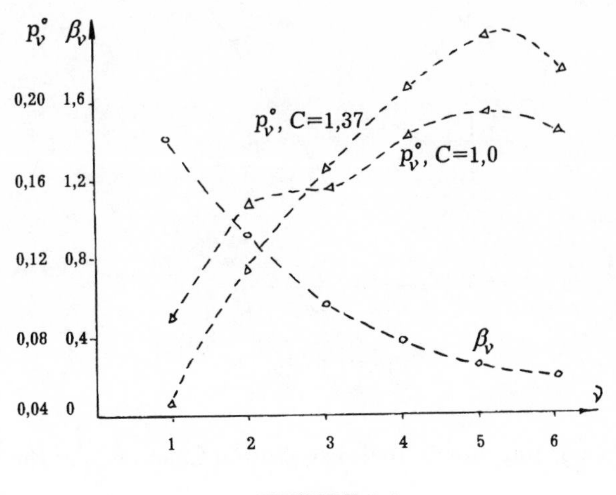

FIGURE 5.4.

According to (27),

$$\left.\frac{\partial \phi}{\partial C}\right|_{C_0} = \sum_{\nu=1}^{6} \widetilde{\beta}_\nu \nu^{\beta_0} e^{-\beta_0 B\nu} = 17.017.$$

$$\left.\frac{\partial \phi}{\partial \beta}\right|_{C_0,\beta_0} = \sum_{\nu=1}^{6} (\widetilde{\beta}_\nu C_0 - 1)\nu^{\beta_0} \ln \nu(-B\nu)e^{-\beta_0 B\nu} = -6.298.$$

Substituting these results into (29), we obtain

$$\beta(C) = 1 + 2.702(C - 1).$$

The graph of this function is shown in fig. 5.5

Thus in the neighborhood of $\beta_0 = C_0 = 1$ the distribution function can be presented in the following form

$$p_\nu^0 = \frac{\nu^{1+2.702(C-1)} \exp[-0.2(1 + 2.702(C - 1))\nu]}{\sum_{\nu=1}^{6} \nu^{1+2.702(C-1)} \exp[-0.2(1 + 2.702(C - 1))\nu]}.$$

Note that this function has the only maximum at the point $\nu_0 = 5$ for any C. Shown in fig. 5.4 are these functions for certain values of C.

Consider a *probabilistic hierarchical structure in forming which one resource can be consumed incompletely.*

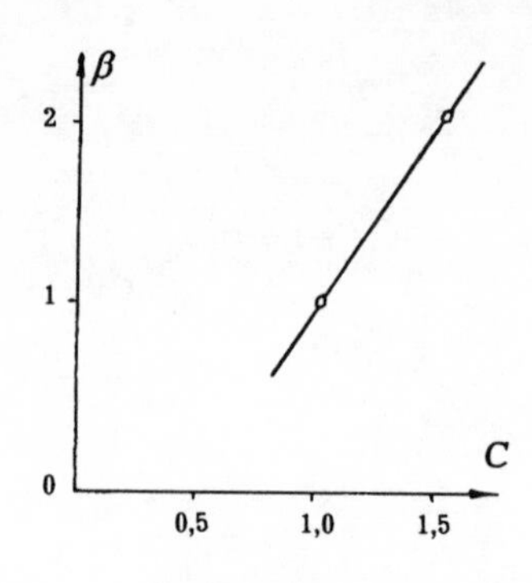

FIGURE 5.5.

The model generating the desired distribution function is of the following form:

$$H(p) = -\sum_{\nu=1}^{q} p_\nu \ln \frac{p_\nu}{a_\nu} \to \max;$$

$$\sum_{\nu=1}^{q} p_\nu = 1, \quad p_\nu \geq 0, \quad \nu \in \overline{1,q}; \tag{30}$$

$$\sum_{\nu=1}^{q} \beta_\nu p_\nu \leq 1, \quad \beta_\nu = \frac{c_\nu}{C}.$$

To solve it, we use the following method.

Consider an auxiliary problem

$$\max H(p), \quad \sum_{\nu=1}^{q} p_\nu = 1, \quad p_\nu \geq 0, \quad \nu \in \overline{1,q}.$$

Its solution p_ν^0 is as in (9). Now it is necessary to check whether the inequality holds in (30). If

$$\sum_{\nu=1}^{q} \beta_\nu p_\nu^0 < 1$$

then the solution of problem (30) coincides with p_ν^0 (9). If

$$\sum_{\nu=1}^{q} \beta_\nu p_\nu^0 \geq 1$$

then the solution of problem (30) coincides with the solution (13) of problem (12), which can be obtained by means of algorithm (14).

Example 2. Industrial production of many products uses socalled flexible technologies (Hatvany, 1981) which, in particular, can have linear structure. Such technologies represent a unidirectional chain without cycles formed by the successive linkage of aggregates A_i, each processing the intermediate products obtained from the previous aggregate A_{i-1} and an external product K_i. Technological flexibility allows such a linear technological chain to produce a wide range of products, i.e. to process external products with quality and cost varying within a wide range.

In developed countries, a great attention is now paid to create such technological lines (usually equipped with programmable robots). One of the main features of flexible technologies is that they can react more efficiently to random fluctuations of the demand.

The assembling line is a typical example of flexible technology. It consists of working sites where certain assembling operations are performed. A unit and additional parts which should be connected to it are fed to the working site from the previous site and the stock, respectively. Every working site A_i processes a particular range of additional parts $\mathcal{D}_1, \ldots, \mathcal{D}_{s_i}$, where s_i is the number of types of these parts for the site A_i $(i \in \overline{1,q})$. A list of required additional parts is defined depending on the requirements on the type of final product. Let ν be a type of final product $(\nu \in \overline{1,R})$; then the list is $D_\nu = \{\mathcal{D}_{11}^\nu, \ldots, \mathcal{D}_{1s_1}; \ldots; \mathcal{D}_{q1}^\nu, \ldots, \mathcal{D}_{qs_q}^\nu\}$.

A set of lists $D_1, \ldots, D_R$ is a characteristic of given conveyer and defines its "flexibility", i.e. the possibility of assembling various products.

Since the demand on the final production varies randomly, realization of each particular list is a stochastic event. Observing the conveyer functioning during a period T in which N products were produced, among these we can find N_1 products of type 1 (obtained by means of assembling the parts from the list D_1), N_2 products of type 2 (the list D_2),..., and N_R products of the type R (the list D_R).

The values $N_1, \ldots, N_R$ are random because of random demand.

Thus it is easy to note an analogy between a general scheme of forming a probabilistic hierarchy and flexible technology realized on the assembling conveyer. In this case, the working sites of the conveyer are the hierarchy levels, and the ensemble of hierarchical structures is formed from various lists of additional parts D_ν required for assembling the products of type ν.

ν	1	2	3	4	5	6	7	8	9	10	11	12
c_ν	620	840	900	980	1200	1290	1380	1420	1500	1600	1810	2000
$P_d(\nu)$	0.09	0.26	0.17	0.14	0.14	0.04	0.07	0.11	0.07	0.04	0.02	0.01

TABLE 5.5.

In designing a flexible technology, it is necessary to take into account that, on the one hand, it should ensure the production of all types of products from 1 to R, and, on the other hand, it should be the most adaptable for producing the products the demand on which is expected to be the most probable. Hardware, software, and management support of a flexible technology depend on this and, in turn, define its economical characteristics.

Instead of the set $\{N_1, \ldots, N_R\}$, it is more convenient to consider a set of relative values $p_\nu = N_\nu / N$ and the distribution function of product types $P(\nu)$ with the values p_ν $(\nu \in \overline{1, R})$. If N is large enough, the function $P(\nu)$ characterizes the probability distribution of assembling different types of product on the given conveyer and therefore defines indirectly the necessary "flexibility" of the technology.

The manufacturing cost is another important characteristic.

The cost of each type of final product consists of the main cost (equipment, construction, etc) and that of additional parts. The main cost is supposed to be known. Therefore, knowing the prices c_{ij} of additional parts, we can calculate the cost c_ν of the final product.

If the prices c_{ij} do not depend on how often the additional parts are used, their values are constant for the given set D_ν. Therefore the average cost of producing all types of final product equals $\sum\limits_{\nu=1}^{R} c_\nu p_\nu$. This value should not exceed the planned average cost C.

Finally, in designing a line, the demands on its products are usually predicted. This means that there is a "predictive" distribution function of probabilities of the demands $P_d(\nu) = N_\nu^d / N$, where N_ν^d is a predictive number of products of the type ν, $\nu \in \overline{1, R}$.

Consider a technological line for welding car bodies designed for producing $R = 12$ models. The costs of bodies for each model and the predictive demand probabilities $P_d(\nu)$ are given in Table 5.5.

C	650	700	750	800	850	900	950	1,000	1,050	1,100	1,150
s	2	2	2	2	2	2	2	2	2	1	1
p_1^0	0.849	0.656	0.488	0.356	0.262	0.196	0.150	0.116	0.090	0.078	0.078
p_2^0	0.111	0.225	0.301	0.338	0.342	0.327	0.301	0.272	0.241	0.224	0.224
p_3^0	0.031	0.082	0.129	0.163	0.180	0.184	0.179	0.168	0.155	0.147	0.147
p_4^0	0.008	0.031	0.061	0.090	0.111	0.124	0.129	0.129	0.124	0.121	0.121
p_5^0	0	0.004	0.013	0.029	0.050	0.071	0.090	0.105	0.116	0.121	0.121
p_6^0	0	0	0.002	0.005	0.016	0.016	0.022	0.027	0.032	0.034	0.034
p_7^0	0	0	0.002	0.006	0.013	0.023	0.033	0.044	0.054	0.060	0.060
p_8^0	0	0	0.002	0.008	0.018	0.032	0.049	0.067	0.085	0.095	0.095
p_9^0	0	0	0.001	0.003	0.009	0.017	0.027	0.039	0.053	0.060	0.060
p_{10}^0	0	0	0	0.001	0.003	0.007	0.013	0.021	0.029	0.034	0.034
p_{11}^0	0	0	0	0	0.001	0.002	0.005	0.008	0.014	0.017	0.017
p_{12}^0	0	0	0	0	0	0.001	0.002	0.004	0.006	0.009	0.009

TABLE 5.6.

The annual production program of the enterprise is $138 million, with the production of 120,000 car bodies. Thus an "average" car body costs $1,150.

The calculations for $C = 1{,}150$ are given in Table 5.6 (the last column). Note that the recourse constraint always holds (strict inequality) for this value of "average cost". The value $s = 1$ means that $\sum \beta_\nu p_\nu < 1$, and $s = 2$ means that $\sum \beta_\nu p_\nu = 1$. In this situation, the second list ($p_2^0 = 0.224$) is the most probable. The distribution function $P_d(\nu)$ has another maximum corresponding to the eighth list ($p_8^0 = 0.095$).

We analyzed how the "average cost" of a car body influenced the form of $P_d(\nu)$. The calculations are given in Table 5.6. It is seen from the table that the resource constraint is active ($s = 2$) for $650 \le C \le 1{,}050$.

Now let us consider the model of general type (3), (4) and (6), where we denote

$$\Psi_s(p) = \sum_{\nu=1}^{q} \phi_s(\nu) P(\nu).$$

Let us rewrite it in the following form:

$$H(p) = -\sum_{\nu=1}^{q} p_\nu \ln \frac{p_\nu}{a_\nu} \to \max,$$
$$\sum_{\nu=1}^{q} p_\nu = 1. \tag{31}$$
$$\Psi_s(p) \leq q_s, \quad s \in \overline{1,r}; \qquad p_\nu \geq 0, \quad \nu \in \overline{1,q}.$$

The functions Ψ_s characterizing the consumption of the s-th resource in forming a probabilistic hierarchy are supposed to be monotonously increasing, convex and continuously differentiable.

By excluding the constraint equality, we transform the problem (31) to the form:

$$\widetilde{H}(p_1, \ldots, p_{q-1}) \to \max;$$
$$\widetilde{\Psi}_s(p_1, \ldots, p_{q-1}) \geq 0; \quad s \in \overline{1, r+1}; \tag{32}$$
$$p_\nu \geq 0, \quad \nu \in \overline{1, q-1},$$

where

$$\widetilde{H}(p) = -\sum_{\nu=1}^{q-1} p_\nu \ln \frac{p_\nu}{a_\nu} - \left(1 - \sum_{\nu=1}^{q-1} p_\nu\right) \ln \frac{1 - \sum_{\nu=1}^{q-1} p_\nu}{a_q}; \tag{32'}$$

$$\widetilde{\Psi}_s(p) = q_s - \Psi_s(p_1, \ldots, p_{q-1}), \quad s \in \overline{1,r}; \tag{32''}$$

$$\widetilde{\Psi}_{r+1}(p) = 1 - \sum_{\nu=1}^{q-1} p_\nu.$$

Consider the Lagrange function

$$L(p, \lambda) = \widetilde{H}(p) - \sum_{j=1}^{r+1} \lambda_j \widetilde{\Psi}_j(p)$$

and the components of its gradient

$$\frac{\partial L}{\partial p_k} = \frac{\partial \widetilde{H}}{\partial p_k} - \sum_{j=1}^{r+1} \lambda_j \frac{\partial \widetilde{\Psi}_j(p)}{\partial p_k}, \quad k \in \overline{1,q};$$

$$\frac{\partial L}{\partial \lambda_j} = -\widetilde{\Psi}_j(p), \quad j \in \overline{1, r+1}.$$

To find a saddle-point of the Lagrange function, we can use the appropriate multiplicative algorithm (see Chapter 4):

$$p^{s+1} = p^s \otimes G(p^s, \lambda^s), \quad \lambda^{s+1} = \lambda^s \otimes Q(p^s), \tag{33}$$

where $\otimes$ is the coordinate-wise multiplication;

$$\begin{aligned} Q(p) &= 1 - \gamma L_\lambda(p); \\ G(p, \lambda) &= 1 + \gamma L_p(p, \lambda); \\ L_p(p) &= \left\{ \frac{\partial L}{\partial p_1}, \ldots, \frac{\partial L}{\partial p_{q-1}} \right\}; \\ L_\lambda(p) &= \left\{ \frac{\partial L}{\partial \lambda_1}, \ldots, \frac{\partial L}{\partial \lambda_{r+1}} \right\}; \end{aligned} \tag{34}$$

and γ is a step coefficient.

Example 3. Let us consider the technological line for producing car bodies, which has been studied in example 2.

Different materials are used in producing a car body, with their costs being the components of final product cost. Some kinds of them are limited. In particular, these are noiseproof plastics, some kinds of non-ferrous metals and anti-corrosion coatings (see Inaba, 1981). Calculated with respect to an "average" car body, the limited costs of five materials are $C = \$110$, $C = \$150$, $C = \$140$, $C = \$165$, and $C = \$115$.

In the previous example, the cost of a car body of type ν was supposed to be constant for any output. However, if the number of produced cars of type ν is quite large, the production cost decreases. It is often convenient to approximate the dependence of the cost on the output by the exponential law, i.e.

$$N_\nu c_\nu(N_\text{v}) = \alpha_\nu + \beta_\nu \exp(-\gamma_\nu N_\nu),$$

where N_ν is the number of products of type ν.
Replacing $N_\nu = p_\nu N$, we obtain

$$N p_\nu c_\nu(p_\nu) = \left[\alpha_\nu + \beta_\nu \exp(-\gamma_\nu p_\nu)\right],$$

where $\gamma_\nu = \widetilde{\gamma}_\nu N$.

The exponential law of changing the car body cost can also be applied to the limited materials used therein. Given in Table 5.7 are the parameters α_ν,

ν	1	2	3	4	5	6	7	8	9	10	11	12
α_ν^1	2.5	2.5	4.0	4.0	4.0	6.1	6.3	6.3	6.3	6.6	7.0	9.5
α_ν^2	5.1	5.6	6.1	6.8	7.0	7.5	8.0	8.2	9.6	9.6	9.6	10.0
α_ν^3	3.6	5.7	6.4	6.8	6.8	6.8	7.5	7.5	7.5	8.0	9.0	11.0
α_ν^4	1.6	5.1	6.0	7.5	8.0	8.0	8.3	9.0	9.7	11.4	11.4	13.2
α_ν^5	1.0	1.5	1.5	2.0	2.0	3.0	4.0	6.0	6.5	7.0	8.5	11.5
α_ν	13	18	30	48	50	55	76	89	90	92	96	100
a_ν	.418	.372	.325	.046	.046	.005	.009	.046	.009	.019	.009	.005
β_ν^1	1.5	2.5	1.5	2.8	3.3	3.3	3.9	4.5	4.5	5.0	5.9	6.9
β_ν^2	3.9	2.4	3.9	4.2	4.5	4.5	5.4	6.0	5.4	5.4	5.4	7.5
β_ν^3	2.9	3.8	3.6	4.2	4.2	4.2	4.5	4.5	4.5	5.3	6.0	8.4
β_ν^4	2.4	3.9	6.0	5.5	5.0	5.0	5.7	6.0	6.3	6.6	6.6	7.8
β_ν^5	2.0	3.5	2.6	2.4	3.5	5.3	5.7	5.7	6.6	6.9	7.5	9.0
β_ν	18	25	27	29	36	38	41	42	45	48	54	60
p_ν^0	.132	.149	.146	.061	.078	.028	.045	.093	.053	.078	.070	.067

TABLE 5.7.

β_ν and a_ν for a car body and five types of limited materials. The parameters γ_ν are supposed to be the same for all types of consumed materials and car bodies and equal to $\gamma = 6.25$. This value is characterized by 30% cost decrease for the output exceeding 5,000 units.

Thus the model (31) for the given problem is of the following form:

$$H = -\sum_{\nu=1}^{12} p_\nu \ln \frac{p_\nu}{a_\nu};$$

$$\sum_{\nu=1}^{12} p_\nu = 1;$$

$$\sum_{\nu=1}^{12} \alpha_\nu^s + \beta_\nu^s e^{-\gamma p_\nu} \leq C_s, \quad s \in \overline{1,5};$$

$$\sum_{\nu=1}^{12} \alpha_\nu + \beta_\nu e^{-\gamma p_\nu} \leq C; \qquad p_\nu \geq 0, \quad \nu \in \overline{1,12}.$$

Calculations of the desired distribution function for the probabilities of producing the products of ν types are given in Table 5.7.

5.4 Models of mechanisms for forming hierarchical structures

Formation of hierarchical chains consists in random and independent entering the subordination levels by their elements.

Let us denote by w_{ij} the probability of the element with the number i to enter the level j. If in forming the hierarchical chains all the elements are distributed among the subordination levels then

$$\sum_{i=1}^{m} w_{ij} = 1, \quad j \in \overline{1,k}. \tag{1}$$

Similarly, if there are no empty subordination levels, i.e. the element having the number i always enters a subordination level, then

$$\sum_{j=1}^{k} w_{ij} = 1, \quad i \in \overline{1,m}. \tag{2}$$

Situations occurs where both assumptions hold unstrictly. Then, instead of (1) and (2), the inequality condition arise, namely

$$\sum_{i=1}^{m} w_{ij} \leq 1, \quad j \in \overline{1,k}; \tag{3}$$

$$\sum_{j=1}^{k} w_{ij} \leq 1, \quad i \in \overline{1,m}. \tag{4}$$

The matrix $W = [w_{ij}]$ is a characteristic of forming the probabilistic hierarchical structure. A model of forming mechanism consists in reproducing this characteristic by mathematical description of its phenomenology.

The phenomenology of forming mechanism consists of the procedure of stochastic distribution of elements among the subordination levels, under environmental constraints influencing this procedure.

To describe a phenomenological scheme of stochastic distribution, let us consider an abstract system consisting of m elements and k subordination levels. We regard the states of given system as the pairs (i, j), where i is the element number, j is the number of level which it enters randomly. Since the system contains discrete objects, the number of states which can be realized is finite.

Thus a set S of states of the abstract system considered consists of $r = mk$ states. Recall that the state s is "the element i on the level j", i.e. the numeration of states s corresponds to that of the pairs (i, j)

$$s \Rightarrow (i, j); \quad s \in \overline{1, r}; \quad i \in \overline{1, m}; \quad j \in \overline{1, k}. \tag{5}$$

Therefore the "stochastic realization of the state s" and "the element i enters randomly the level j" are equivalent stochastic events.

Using this equivalence, we represent the random realization of the states from the set $\mathfrak{S}$ as a sequence of mk independent series of the Bernoulli trials with return.

Recall that the classical Bernoulli scheme considers a sequence of independent trials where two results, namely "successful" and "unsuccessful", are realized with the probabilities u and $\nu = 1 - u$, respectively. If we have M trials with ρ "successes" and $M - \rho$ "failures" then the probability of such an event is

$$P(\rho, M - \rho) = \frac{M!}{\rho!(M - \rho)!} u^{\rho} \nu^{M-\rho}. \tag{6}$$

Now let us consider the set $\mathfrak{S}$ in which the states are realized randomly and independently; the Bernoulli scheme can be applied to any of these states.

Let us represent the set $\mathfrak{S}$ as an integration of two subsets $\mathfrak{S}_1$ and $\mathfrak{S}_{r-1}$:

$$\mathfrak{S} = \mathfrak{S}_1 \cup \mathfrak{S}_{r-1}, \tag{7}$$

where $\mathfrak{S}_1$ contains one state with the number 1, while $\mathfrak{S}_{r-1}$ contains the other $(r - 1)$ states.

Let us consider a sequence of Bernoulli trials with these two subsets and let us assume that a subset is "successful" if state 1 is realized in it ($\mathfrak{S}_1$) and "unsuccessful" if otherwise ($\mathfrak{S}_{r-1}$). Denote the probability of "success" in every trial by u_1, the number of "successful" trials by ρ_1, and the number of "unsuccessful" trials by $\overline{\rho}_1$. Then the probability of state 1 (the hierarchical chain having the number 1) being realized ρ_1 times and any of states $2, 3, \ldots, r - 1$ (hierarchical chains $2, 3, \ldots, r - 1$) being realized $\overline{\rho}_1$ times in M independent trials is

$$P(\rho_1, \overline{\rho}_1) = \frac{M!}{\rho_1! \overline{\rho}_1!} u_1^{\rho_1} \nu_1^{\overline{\rho}_1}. \tag{8}$$

Note that in this expression (as in (6)) the value $\nu_1^{\overline{\rho}_1}$ is the probability of realizing the subset $\mathfrak{S}_{r-1}$, i.e.

$$\nu^{\overline{\rho}_1} = \mathrm{Prob}(\mathfrak{S}_{r-1}). \tag{9}$$

Let us consider the subset $\mathfrak{S}_{r-1}$ and represent it as an integration of two subsets $\mathfrak{S}_2$ and $\mathfrak{S}_{r-2}$:

$$\mathfrak{S}_{r-1} = \mathfrak{S}_2 \cup \mathfrak{S}_{r-2}, \tag{10}$$

where $\mathfrak{S}_2$ contains one state with the number 2, while $\mathfrak{S}_{r-2}$ contains the other $r-2$ states.

According to the previous operation with the subset $\mathfrak{S}_{r-1}$, $\overline{\rho}_1$ independent trials were performed. Assume that ρ_2 "successes" (i.e. state 2 (subset $\mathfrak{S}_2$)) were realized with the probability u_2, and $\overline{\rho}_2$ "failures" (i.e. states $3, \ldots, r$ (subset $\mathfrak{S}_{r-2}$)) were realized with the probability ν_2.

The probability of $\mathfrak{S}_2$ being realized ρ_2 times and $\mathfrak{S}_{r-2}$ being realized $\overline{\rho}_2$ times in $\overline{\rho}_1$ trials is

$$P(\rho_2, \overline{\rho}_2) = \frac{\overline{\rho}_1!}{\rho_2! \, \overline{\rho}_2!} u_2^{\rho_2} \nu_2^{\overline{\rho}_2}, \quad \overline{\rho}_1 = \rho_2 + \overline{\rho}_2. \tag{11}$$

But, by virtue of (10) and (9), we have

$$P(\rho_2, \overline{\rho}_2) = \mathrm{Prob}(\mathfrak{S}_{r-1}) = \nu^{\overline{\rho}_1}.$$

Therefore, using the equality (9), we obtain from (8) and (11)

$$P(\rho_1, \overline{\rho}_1) = P(\rho_1, \rho_2, \overline{\rho}_2) = \frac{M!}{\rho_1! \, \rho_2! \, \overline{\rho}_2!} u_1^{\rho_1} u_2^{\rho_1} \nu_2^{\overline{\rho}_2}. \tag{12}$$

Continuing the process (7), (10), we obtain that the hierarchical chain 1 is realized ρ_1 times, the hierarchical chain 2 is realized ρ_2 times, ..., and the hierarchical chain r is realized ρ_r times in M Bernoulli trials with the probability

$$P(\rho_1, \ldots, \rho_r) = \frac{M!}{\rho_1! \ldots \rho_r!} u_1^{\rho_1} \ldots u_r^{\rho_r}, \tag{13}$$

where u_i is the apriori probability of the i-th hierarchical chain being realized.

The distribution function (13) is the well-known polynomial distribution (Foeller, 1957). The investigation of morphological properties of this function shows that it has a unique maximum, with its "pick" (see Chapter 2) increasing together with the number of trials M.

In other words, the more times the probabilistic hierarchical structure is realized, the more often the distribution is realized that corresponds to the maximum of the function $P(\rho_1, \ldots, \rho_r)$ (13), i.e.

$$\max P(\rho_1, \ldots, \rho_r) = \{\rho_1^0, \ldots, \rho_r^0) = P^0. \tag{14}$$

According to the general macrosystems concept (Chapter 2), it is more convenient to pass from the distribution function (13) to the entropy of Bernoulli trials. Finding the logarithm of (13) and using the Stirling approximation, we obtain

$$\widetilde{H}(\rho_1, \ldots, \rho_r) = -\sum_{s=1}^{r} \rho_s \ln \frac{\rho_s}{u_s e}. \tag{15}$$

Let us introduce the notation

$$\frac{\rho_s}{M} = w_s \to w_{ij}; \quad u_s \to a_{ij}, \tag{16}$$

where a_{ij} is the apriori probability of the element i entering the level j.

Then (15) can be transformed to the following form neglecting a positive constant and positive multiplier:

$$H(W) = -\sum_{i=1}^{m}\sum_{j=1}^{k} w_{ij} \ln \frac{w_{ij}}{a_{ij} e}. \tag{17}$$

This defines the generalized information entropy of Boltzmann expressed in the probabilities of distributing the elements of a probabilistic hierarchical structure among the subordination levels. The apriori probabilities a_{ij} (16) are important parameters here.

Recall that an ensemble of hierarchical chains results from random distribution of the elements among the subordination levels. Using the technique proposed in Section 3, we can define the required probability distribution function $P^0(\nu)$. The formation of chains is characterized by the probability matrix W. Therefore the desired characteristics of the ensemble of hierarchical chains, i.e. the type of the desired probability distribution function $P(\nu)$, should be taken into account in a model generating the matrices W.

Because of the described mechanism of forming the chains, there exists an apparent link (if this process is regarded as independent) between $P^0(\nu)$

and the probabilities characterizing the element entering the level of subordination. In particular, such a link can be realized between $P^0(\nu)$ and the apriori probabilities $[a_{ij}]$.

The chain having the number ν consists of the elements $\mu_1, \ldots, \mu_{s_1}$ on the first subordination level, the elements $\alpha_, \ldots, \alpha_{s_1}$ on the second subordination level, ..., and the elements $\omega_1, \ldots, \omega_{s_k}$ on the k-th subordination level. Besides, each element can be on only one level, therefore $\mu \neq \alpha \neq \ldots \neq \omega$.

Now we consider external constraints characterizing the formation of the probabilistic hierarchical structure. The first type of these constraints is related to the consumption of resources accompanying the process. Here the situation is similar to that considered in Section 3 in modeling the desired characteristics of an ensemble of hierarchical structures.

We characterize the consumption of resources by the functions $\phi_1(i,j), \ldots, \phi_r(i,j)$ which are amounts of the resources of the first, second, ..., and the r-th type, respectively, consumed for an element i to enter the subordination level j. Since the elements are distributed randomly, it is useful to characterize this process by a "mean" consumption of each of r resources.

Let us denote by $b_1, \ldots, b_r$ the stocks of r resources, respectively. Then

$$\sum_{j=1}^{k}\sum_{i=1}^{m} \phi_s(i,j) w_{ij} = b_s, \quad s \in \overline{1,r} \tag{18}$$

if the stock is consumed completely in forming the hierarchical chains, and

$$\sum_{j=1}^{k}\sum_{i=1}^{m} \phi_s(i,j) w_{ij} \leq b_s, \quad s \in \overline{1,r} \tag{19}$$

if otherwise.

Thus, following the general macrosystem concept (Chapter 2), a model of mechanism for forming a probabilistic hierarchical structure can be represented as

$$H(W) = -\sum_{i=1}^{m}\sum_{j=1}^{k} w_{ij} \ln \frac{w_{ij}}{a_{ij} e} \to \max; \tag{20}$$

$$\sum_{i=1}^{m} w_{ij} = 1, \quad j \in \overline{1,k}, \tag{21}$$

$$\sum_{j=1}^{k} w_{ij} = 1, \quad i \in \overline{1,m}. \tag{22}$$

$$\sum_{i=1}^{m}\sum_{j=1}^{k} \phi_s(i,j) w_{ij} = b_s, \quad s \in \overline{1,r} \tag{23}$$

or

$$\sum_{i=1}^{m}\sum_{j=1}^{k} \phi_s(i,j) w_{ij} \leq b_s, \quad s \in \overline{1,r}, \tag{24}$$

where the apriori probabilities a_{ij} in (21) are related to $P^0(\nu)$.

Example 1. Now we return to constructing the model of urban system consisting of three subsystems, which was described in detail in Section 3 (example 1). To determine the characteristics of its probabilistic hierarchical structure, we shall use the model (20)–(23) with the cost functions of the following two types:

$$\phi_1(i,j) = a + bi + dj; \tag{25}$$

$$\phi_1(i,j) = a + bi + dj + \alpha \ln(\widetilde{a} + \widetilde{b}i + \widetilde{d}j), \tag{26}$$

where $a, b, d, \widetilde{a}, \widetilde{b}, \widetilde{d}$ and α are constant parameters.

In this case, the ensemble of hierarchical chains consists of 27 chains given in Table 5.8.

First, let us consider the case of linear costs (25). Using the methods of Chapter 2, we can represent the solution of problem (21)–(23) with the function $\phi_1(i,j)$ (25) as

$$w_{ij} = a_{ij} \exp(-\lambda_i - \alpha_j - \beta a - \beta bj - \beta di), \tag{27}$$

where λ_i, α_j and β are the Lagrange multipliers corresponding to the constraint equalities (21, 22, 23).

Given in Table 5.9 are the calculations of w_{ij} for various matrices of apriori probabilities $[a_{ij}]$ and $b_1 = 6.00$.

Now let us consider the case when the cost function is logarithmic and apriori probabilities are constant. In this case, the model (20)–(23) becomes non-linear.

(a) In the general expression for cost function (26), the coefficients are $a = b = d = 0$, $\alpha = 1$, $\widetilde{a} = 1$, $\widetilde{b} = 0.2$ and $\widetilde{d} = 0.3$. Then

$$\phi_1(i,j) = \ln(1 + 0.2i + 0.3j).$$

Chain \ Level	1	2	3	Chain \ Level	1	2	3
1	1,2,3			15	3		1,2
2		1,2,3		16		1,2	3
3			1,2,3	17		1,3	2
4	1,2	3		18		2,3	1
5	1,3			19		1	2,3
6	2,3	1		20		2	1,3
7	1,2		3	21		3	1,2
8	1,3		2	22	1	2	3
9	2,3		1	23	1	3	2
10	1	2,3		24	2	1	3
11	2	1,3		25	2	3	1
12	3	1,2		26	3	1	2
13	1		2,3	27	3	2	1
14	2		1,3				

TABLE 5.8.

Version \ Variables	$[a_{ij}]$			$[w_{ij}]$		
1	0.8	0.1	0.1	0.7998	0.1000	0.1000
	0.1	0.8	0.1	0.1000	0.8000	0.1000
	0.1	0.1	0.8	0.1000	0.1000	0.8000
2	0.1	0.6	0.5	0.0677	0.4656	0.4667
	0.9	0.2	0.2	0.5355	0.1364	0.3281
	0.8	0.7	0.3	0.3968	0.3980	0.2052
3	0.05	0.5	0.45	0.0500	0.5000	0.4500
	0.70	0.1	0.20	0.7001	0.1000	0.1999
	0.25	0.4	0.35	0.2500	0.3999	0.3500

TABLE 5.9.

i \ j	1	2	3
1	0.41	0.59	0.47
2	0.53	0.69	0.83
3	0.64	0.79	0.92

TABLE 5.10.

β	b_1	W
0.01000	2.04615	.33339.33333.33328 .33333 33333 33334 .33328.33334.33338
0.05000	2.04614	.33361.33331.33308 .33331.33334.33335 .33307.33336.33357
0.25000	2.04607	.33474.33320.33207 .33324.33335.33341 .33202.33346.33452
1.25000	2.04575	.34037.33263.32700 .33284.33343.33373 .32679.33394.33927
6.25000	2.04413	.36864.32950.30186 .33052.33435.33513 .30084.33615.36301

TABLE 5.11.

The values of this function are given in Table 5.10.

Results of calculation of the dependence of β on b_1 in case $a_{ij} = \text{const} = 0.3333$ are given in Table 5.11.

It is seen from Table 5.11 that β (the Lagrange multipliers corresponding to the resource constraint) vary in a wide range when the resource stock b_1 varies negligibly. In this case, variations of the elements of W are of the same order as those of b_1.

(b) Let us consider the cost function with a linear and logarithmic component

$$\phi_1(i,j) = 1 + 0.2i + j - \ln(i+j). \tag{28}$$

We estimate the influence of average cost b_1 on the matrix of probabilities W for given function of specific costs.

Calculations are given in Table 5.12.

It is seen from the table that β rises significantly and the matrix of posterior probabilities differs more from that having equal probabilities when b_1 decreases.

Given in Tables 5.13–5.15 are the calculations of matrices W for the matrices of apriori probabilities given in Table 5.9.

β	b_1	W
11.0	1.91324	.08728.34617.56655 .34616.36490.28893 .56654.28893.14453
20.0	1.88832	.01987.30633.67380 .30631 43482 25886 .67373.25885.06742
40.0	1.86790	.00053.19365.80582 .19277.61631.19092 .79855.19006.01139
60.0	1.85677	.00001.11661.88337 .11639.76495.11865 .88000.11843.00157
80.0	1.85018	.00000.06568.93432 .06555.86514.06931 .93062.06918.00021
100.0	1.84628	.00000.03524.96476 .03516.92558.03926 .96079.03918.00003

TABLE 5.12.

β	b_1	W
30.0	1.87951	.05465.15061.79474 .15058.77562.07380 .79422.07377.13201
40.0	1.85999	.00859.11111.88029 .11110.82695.06196 .88002.06195.05803
50.0	1.85242	.00119.07638.92243 .07635.87145.05220 .92183.05218.02598
60.0	1.85081	.00012.04435.95554 .04388.90538.05075 .93432.05015.01553
70.0	1.84682	.00002.03393.96605 .03380.93007.03613 .95887.03599.00514

TABLE 5.13.

β	b_1	W
30.0	1.88487	.00034.30056.69910 .36488.35181.28331 .63475.34762.01763
40.0	1.87743	.00006.25049.74946 .30758.55867.24375 .69232.30084.00684
50.0	1.87077	.00001.20442.79557 .25334.54476.20190 .74660.25082.00259
60.0	1.86485	.00000.16321.83679 .20380.63388.16231 .79614.20290.00096
70.0	1.85973	.00000.12770.87230 .16042.71215.12743 .83950.16016.00035

TABLE 5.14.

β	b_1	W
30.0	1.89308	.00021.27715.72264 .49668.27416.22916 .50310.44868.04821
40.0	1.88432	.00003.22741.77255 .42923.36328.20749 .57072.40930.01998
50.0	1.87689	.00001.18541.81458 .36392.45848.17760 .63606.35610.00784
60.0	1.87030	.00000.14895.85104 .30143.55258.14600 .69856.29846.00298
70.0	1.86448	.00000.11756.88244 .24373.63979.11648 .75625.24265.00111

TABLE 5.15.

CHAPTER

SIX

MODELLING AND ANALYSIS OF INTERREGIONAL EXCHANGE OF CONSTRUCTION MATERIALS

6.1 Interregional exchange of construction materials as object of analysis

Industry of construction materials (CMI) meets the demands of civil engineering in row materials, fabricates and final goods. CMI provides 45 percent of material resources needed in construction. Close connection of CMI with civil engineering causes a number of its specific properties in producing and distributing products.

First of all, since the construction can be everywhere, the products of CMI should be very "mobile". This explains its high freight demands, CMI production occupies 25 percent of railway freight traffic.

Another specific feature of this industry and related product exchange are numerous intra- and interconnections, for example, almost 50 percent of cement produced in former USSR were consumed in production of asbestine, ferro-concrete panels and wall panels. In general, CMI consumes 20 percents of its own production.

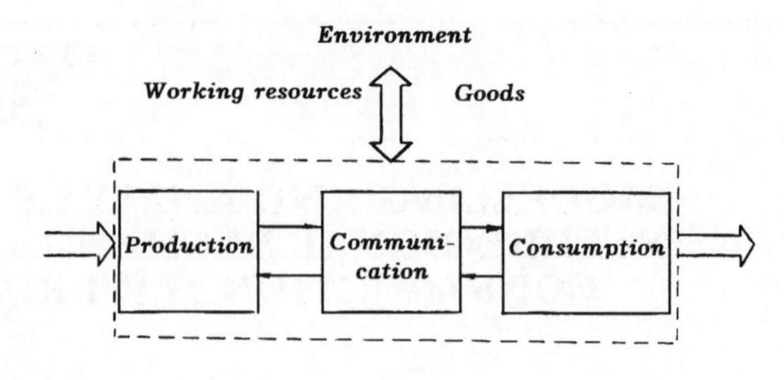

FIGURE 6.1.

Formation and development of these connections are significantly influenced by the geographical location of the functional elements of industry. The geographical factor arises because CMI has features of both mining and manufacturing industries. Therefore geological and other natural conditions cause significantly the geographical location and economics of a number of CMI branches. Chemical and thermodynamic processes are involved in other branches, and the latter, as a result, are attracted to the regions with large energy resources.

Thus the distinguishing feature of exchange and distribution of CMI products is inhomogeneity. These involves economical, technical, natural and other elements.

There are many qualitative indices of product exchange. Some of them can be quantified, for example, production volumes, transport approachability, technological parameters and so on. Other are only qualitative.

A collection of indices mentioned above characterize a state of the system, while their changes in time characterize its development. More exactly, they reflect the results of processes in the system and outside it. Internal processes are caused by interactions of elements which are economically connected in exchanging the resources. These elements can be united in subsystems. For this purpose, a functional principle is used (production – communication – consumption). Applying this, we can obtain three main subsystems (fig. 6.1).

The first subsystem, or the industry of construction materials, involves producers providing necessary resources for industry itself and for the civil engineering.

The second subsystem is the civil engineering. It involves various construction objects and unites consumers of the construction materials.

The transport subsystem differs from the others. It is a factor forming the structure of product exchange between the first two subsystems.

At the same time, the industry cannot exist out of the external environment, since they exchange products and working resources.

Structure analysis of interregional transport-economical connections is an important problem arising in study of product exchange system. Such an analysis allows one to estimate the rationality of interregional connections of elements in economic system, possibility of decreasing the costs in exchange and congestion of transportation system.

Because of this, it is necessary to study logical properties of product exchange system existing in reality. To do this, we can use a mathematical model describing the internal structure of product exchange system, behavior of its constituents and interactions between them in maximum detail.

6.2 Characteristics of state of an interregional exchange system

Let us consider an area R (fig. 6.2) divided into the regions R_i so that $\bigcup_{i=1}^{n} R_i = R$, where n is the number of regions. Let r types of products be produced and let s types be consumed in given area. Denote by $k_1(i), \ldots, k_{l_i}(i)$ the types of products produced in the region R_i, by $s_1(i), \ldots, s_{q_i}(i)$ the types of products consumed in this region, by l_i and q_i the numbers of types of produced and consumed products, respectively, and by $P_i^{k_\mu(i)}$ and $Y_i^{s_\nu(i)}$ the whole production and consumption of the products of types $k_\mu(i)$ and $s_\nu(i)$, respectively. The situations are possible where both $k_\mu(i) \neq s_\nu(i)$, i.e. the produced and consumed products are different, and $k_\mu(i) = s_\nu(i)$, i.e. a product is both consumed and produced in the region R_i, besides

$$\sum_{i \in I_{k_\mu}} P_i^{k_\mu} = \sum_{j \in J_{k_\mu}} Y_j^{k_\mu} = N^{k_\mu}, \quad \text{for all} \quad k_\mu \in \overline{1,r}, \tag{1}$$

where I_{k_μ} and J_{k_μ} are the sets of indices of regions producing and consuming the product of type k_μ.

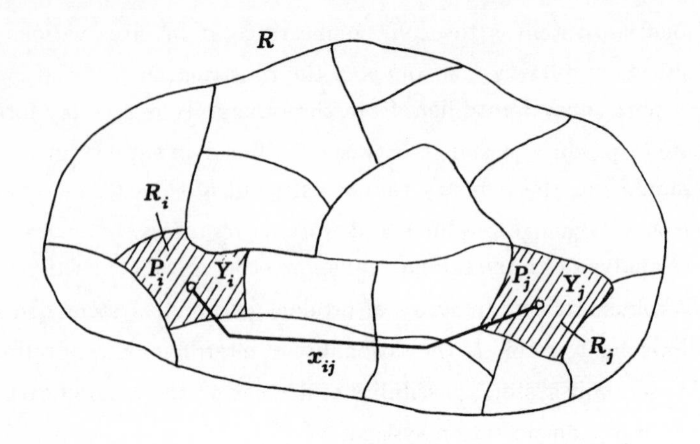

FIGURE 6.2.

The equality (1) means that the exchange system is balanced with respect to the consumption and production.

The technology of production and consumption is important in the product exchange system.

First, let us consider the structure of this technology in an arbitrary region R_i, assuming that all types of products can be produced and consumed in this region.

The consumption of product of type k is considered as its use in producing other products and the k-th product itself (productive consumption) and its use for social and individual needs, savings, and export (nonproductive consumption).

Table 6.1 Illustrate the economical structure of a region.

In the table, P^{ks} is the amount of the k-th product consumed in producing the s-th product. It follows from the table that

$$P^k = \sum_{s=1}^{r} P^{ks} + Y^k, \quad k \in \overline{1,r}. \tag{2}$$

Denote

$$P^{ks} = a^{ks} P^s, \tag{3}$$

where a^{ks} is the part of the k-th product used in producing the s-th product.

Type of product	Output	Productive consumption $1 \dots k \dots r$	Nonproductive consuption
1	P^1	$P^{11} \dots P^{1k} \dots P^{1r}$	Y^1
$\vdots$	$\vdots$	$\vdots \quad \vdots \quad \vdots$	$\vdots$
k	P^k	$P^{k1} \dots P^{kk} \dots P^{kr}$	Y^k
$\vdots$	$\vdots$	$\vdots \quad \vdots \quad \vdots$	$\vdots$
r	P^r	$P^{r1} \dots P^{rk} \dots P^{rr}$	Y^r

TABLE 6.1.

The coefficients a^{ks} are called technological, since they are related to the technology of transforming one product into another. Assume that the technological coefficients are constant. Then the equalities (2) can be represented as

$$P^k = \sum_{s=1}^{r} a^{ks} P^s + Y^k, \quad k \in \overline{1,r} \tag{4}$$

or

$$(-A + E)P = Y, \tag{5}$$

where

$A = [a^{ks}]$ is the matrix of technological coefficients,

E is the unit matrix,

$P = \{P^1, \dots, P^r\}$ is the output vector, and

$Y = \{Y^1, \dots, Y^r\}$ is the vector of nonproductive consumption.

The equations (4) and (5) describe the technological connections between the production and consumption, with their structure assuming the same for all regions. However, the parameters of these connections can vary.

The products produced in the regions are exchanged through the transportation network.

Denote by $x_{ij}^{k_\mu}$ the amount of product of type k_μ transfered from the region R_i to R_j during the fixed time interval τ (freight flow). It is convenient to measure the product mass as an amount of product portions having a definite mass. Then $x_{ij}^{k_\mu}$ is an integer meaning the amount of portions of the product k_μ transfered from R_i to R_j during the time τ.

A state of product exchange system is characterized by the freight flow matrix

$$X^{k_\mu} = \left[\; x_{ij}^{k_\mu} \; \right], \quad k_\mu \in \overline{1,r}.$$

Shipments of products require expenditures depending both on the regions R_i and R_j and on the type and amount of product.

In general, this dependence is nonlinear, namely $\Phi(x_{ij}^{k_\mu})$. The nonlinearity is the most significant in dependence of Φ on the regions R_i and R_j and on the type k_μ of the product.

The dependence of expenditures on the amount of the product transfered can be assumed linear. Then the expenditures of the freight flow $x_{ij}^{k_\mu}$ can be estimated as $c_{ij}^{k_\mu}\, x_{ij}^{k_\mu}$, where $c_{ij}^{k_\mu}$ are the specific costs of transferring one portion of the k_μ-th product from R_i to R_j .

In this case, the general expenditures in transferring the k_μ-th product in the product exchange system are $\sum_{ij} c_{ij}^{k_\mu}\, x_{ij}^{k_\mu}$. In general, these values are compared with the acceptable values of the general expenditures C^{k_μ} ($k_\mu \in \overline{1,r}$) which are important parameters of the exchange system.

6.3 Macrosystem model of single-product interregional exchange

A mathematical model of the single-product interregional exchange is based on the assumption that the portions of produced single product enters the communications randomly and independently with a known apriori probability.

Some considerations can be given in favor of this hypothesis. The first is associated with the random nature of the consumption and choice of suppliers.

These factors are important even in the plan economy of former USSR. In 1980–1985, according to the statistics, 45 percent of the consumers have used nonplaned supplies of cement, 31 percent have used nonplaned supplies of enforced concrete and 65 percent have used nonplaned supplies of bricks.

The second consideration is associated with various contingencies accompanying the transportation, namely the lack of wagons, emergencies and meteorological conditions.

Assumption on random forming the freight flows allows us to apply the concept of macrosystem stationary state model.

According to this, the product exchange system is considered as consisted of a set of elements (the portions of homogeneous product) and a set of states. The communication (i, j) is considered as a state. Since the transportation network is multi-purpose, its capacity with respect to the shipments of any fixed product can be considered substantially larger than the amount of product transfered.

Therefore average occupation numbers (freight flows) are smaller than the potential capacity of the states (communications). This allows us to apply the MSS of Boltzmann type:

$$H(X) = -\sum_{i,j=1}^{n} x_{ij} \ln \frac{x_{ij}}{\rho_{ij}} \Rightarrow \max; \tag{1}$$

$$\left.\begin{aligned} \sum_{i=1}^{n} x_{ij} &= Y_j, \quad j = \overline{1,n}; \\ \sum_{j=1}^{n} x_{ij} &= P_i, \quad i = \overline{1,n}; \end{aligned}\right\} \tag{2}$$

$$\sum_{i=1}^{n} c_{ij} x_{ij} = C. \tag{3}$$

The equalities (2) characterize the balances with respect to the consumption and production of the product, while (3) characterize the generalized transportation costs.

The parameters Y_j, P_i $(i, j \in \overline{1,n})$ and C of the model can be measured directly, while the apriori probabilities ρ_{ij} of a product portion to enter the communication (i, j), do not. They depend on many various factors, namely economical, geographical, social, technical, etc. If a quite objective indexation of them could be developed then, probably, we could find quantitative relations between ρ_{ij} and the parameters characterizing these factors. One of the ways to realize such an approach is based on the usefulness functions (Wilson, 1978).

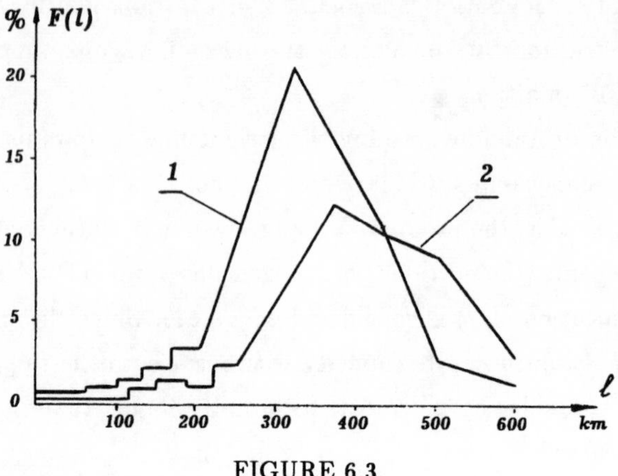

FIGURE 6.3.

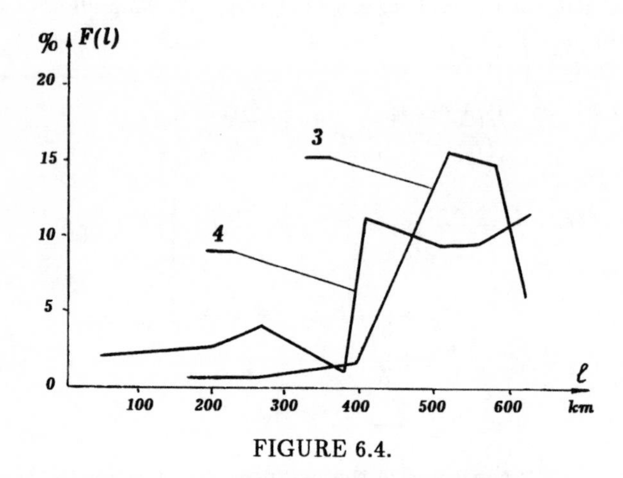

FIGURE 6.4.

Another approach is reduced to the search for invariant (lightly changing in time) characteristics of the transportation process which would take into account all the factors mentioned above in an aggregated form.

The function $F(l)$ of distribution of the freight flow with respect to the distance is one of such invariants. It determines a relative (with respect to the general shipment volume) amount of a product transfered on the distances less than l (an integral distribution function). Realization of such an approach is based on the analysis of real exchanges in the industry of construction materials transfered by the Soviet railways in 1972–1983's (Kotel'nikov,

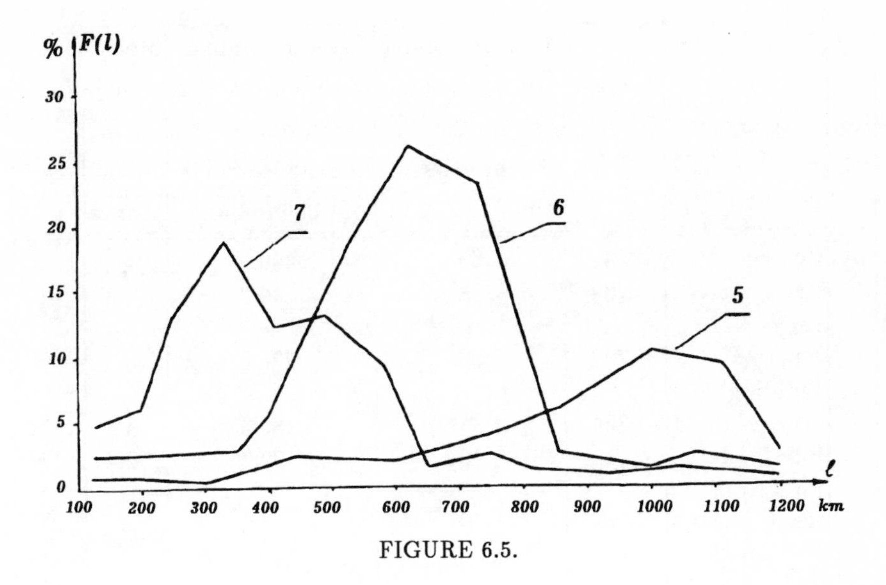

FIGURE 6.5.

1985). Data about the freight volumes with respect to the distance ranges were collected for the following groups of construction materials: (1) building bricks, (2) fire-bricks, (3) fire- and acid-proof materials, (4) fire-proof materials and products, except fire-bricks, (5) water-proof and rouleaux roof materials, (6) building stones, and (7) plaster, lime and chalk.

Distributions of the freight flows among the distance ranges are shown in figures 6.3–6.5.

The study of stability and invariantness of the distributions is quite interest.

A product portion entering the given distance range Δ_k ($k \in \overline{1,k}$) is considered as a random value. The retrospective data on the distribution of freight flows among the distance ranges are considered as samples from the general ensemble, with Δ_1 observed n_1 times, Δ_2 observed n_2 times, Δ_k observed n_k times, and $\sum_k n_k = n$ being the volume of each sample. Then the distributions shown in figures 6.3–6.5 are statistical distributions of the samples with respect to the years of the period considered (1972–1983's).

In studying the distribution invariantness, the samples were taken for each material for the most congested months, namely april and october, and the statistical analysis was done, with the results presented in Table 6.2. Next,

No.	Cargo	Average dist.	Average volume of shipments (%) on the distances less than 1,000 km	Average volume of shipments with with respect to a distance range (1,000 tonnes)	Mean quadr. deviation	
					1,000 tonnes	%
1.	Cement	547	85	400.0	19.4	5
2.	Keramsite	463	88	40.3	4.6	11
3.	Ferro-concrete	702	80	326.6	18.4	6
4.	Building bricks	354	94	183.3	9.8	5
5.	Fire-bricks	1177	59	22.4	2.4	11
6.	Fire-proof row materials	668	79	69.0	12.5	21
7.	Fire-proof materials and products	1262	60	18.5	2.5	14
8.	Water-proof and rouleaux roof materials	1140	66	10.5	1.4	14
9.	Building stones	335	96	1384.0	55.4	4
10.	Plaster, lime chalk	618	84	50.5	4.6	9

TABLE 6.2.

normed correlation matrices were constructed for the corresponding samples ("april" and "october") with respect to the whole retrospective period (12 years). Analyzing them, we saw that the distributions of the freight flows among the distance ranges are almost invariant with respect to changes of the production volumes P_i and consumptions Y_i and practically do not change in time. Hence these distributions can be predicted quite accurately for the practical purposes, which is important in modeling the perspective interregional product exchange.

The study of homogeneity of the statistical distributions is also important. The latter was estimated by the Pirson criterion.

Processing empiric data, we can conclude that all the samples (except for enforced concrete) are homogeneous with respect to the significance level accepted, and consequently they are chosen from the same general ensemble.

Thus there exists the function $F(l)$ of distribution of the freight flows among the distance ranges for each construction material. This function is a stable aggregated characteristic of the spatial interactions of economical system elements which exchange the products through real transportation communications.

The existence of such a stable characteristic shows that the distance factor is essential in random choosing the pair "producer – consumer" (or, in other words, the communication (i, j)).

Therefore, in model (1–3), it is logical to connect the apriori probabilities ρ_{ij} characterizing this choice with the distances, taking into account the distribution of the freight flows among the distance ranges and real characteristics of the transportation network.

Assume that the economic resources are both exchanged and distributed among the objects through the shortest paths (connections). Denote by L $\{l_{ij};\ i, j \in \overline{1,n}\}$ the matrix of shortest paths between all regions.

The following ways for determining the parameters ρ_i can be proposed if we have this matrix and the apriori distribution $F(l)$.

Consider the function $F(l)$ defined on the interval $(l^{\min}, l^{\max})$. Let us divide the distance axis between $l^{\min}$ and $l^{\max}$ into K adjacent segments Δ_k, $k \in \overline{1,K}$ (fig. 6.6(a)) and consider l_{ij} as being undistinguishable inside the segments (Shmulyan, Imelbayev, 1978).

We introduce the indicator functions of the segments Δ_k (fig. 6.6(b)):

$$\lambda_k(l) = \begin{cases} 1 & (l \in \Delta_k), \\ 0 & (l \notin \Delta_k). \quad k \in \overline{1,K} \end{cases}$$

Using a real transportation network, we can determine the number n_k of connections with $l_{ij} \in \Delta_k$ (see fig. 6.6 (c)), namely

$$n_k(l) = \sum_{i,j \in \Omega_k} \lambda_k(l_{ij}), \quad k \in \overline{1,K},$$

where

$$\Omega_k = \{(i,j) :\ l_{ij} \in \Delta_k\}, \quad k \in \overline{1,K}.$$

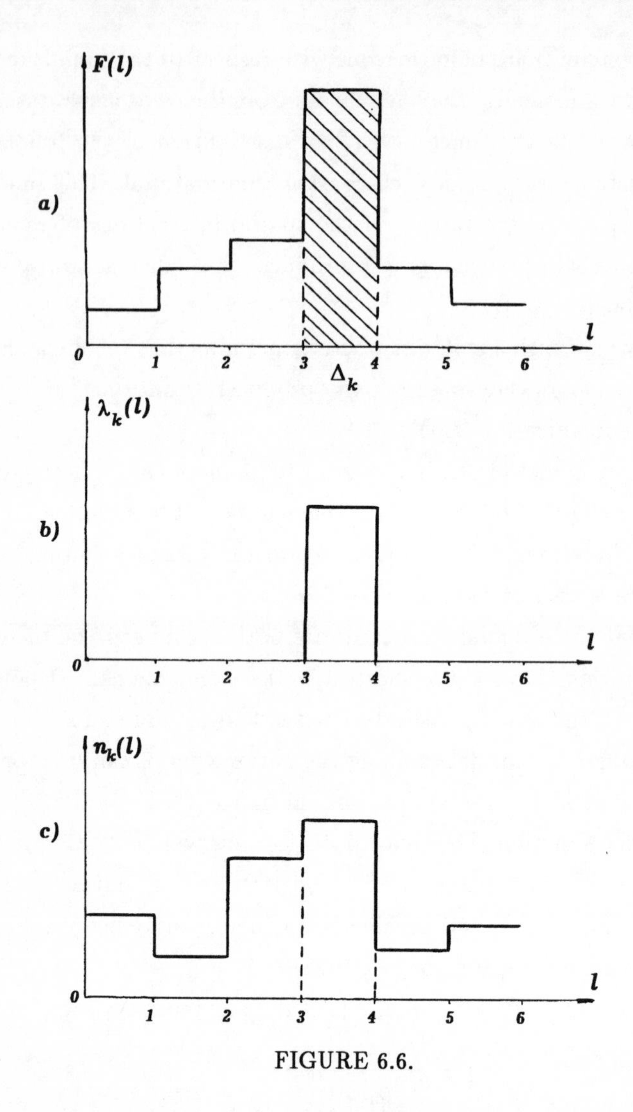

FIGURE 6.6.

Apriori information on the freight volumes F_k with respect to the distance intervals (ranges) Δ_k, $k \in \overline{1, K}$, allows us to calculate the apriori probabilities

$$\rho_{ij} = F_k / N, \quad (i, j) \in \Omega_k,$$

where N is the overall volumes of economic resource freighted.

Assuming the "weights" of elements to be the same inside the subset Ω_k, we obtain

$$\rho_{ij} = \frac{F_k}{N n_k}. \tag{4}$$

Thus the equality obtained characterizes the way the apriori probabilities depend on the "transportation" characteristics of communications.

However, the communication characteristics are not the only parameters influencing the spatial interaction between the regions. So-called "economical capacities" of the regions are another group of factors stimulating the interregional exchanges.

To measure this, many indices exist, for example, the resource consumption in given region. Denote this characteristic of region by w_i; then the apriori probabilities ρ_{ij} can be presented as a product of two terms, one describing the influence of "transportation" characteristics of communications, while the other describing the consumption level in given region, namely

$$\rho_{ij} = \left(\frac{F_k}{Nn_k}\right) \frac{w_j}{\sum_j w_j}. \tag{5}$$

Analyzing real information on the cement consumption in 125 regions of Russia during 12 years, Kotel'nikov (1985) shows that this characteristic is "stable".

6.4 Modeling and analysis of structure of interregional transport and economical connections in cement industry (single-product model)

Cement industry is an important branch of material production and determines the economical potential and level of technical development in a country. The rate of cement production growth and its quality essentially influence civil engineering, its efficiency and technical base.

Location of raw materials influences mainly the allocation and development of the cement industry factories. Cement industry cannot be distributed among the country regions according to their demands, because the raw materials are not located uniformly. Therefore the cement industry generates high transportation demands. The main part of this product is carried by the railway communications.

No.	Economical regions	Production versus consumption (%)		
		1970	1975	1980
1.	North-West	99.8	83.5	55.1
2.	Central	73.6	66.2	71.8
3.	Volgo-Vyatsky	63.1	84.9	77.5
4.	Central-Chernozemny	135.7	175.9	207.4
5.	Volga region	133.5	111.3	110.3
6.	North Caucasian	78.5	87.4	100.2
7.	Ural	161.7	186.6	184.2
8.	West-Siberian	115.8	108.4	99.8
9.	East-Siberian	98.8	87.0	91.3
10.	Far-East	67.7	64.6	102.8
11.	Donetsk-Dnepr	137.1	128.9	127.4
12.	South-West	80.4	104.1	116.3
13.	South	71.5	61.5	65.7
14.	Baltic	86.6	113.2	136.0
15.	Trans-Caucasian	82.0	97.6	83.4
16.	Central Asia	87.4	78.8	90.9
17.	Kazakhstan	93.5	94.2	95.4
18.	Belorussia	65.2	51.9	47.3
19.	Moldavia	65.0	71.1	77.3

TABLE 6.3.

Characteristic feature of national cement industry is its growing concentration. However, growing demands on the cement and amount of places of its consumptions causes rather large transportation costs. Table 6.3 shows rough relation between the consumption and production of cement for economical regions at the ends of 5-years plan periods (see the reports "Cement industry", 1971, 1976, 1981).

Estimation of structure of interregional connections developed in producing and distributing this important resource is more complex. But such an estimation is necessary, because it is the structure of connections that determines the congestion of the national transportation subsystem and produces the important index of exchanges, namely the costs on delivering the products from the producer to the ultimate consumer. Analyzing the structure of interregional connections, we can conclude about how effectively is located the industry and organized the interregional product exchange. The information about the structure is contained in the matrix $X = \{x_{il}\}$ of interregional flows of the product. The model (3.1)–(3.3) can be used to obtain this matrix under fixed location of the factories. The initial information and computer realization of the model assume that the territory of former USSR is divided into quite small calculation regions. In dividing the territory, each calculation region has been chosen so that it corresponded to at least one railway station, with the distance between the neighbor station being less than 100 km, and its boundaries coincided with the existing administrative structure.

As a result, the territory of former USSR was divided into 691 calculation regions (Kamchatka, Magadan, Sakhalin and Tuva were not included, because of lack of railways connected with the common network of USSR).

Such a division allows us to analyze the structure of interregional connections in detail. However, this is not convenient for aggregated calculations and estimation of variants. Therefore, the following aggregation levels were introduced, with their boundaries coinciding with those of the calculation regions.

(1) Aggregation into administrative regions.

(2) Aggregation into economical regions.

Specialists consider the latter to be preferable in solving practical problems. Aggregation is also caused by the real information about the model parameters, namely volumes of production and consumption, which are presented for the regions of country. Table 6.4 illustrate the division of the former USSR territory.

The parameters ρ_{ij} of model (3.1)–(3.3) were determined via the distribution of the freight flows among the distance ranges, according to the method given in Section 3. Since this characteristic of interregional exchange is stable, it can be approximated appropriately, and the elements ρ_{ij} of the matrix of apriori probabilities can be calculated by using an approximating distribution.

The Gamma-distribution is such an "appropriate" one:

$$\Phi(l, \alpha, \beta) = \frac{\beta^{\alpha}}{\Gamma(\alpha)} l^{\alpha-1} e^{-\beta l} = \frac{F(l)}{N}, \tag{1}$$

where

$$\Gamma(\alpha) = \int_0^{\infty} y^{\alpha-1} e^{-y}\, dy \tag{2}$$

is the gamma function, while α and β are the distribution parameters.

No.	Economical regions	Number of regions		Number of aggregates	
		admini-strative	calcu-lation	produc-tion	consump-tion
1.	North-West	8	55	10	8
2.	Central	12	52	9	12
3.	Volgo-Vyatsky	5	21	3	5
4.	Central-Chernozemny	5	24	6	5
5.	Volga region	9	40	11	9
6.	North Caucasian	7	31	6	7
7.	Ural	6	37	12	6
8.	West-Siberian	6	41	7	6
9.	East-Siberian	4	31	7	4
10.	Far-East	4	29	6	4
11.	Donetsk-Dnepr	8	34	9	8
12.	South-West	13	36	11	13
13.	South	4	17	6	4
14.	Baltic	4	31	7	4
15.	Trans-Caucasian	3	24	10	3
16.	Central Asia	14	52	14	4
17.	Kazakhstan	19	103	9	19
18.	Belorussia	6	24	4	1
19.	Moldavia	1	9	2	1

TABLE 6.4.

The mean value of Gamma-distribution is

$$\bar{l} = \int_0^\infty l\Phi(l, \alpha, \beta)\, dl = \frac{\alpha}{\beta}. \tag{3}$$

For $0 < \alpha < 1$, $\Phi(l, \alpha, \beta) = \infty$,
for $\alpha = 1$,

$$\Phi(l, \alpha, \beta) = \beta e^{-\beta l}$$

corresponds to the exponential distribution, and for $\alpha > 1$, $\Phi(l, \alpha, \beta) = 0$. The mode of $\Phi(\cdot)$ is

$$l_{\Phi_{\max}} = \frac{\alpha - 1}{\beta} = \bar{l} - \frac{1}{\beta}. \tag{4}$$

Thus the maximum of $\Phi(\cdot)$ is shifted to the left from the mean value, with the difference $1/\beta$.

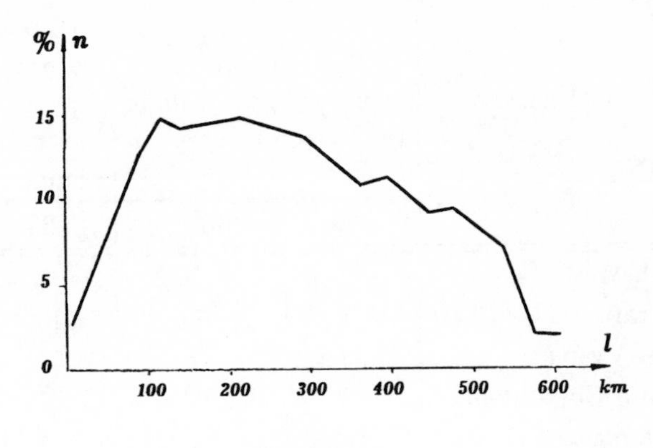

FIGURE 6.7.

The Gamma-distribution reflects qualitative features of empirical distributions of the freight flows among the distance ranges for CMI and, in particular, for cement (see figures 6.3–6.5).

Note that only two parameters α and β are necessary to predict the distribution (1), with the former determining the mean, while the latter so doing the mode of the distribution. According to (3.4),

$$\rho_{ij} = f_k / n_k,$$

where

$$f_k = \int_{l_{ij} \in \Delta_k} \Phi(l, \alpha, \beta)\, dl$$

and $L = \left[l_{ij} \right]$ is the matrix of the shortest paths.

The matrix of the shortest paths is formed according to the existing transportation network.

Two states of the network were considered, namely in 1980 and 1990 years. The quantitative characteristics of the transportation network graph are: (1) up to 700 nodes and (2) up to 3,000 edges (the edges are bidirectional, so that the graph has 6,000 arcs). The distribution of the connection numbers $n(l)$ among the distance ranges illustrate the transportation network (fig. 6.7).

By using the information on the structure of interregional connections and average distance of cement shipments in economical regions and former Soviet republics, the adequacy of the macrosystem models was analyzed in two ways.

Economical regions	Relative error of export/import include the economical region		Relative error with respect to average freight distance in repablics and economical regions	
	123	691	123	691
1. North-West	9	7	7	7
2. Central	1	10	15	17
3. Volgo-Vyatsky	53	37	54	32
4. Central-Chernozemny	15	39	7	20
5. Volga region	16	2	4	18
6. North Caucasian	12	15	13	10
7. Ural	4	4	17	1
8. West-Siberian	0	3	20	8
9. East-Siberian	6	2	17	11
10. Far-East	10	8	24	24
11. Donetsk-Dnepr	14	17	21	20
12. South-West	3	13	5	17
13. South	54	29	57	71
14. Baltic	78	13	25	12
15. Trans-Caucasian	1	4	7	18
16. Central Asia	12	4	44	3
17. Kazakhstan	21	17	20	25
18. Belorussia	99	35	41	13
19. Moldavia	95	74	21	7

TABLE 6.5.

(1) The country is divided into 123 calculation regions. The consumption and production of cement are contracted into one point, namely the administrative center.

(2) The country is divided into 691 calculation regions. The production is located in real factories, while the consumption is aggregated with respect to the administrative regions (according to the available information).

The railway network is characterized by the matrix of shortest paths $L = \{l_{ij},\ i, j \in \overline{1, n}\}$. The distance axis was divided into 1,000 intervals (distance ranges), each being 10 km length. The relative capacities ρ_{ij} were calculated

Economical regions	Relative error of export/import include the economical region		Relative error with respect to average freight distance in repablics and economical regions	
	123	691	123	691
1. North-West	17	15	3	1
2. Central	13	18	7	12
3. Volgo-Vyatsky	72	48	49	28
4. Central-Chernozemny	9	29	26	12
5. Volga region	26	14	4	0
6. North Caucasian	23	20	0	4
7. Ural	11	5	30	17
8. West-Siberian	15	17	11	2
9. East-Siberian	3	6	14	11
10. Far-East	8	7	20	27
11. Donetsk-Dnepr	14	13	15	19
12. South-West	8	10	15	14
13. South	0	0	25	27
14. Baltic	67	11	100	31
15. Trans-Caucasian	3	5	1	19
16. Central Asia	17	4	30	2
17. Kazakhstan	41	26	11	14
18. Belorussia	76	42	98	7
19. Moldavia	0	0	99	53

TABLE 6.6.

by using the Gamma-distribution with the parameters $\alpha = 454$ km (the mean shipment distance) and $\beta = 100$ km (the distribution mode). In each variant the calculations were made with and without constraints on the interregional connections.

The results of modeling are given in Table 6.5 (for the model without the constraints) and in Table 6.6 (for the models with the constraints).

Analyzing the results, we can see that relative errors of import/export with respect to an average economical region are 27% in the 1-st variant and 18% in the second; relative errors are 22% and 15%, respectively; relative errors with respect to the whole matrix of interregional connections are 23% and 21%, respectively. Average relative errors with respect to the average shipment distance in economical regions and republics are 30% in the 1-st variant and 16% in the second.

Thus the results obtained allows us to conclude that the model describes quite accurately (for practical purposes) real interregional product exchanges and can be used for predictions. Note that, taking into account existing information on the interregional connections and increasing the number of calculation regions, we can calculate more accurately.

6.5 Multi-product macrosystem models of interregional exchange

Thus far we considered the models for independent flows of the products. In reality, however, an economical system of product exchange is a more complex system in which the resources are exchanged and distributed to meet the production needs of various functional subsystems of economy and ultimate demands.

We consider the set of regions R_i, $i \in \overline{1,n}$, again. Let the objects of various functional subsystems producing and distributing the resources (raw materials, half-ready products, final products) be situated in the regions. Using the functional principle we can distinguish three main subsystems:

(a) *Production.* This involves the objects of industry branches and agriculture producing various product groups.

(b) *Consumption.* This involves the objects of material production and nonproductive branches, savings, social and individual consumption.

(c) *Transport.* As above, we describe the transport subsystem by the matrix $L = \{l_{ij}\}$ of shortest ways between the regions.

According to Leontieff, Strout, 1963, we assume that the final distribution of products is not of interests for the producers, while the product origins does so for the consumers. Hence all the products produced in the region i are collected in a common supply fund, while all the products consumed

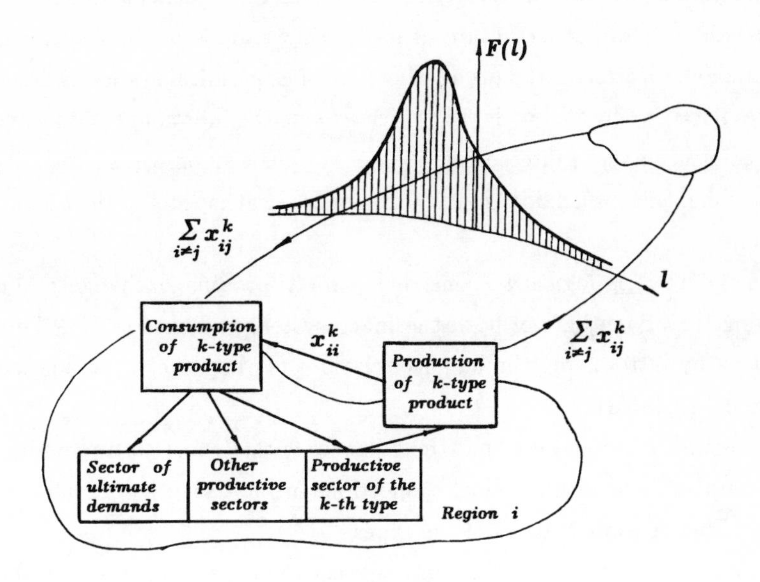

FIGURE 6.8.

in the region j are extracted from a common consumption fund. The flows of such a system are shown in fig. 6.8. The notation from the singleregional model is used, with regional indices added.

Thus let x_{ij}^k be a flow of the k-th product carried from the region i to j by the shortest way, and let N be the whole production of the r-th product in the system. Now each region should satisfy the balance relation like (2.5). We see from fig. 6.8 that the corresponding equation system is (Wilson, 1970)

$$\sum_j x_{ji}^k = \sum_s a_i^{ks} \sum_j x_{ji}^s + y_i^k, \qquad i \in \overline{1,n}, \quad k \in \overline{1,r}. \tag{1}$$

where $\sum_j x_{ji}^k$ is the consumption of final demand in the i-th region by the sector;

$\sum_s a_i^{ks} \sum_j x_{ji}^s$ is the amount of product k consumed by other sectors and

a_i^{ks} is the matrix of technological coefficients in the region i.

A real economical system has two linked features of its operation, namely productive and spatial. The latter is associated with the exchange and distribution of produced resources. It is described quite constructively by a macrosystem model of interregional exchange. However, it is difficult to describe the economical structure of interacting regions in the frameworks of this model, vice versa, the productive part of economical system is described quite effectively by the model (1) which reflects the functional and technological connections and interconnections of various economical sectors. At the same time, this model does not involve real spatial aspects of the economical system.

Therefore it is logical to construct a multi-product exchange model involving the advantages of both the macrosystem and models (1). This can be done by introducing the balance relations (1) into the constraint system of model (3.1–3.3).

It should be noted here that the balances with respect to the production, consumption and technological connections are not satisfied for all product types, but for some subsets of the types.

Then the general form of a macrosystem multi-product model is the following

$$H(X) = \sum_{i,j=1}^{n} \sum_{k=1}^{r} x_{ij}^{k} \ln \frac{\rho_{ij}^{k}}{x_{ij}^{k}} \Rightarrow \max; \tag{2}$$

$$\sum_{j} x_{ij}^{k} = P_{i}^{k}, \qquad i = \overline{1,n}, \quad k \in M_1, \tag{3}$$

$$\sum_{i} x_{ij}^{k} = Y_{j}^{k}, \qquad j = \overline{1,n}, \quad k \in M_2, \tag{4}$$

$$\sum_{j=1}^{n} x_{ji}^{k} = \sum_{s=1}^{r} a_{i}^{ks} \sum_{j=1}^{n} x_{ji}^{s} + y_{i}^{k}, \qquad i \in \overline{1,n}, \quad k \in M_3 \tag{5}$$

$$x_{ji}^{k} \geq 0,$$

where M_1, M_2 and M_3 are the subsets of the product types for which the balances are satisfied with respect to the production, consumption and technology, respectively. Maximizing the entropy criterion, under appropriate constraints, we can determine the state of macrosystem which realization probability is maximum. The model parameters, namely the apriori probabilities, are determined according to the technique given in section 3.

The model (2–5) differs from those described, for example, in Wilson (1970) in the fact that (2–5) allows us to take into account the characteristics of real exchange process and information on how the freight flows are distributed among the distance ranges. This provides more model adequacy with respect to the phenomenon studied.

Let us classify the set of economic resources with respect to the existence of information about the regional production and consumption of a product. We obtain four groups (subsets) of economic resources. The first (1) includes the products with no information on their consumption or production, for the second (2) group we know only the production volumes, for the third (3) group we know only the consumption volumes, and, finally, we have all information about the products of the fourth (4) group.

Model of subsystem without constraints on production and consumption (1). Here we assume that there are no independent estimates P_i^r and Y_j^r, i.e. the constraints (3) and (4) are absent.

Hence the macrosystem multi-product model is of the form

$$H(X) = \sum_{i,j=1}^{n} \sum_{k=1}^{r} x_{ij}^k \ln \frac{\rho_{ij}^k}{x_{ij}^k}, \tag{6}$$

$$\sum_j x_{ji}^k = \sum_{s=1}^{r} a_i^{ks} \sum_{j=1}^{n} x_{ji}^s + y_i^k, \qquad i \in \overline{1,n}, \quad k \in \overline{1,r} \tag{7}$$

$$x_{ji}^k \geq 0.$$

Let us construct the Lagrange function for this problem:

$$L = H(X) + \sum_{i=1}^{n} \sum_{k=1}^{r} \gamma_i^k \left(y_i^k + \sum_{s=1}^{r} a_i^{ks} \sum_{j=1}^{n} x_{ij}^s - \sum_{j=1}^{n} x_{ji}^k \right). \tag{8}$$

The stationary point of the Lagrange system is determined by the solution of equation system

$$\frac{\partial L}{\partial x_{ij}^k} = \ln \frac{\rho_{ij}^k}{x_{ij}^k} - 1 - \gamma_k^k + \sum_{s=1}^{r} \gamma_i^s a_i^{ks} = 0, \quad i \in \overline{1,n}; \quad k \in \overline{1,r}.$$

Hence the solution is of the form

$$\overset{*}{x}{}_{ij}^{k} = \rho_{ij}^k \exp\left(\sum_{s=1}^{r} \gamma_i^s a_i^{ks} - \widetilde{\gamma}_j^k \right), \quad i,j \in \overline{1,n}; \quad k \in \overline{1,r}, \tag{9}$$

where $\widetilde{\gamma}_j^k = 1 + \gamma_j^k$.

Denote

$$\delta_i^k = \exp\left(\sum_{s=1}^{r} \gamma_i^s a_i^{ks}\right),$$
$$\varepsilon_j^k = \exp(-\widetilde{\gamma}_j^k). \tag{10}$$

Then (9) can be rewrite as follows:

$$\overset{*}{x}{}_{ij}^{k} = \rho_{ij}^k \delta_i^k \varepsilon_j^k. \tag{11}$$

To determine the Lagrange multipliers ε_i^k, we have the following equation system:

$$\Phi_i^k(\varepsilon) = \frac{1}{y_i^k}\left[\varepsilon_i^k \sum_{j=1}^{n} \rho_{ji}^k \delta_j^k - \sum_{s=1}^{r} a_i^{ks} \delta_i^s \sum_{j=1}^{n} \rho_{ji}^k \varepsilon_j^s\right] = 1. \tag{12}$$

Hence we obtain

$$\varepsilon_i^k = \frac{y_i^k + \sum_{s=1}^{r} a_i^{ks} \delta_i^s \sum_{j=1}^{n} \rho_{ji}^k \varepsilon_j^s}{\sum_{j=1}^{n} \rho_{ji}^k \delta_j^k}, \quad i \in \overline{1,n}; \quad k \in \overline{1,r},$$

where

$$\delta_j^k = \prod_{s=1}^{r} (\varepsilon_i^k)^{-a_i^{ks}}. \tag{12'}$$

Thus the freight flows x_{ij}^k are determined by the system (9)–(12). The results of modeling are the volumes of production and consumption of products distributed among the regions, namely

$$P_i^k = \delta_i^k \sum_j \rho_{ji}^k \varepsilon_j^k \qquad i \in \overline{1,n}; \quad k \in \overline{1,r},$$
$$Y_j^k = \varepsilon_j^k \sum_i \rho_{ji}^k \delta_i^k \qquad j \in \overline{1,n}; \quad k \in \overline{1,r}. \tag{13}$$

Multiplicative algorithms are used to solve the equations (12) and (12′) (see Chapter 4).

Model of subsystem with constraints on the production (2). Let the estimates P_i^k of product productions be available for the regions. Then the macrosystem multi-product model is of the form

$$H(X) = \sum_{i,j=1}^{n} \sum_{k=1}^{r} x_{ij}^{k} \ln \frac{\rho_{ij}^{k}}{x_{ij}^{k}}, \tag{14}$$

$$\sum_{j=1}^{n} x_{ij}^{k} = P_{i}^{k}, \qquad i \in \overline{1,n}, \quad k \in \overline{1,r} \tag{15}$$

$$\sum_{j=1}^{n} x_{ji}^{k} = \sum_{s=1}^{r} a_{i}^{ks} P_{i}^{s} + y_{i}^{k}, \qquad i \in \overline{1,n}, \quad k \in \overline{1,r}, \tag{16}$$

$$x_{ji}^{k} \geq 0.$$

The Lagrange function of the problem (14)–(16) is

$$L = H(X) + \sum_{i=1}^{n} \sum_{k=1}^{r} \lambda_{i}^{k} \left(P_{i}^{k} - \sum_{j=1}^{n} x_{ji}^{k} \right) +$$

$$+ \sum_{i=1}^{n} \sum_{k=1}^{r} \gamma_{i}^{k} \left(y_{i}^{k} + \sum_{s=1}^{r} a_{i}^{ks} P_{i}^{s} - \sum_{j=1}^{n} x_{ji}^{k} \right).$$

The stationary point of the Lagrange function is determined by the equation system:

$$\frac{\partial L}{\partial x_{ij}^{k}} = \frac{\rho_{ij}^{k}}{x_{ij}^{k}} - 1 - \gamma_{j}^{k} - \lambda_{i}^{k} = 0,$$

$$i, j \in \overline{1,n}; \quad k \in \overline{1,r}.$$

Hence

$$\overset{*}{x}{}_{ij}^{k} = \rho_{ij}^{k} \exp(-1 - \gamma_{j}^{k} - \lambda_{i}^{k}), \qquad i, j \in \overline{1,n}; \quad k \in \overline{1,r}. \tag{17}$$

Denote

$$a_{i}^{k} = \exp(-1 - \lambda_{i}^{k}), \qquad \varepsilon_{j}^{k} = \exp(-\gamma_{j}^{k}). \tag{18}$$

Then (17) can be rewrite as follows:

$$\overset{*}{x}{}_{ij}^{k} = \rho_{ij}^{k} a_{i}^{k} \varepsilon_{j}^{k}. \tag{19}$$

Substituting (19) in (15) and (16), we obtain the equation for determining the Lagrange multipliers:

$$\Psi_i^k(\alpha,\varepsilon) = \frac{\alpha_i^k}{P_i^k}\sum_{j=1}^{n}\rho_{ji}^k\varepsilon_j^k = 1;$$

$$\Omega_i^k(\alpha,\varepsilon) = \frac{\varepsilon_i^k}{y_i^k\sum_{s=1}^{r}a_i^{ks}P_i^s}\sum_{j=1}^{n}\rho_{ji}^k\alpha_j^k = 1;$$

$$i\in\overline{1,n};\quad k\in\overline{1,r}.$$

To solve this system, multiplicative algorithms can be used (see Chapter 4).

Consumption volumes for the product k are the results of modeling.

$$Y_j^k = \varepsilon_j^k\sum_i \rho_{ji}^k\alpha_i^k \qquad j\in\overline{1,n};\quad k\in\overline{1,r} \tag{21}$$

Model of subsystem with constraints on the consumption (3). Let independent consumption estimates Y_j^k be available for the regions. Then the macrosystem multi-product model is of the following form:

$$H(X) = \sum_{i,j=1}^{n}\sum_{k=1}^{r}x_{ij}^k\ln\frac{\rho_{ij}^k}{x_{ij}^k} \Rightarrow \max; \tag{22}$$

$$\sum_i^n x_{ij}^k = Y_j^k \qquad j=\overline{1,n},\quad k\in\overline{1,r} \tag{23}$$

$$Y_i^k = \sum_{s=1}^{r}a_i^{ks}\sum_{j=1}^{n}x_{ji}^s + y_i^k, \qquad i\in\overline{1,n}\quad k\in\overline{1,r} \tag{24}$$

$$x_{ji}^k \ge 0,$$

The Lagrange function for this model is:

$$L = H(X) + \sum_{j=1}^{n}\sum_{k=1}^{r}\mu_j^k\left(Y_j^k - \sum_{i=1}^{n}x_{ij}^k\right) +$$

$$+\sum_{i=1}^{n}\sum_{k=1}^{r}\gamma_i^k\left(y_i^k - Y_i^k + \sum_{s=1}^{r}a_i^{ks}\sum_{j=1}^{n}x_{ji}^s\right).$$

We have

$$\frac{\partial L}{\partial x_{ij}^k} = \ln \frac{\rho_{ij}^k}{x_{ij}^k} - 1 - \mu_j^k + \sum_{s=1}^{r} \gamma_i^s a_i^{ks} = 0,$$

$$i, j = \overline{1,n}; \quad k \in \overline{1,r}.$$

Hence we obtain the solution

$$\overset{*}{x}{}_{ij}^{k} = \rho_{ij}^k \exp\left(\sum_{s=1}^{r} \gamma_i^s a_i^{ks} - 1 - \mu_j^k \right), \qquad (25)$$

$$i, j \in \overline{1,n}; \quad k \in \overline{1,r}.$$

Denote

$$\delta_i^k = \exp\left(\sum_{s=1}^{r} \gamma_i^s a_i^{ks} \right),$$

$$\beta_j^k = \exp(-1 - \mu_j^k),$$

and rewrite (25) as

$$\overset{*}{x}{}_{ij}^{k} = \rho_{ij}^k \delta_i^k \beta_j^k. \qquad (26)$$

Substituting (26) in (23) and (24), we obtain the equations determining the Lagrange multipliers:

$$\frac{\beta_j^k}{Y_j^k} \sum_{i=1}^{n} \rho_{ij}^k \delta_i^k = 1;$$

$$\frac{\delta_i^k}{Y_i^k - y_i^k} \sum_{s=1}^{r} a_i^{ks} \sum_{i=1}^{n} \rho_{ij}^k \beta_i^k = 1; \qquad (27)$$

$$i \in \overline{1,n}; \quad k \in \overline{1,r}.$$

These can be solved by using multiplicative algorithms described in Chapter 4.

We can find the way the production volumes are distributed among the regions for the product groups considered:

$$P_i^k = \delta_i^k \sum_j \rho_{ji}^k B_i^k C_k^k$$

$$j \in \overline{1,n}; \quad k \in \overline{1,r}$$

Model of subsystem with constraints on both the production and consumption with respect to all product ($M_1 = M_2 = M_3$) (4). It is easy to see that in this case the balance constraint (5) is of the form:

$$Y_i^k = \sum_{s=1}^{r} a_i^{ks} P_i^s + y_i^k, \qquad i \in \overline{1,n}, \quad k \in \overline{1,r}. \tag{28}$$

Now x_{ij}^k do not enter (28), and P_i^s and Y_j^k should agree with (28); then the freight flows x_{ij}^k can be determined separately for each product of the group by means of the single-product model considered earlier.

CHAPTER

SEVEN

IMAGE RECONSTRUCTION

7.1 Main notions

The problem of recovering (reconstructing) images arises in various fields of science, technic, medicine and art. This is because the visible information about an image is distorted. These distortions result from both noises accompanying inevitably the observation and the fact that the object under investigation is hidden from the visual perception of a human being.

Before considering the features of the subject matter, we introduce definitions and terminology.

We will consider monochrome two-dimensional images (fig. 7.1). Such an image can be characterized by a density function $\varphi(x, y)$ related to the image through mutually simple correspondence.

We cannot have an ideal image because of various reasons. Usually, instead of this, a two-dimensional function $\mathcal{D}(x, y)$ different from $\varphi(x, y)$ is observed. Here we assume that the observation device allows us to obtain a function of continuous arguments.

The link between the image density function $\varphi(x, y)$ and $\mathcal{D}(x, y)$ is given by the characteristics of observation device. Though, in general, this link is

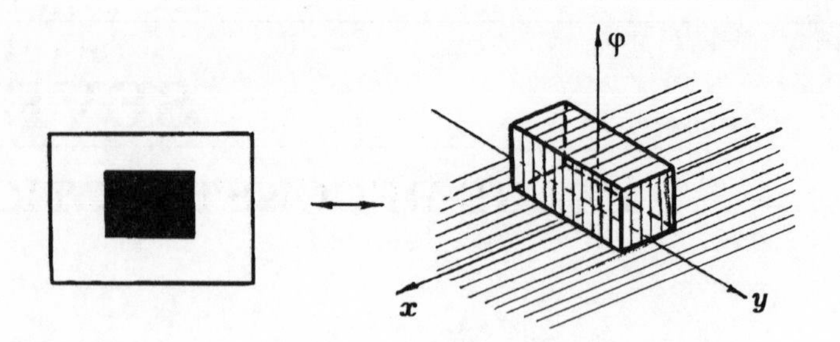

FIGURE 7.1.

nonlinear and dynamic, it can often be approximated with a high accuracy by a linear integral functional

$$\mathcal{D}(x,y) = \mathcal{L}[\varphi(x,y)] = \iint_G \varphi(\xi,\eta)K(x,y,\xi,\eta)\,d\xi\,d\eta, \tag{1}$$

where $K(x,y,\xi,\eta)$ is the functional kernel characterizing the properties of the observation device;

G is the area of image localization.

The devices performing data "fold", namely

$$\mathcal{D}(x,y) = \iint_G \varphi(\xi,\eta)K(x-\xi,y-\eta)\,d\xi\,d\eta, \tag{2}$$

and data projection, namely

$$\mathcal{D}(x,y) = \iint_G \varphi(\xi,\eta)K(x-\xi,\eta)\,d\xi\,d\eta, \tag{3}$$

are the most widespread.

Digital devices for representing and processing the information are used in most modern observation devices. Therefore a real observation device gives a two- or one-dimensional array of numbers:

$$\mathcal{D}(i,j) = \sum_{k,l} \varphi(k,l)K(i,j,k,l) \tag{4}$$

$$\mathcal{D}(i) = \sum_{k,l} \varphi(k,l)K(i-k,l) \tag{5}$$

Errors are inevitable, i.e. the "mean" intensity $\mathcal{D}_d$ measured by the detector involves information about its exact value $\mathcal{D}$ and a noise ξ. Usually, an additive model of measuring is considered where

$$\mathcal{D}_d = \mathcal{D} + \xi. \tag{6}$$

The problem of reconstructing images is in determining the density function $\varphi(x, y)$ (or $\varphi(i, j)$) by measuring the function $\mathcal{D}_d(x, y)$ (or $\mathcal{D}_d(i, j)$)

7.2 Examples

7.2.1 Recovering "blurred" images

This problem arises, for example, in photographing moving objects.

Consider an object moving in a plane with the speed $v(t)$ (fig. 7.2). Define by $\varphi(x, y)$ the image density function of a stationary object and by τ the exposure time.

Since the object changes its position in the plane during the exposure, its photograph turns out to be distorted ("blurred"). Knowing the properties of the film, we can obtain the link between the image density function $\varphi(x, y)$ of the stationary object and the density function $\mathcal{D}(x, y)$ of the photograph obtained in moving the object with the speed $v(t)$.

In a linear approximation, this link is of the form (1.2), namely

$$\mathcal{D}(x, y) = \int_y^{y+\Delta y(\tau)} \int_x^{x+\Delta x(\tau)} \varphi(\xi, \eta) K(x - \xi, y - \eta)\, d\xi\, d\eta, \tag{1}$$

where

$$\begin{aligned} \Delta x(\tau) &= \int_0^\tau v_x(\lambda)\, d\lambda; \\ \Delta y(\tau) &= \int_0^\tau v_y(\lambda)\, d\lambda; \end{aligned} \tag{2}$$

and

$$v_x(t) = v(t) \cos\theta(t); \qquad v_y(t) = v(t) \sin\theta(t).$$

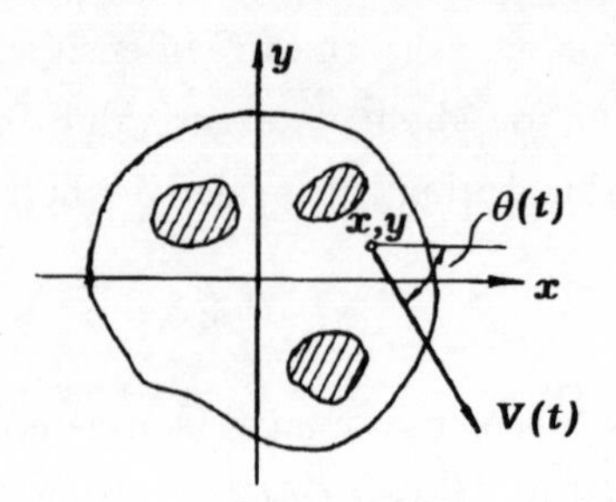

FIGURE 7.2.

The function $K(x,y)$ in (1) depends on the parameters of photographic material, an optical system and the object speed. For example, it can be as follows

$$K(x,y) = \exp\left(-v_x(t) + v_y(t)\right)\Big|_{t=t^*(x,y)}, \tag{3}$$

where $t^*(x,y) = t^*(\sqrt{x^2+y^2})$ is the solution of the equation

$$\omega(t) = \rho = \sqrt{x^2+y^2}; \qquad \omega(t) = \int_0^t |v(\lambda)|\, d\lambda.$$

If the object moves in one of the orthogonal directions ($\theta = 0$) then the expression for the photographic density function (1) is simpler (Bakushinsky, Goncharsky, 1989):

$$\mathcal{D}(x,y) = \int_x^{x+\Delta x(\tau)} \varphi(\xi,y)K(x-\xi)\, d\xi, \tag{4}$$

where

$$K(u) = \frac{1}{v(t)}\Big|_{t=t^*(u)} \tag{5}$$

$t^*(u)$ is the solution of the equation $\omega(t) = u$; and

$$\omega(t) = \int_0^t v(\lambda)\, d\lambda.$$

Finally, in many practical problems, the exposure time is so short that the object speed can be assumed constant during that time, i.e. $v(t) = v_0$. Then

$$\mathcal{D}(x,y) = \frac{1}{v_0} \int_x^{x+v_0\tau} \varphi(\xi, y)\, d\xi. \tag{6}$$

The photograph of the moving object, i.e. $\mathcal{D}(x,y)$, is registered in all the cases considered, and it is necessary to recover the image of the stationary object, i.e. $\varphi(x,y)$.

7.2.2 Recovering defocused images

The image of a spatially-distributed object in a plane is of non-uniform sharpness, i.e. it is defocused in a sense. To explain this situation, consider the geometry of arising the defocused areas in photographic or television images.

Let an initial (flat) object be situated in a plane A and let the film on which the image is recorded be in a plane B (fig. 7.3). The light reflected from the object attains the film through the lens Λ of the objective, as shown in fig. 7.3. Since the film is in the plane B, which is at a distance Δ from the plane B_0 (the focal distance of the lens), any object point is mapped into the circle K, with the diameter ρ being proportional to Δ (Bakushinsky, Goncharsky, 1989).

Image blurring is characterized by the diameters of these circles. It is seen from the figure that the image blurring is zero only if B coincides with B_0. But note that it is true if the initial object is in the plane. If it is spatially-distributed and its "depth" is quite great with respect to the focal distance of the lens, image blurring is inevitable even if the planes B and B_0 coincide.

The link between the density function of the registered (recorded) image $\mathcal{D}(x,y)$ and that of the ideal image $\varphi(x,y)$ can be represented in the form (1.2), where

$$K(x,y) = \begin{cases} 0, & \text{for} \quad x^2+y^2 > \dfrac{\rho^2}{r}; \\ \dfrac{4}{\pi\rho^2}, & \text{for} \quad x^2+y^2 \le \dfrac{\rho^2}{r}. \end{cases} \tag{7}$$

A more difficult situation arises when the initial object is in the plane A not parallel to the plane B. Such a situation is shown in fig. 7.4. In this

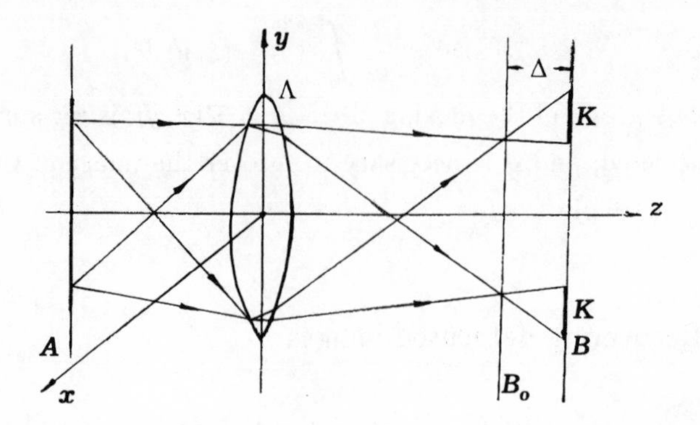

FIGURE 7.3.

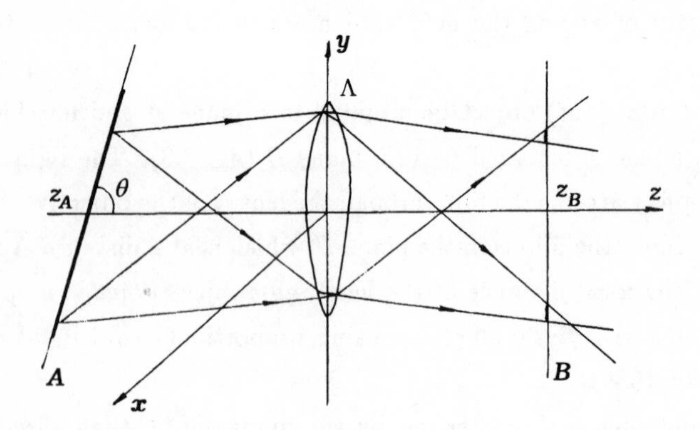

FIGURE 7.4.

case, the link between the functions $\mathcal{D}(x, y)$ and $\varphi(x, y)$ will be described by the following equation

$$\mathcal{D}(x,y) = \iint\limits_{G} \varphi(\xi,\eta) K\big(x - \widetilde{\xi}(\xi,\eta), y - \widetilde{\eta}(\xi,\eta)\big)\, d\xi\, d\eta, \tag{8}$$

where $K(x,y) = \begin{cases} 0, & \text{for } \; x^2 + y^2 > d^2; \\ 1/\pi d^2, \text{for} & x^2 + y^2 \leq d^2; \end{cases}$

$$d = a\left|\left(1 - \frac{z_B}{z} + \frac{z_B}{z_A}\right) + \frac{\widetilde{\eta}\operatorname{ctg}\theta}{z_A}\right|; \tag{9}$$

$$\frac{1}{z} = \frac{1}{z_A} + \frac{1}{z_B};$$

$$\widetilde{\xi} = -\xi \, \frac{z_B}{z_A + \xi \operatorname{ctg}\theta}; \quad \widetilde{\eta} = -\eta \, \frac{z_B}{z_A + \xi \operatorname{ctg}\theta};$$

and a is a parameter of the optical system.

7.2.3 Computer tomography

Most modern methods of diagnostics in technic, medicine and geology are based on the analysis of visual information about an investigated object. However, the object is often hidden and cannot be observed directly. For example, spatial crystalline structure, configuration of human internal organs, structure of star-clusters, geography of mineral deposits, etc.

One of the methods for investigating such objects, namely the methods for obtaining their images, is tomography.

The tomography is a method for obtaining an object image by using the object "shadow" or, more precisely, its projections. Therefore tomographic methods are often considered as the reconstruction of images by means of their projections.

A standard scheme of tomography involves a source S of "radiation" (X-ray, light, neutron, cosmic, etc), an object under investigation O and a detector D (fig. 7.5). Consider the tomographic investigations in more detail, assuming that the source S radiates X-ray photons.

Let the source S radiate a flow of X-ray photons in the direction L. A part of radiated photons are reflected from the object. Some of them, however, penetrate the object and reach the detector D.

It is impossible to predict what will happen with every photon, whether it will be absorbed into the object, reflected from its boundary, or reach the detector. These processes depend on the object properties and photon energy. Both are unpredictable.

A stochastic model is quite convenient for these processes. Its concept is based on the assumption that there exists a probability ρ of transmittance through the object for every value of a photon energy, i.e. the probability of its being captured by the detector. Here it is also supposed that the environment of the object neither reflects nor absorbs X-ray photons.

To define the transmittance probability, a procedure standard for probability theory is usually used. Assume that we can successively irradiate the object by *single* photons having the energy e. Then t_1 of photons will reach the detector in n such experiments. It is clear that $t_1 = t_1(n)$.

Then

$$\rho = \lim_{n \to \infty} \frac{t_1(n)}{n}, \tag{10}$$

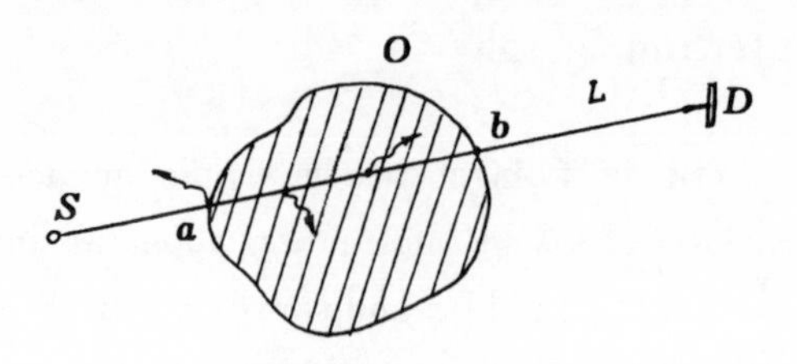

FIGURE 7.5.

where ρ is the probability of a photon reaching the detector.

If we suppose that there is no reflection or dispersion from the object boundaries then the photons not reaching the detector are absorbed in the object. In this case, the probability of each proton to be absorbed in the object is $1 - \rho = \nu$.

Usually, a coefficient of attenuation is introduced to characterize the absorbing ability of the object

$$\mu = -\ln \rho; \qquad \mu \cong k\nu, \tag{11}$$

where k is a proportionality coefficient.

The probability of reaching the detector, as well as the coefficient of attenuation depend on the distance from the source to the detector along the line L. Thus if the detector D could be set inside the object (on the segment a, b), we could fix a change of the coefficient of attenuation along the segment a, b.

Thus the coefficient of attenuation as a characteristic of the "absorbing ability" is a function of points within the object in the corresponding system of coordinates.

Consider a general scheme of tomographic investigation, associated with a system of coordinates (fig. 7.6). The radiation source S can be in any positions along the axis y, while the radiation directions are parallel to the axis x. The detector D is situated on the line $x = x_D$. We assume that there is neither reflection nor dispersion of photons from the object boundaries.

We shall characterize the object under investigation by the coefficient of attenuation μ (11); in this case, this is a function of two variables $\mu(x, y)$ which is zero beyond the object O.

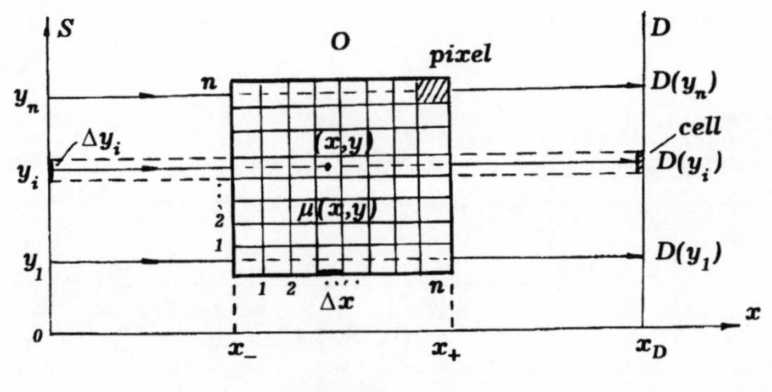

FIGURE 7.6.

Since $\mu(x, y)$ characterizes a relative quantity of photons absorbed at the point (x, y), we see that if a photographic plate were at the place of the objective then its darkening would be proportional to $\mu(x, y)$. Therefore the image of the object on such a photographic plate can be described by the function $\mu(x, y)$.

Now consider a position of the source, for example the point $(0, y_i)$. If a photographic plate is the detector D, its darkening $D(y_i)$ at the point (x_D, y_i) will be proportional to the number of photons passed through the object. According to (11),

$$\mathcal{D}(y_i) = k_0 \exp\left(-\int_{x_-}^{x_+} \mu(x, y_i)\, dx\right), \tag{12}$$

where k_0 is a proportionality coefficient.

In practical tomographs, the irradiation is realized by a beam which is narrow but still spatially-distributed, and the information is processed in a digital form. Therefore the object is described by a set of pixels and the detector is described by a set of cells. The functions $\mu(x, y)$ and $\mathcal{D}(y)$ are assumed to be constant inside the pixels and cells. As a result of such discretization, the equality (12) is transformed to the following form:

$$\mathcal{D}(i) = k_0 \exp\left(-\sum_{j=1}^{n} \widetilde{\mu}(i, j)\Delta x\right), \tag{13}$$

where

$$\widetilde{\mu}(i, j) = \int_{y_i}^{y_i+\Delta y} \mu(x_i, y)\, dy. \tag{13'}$$

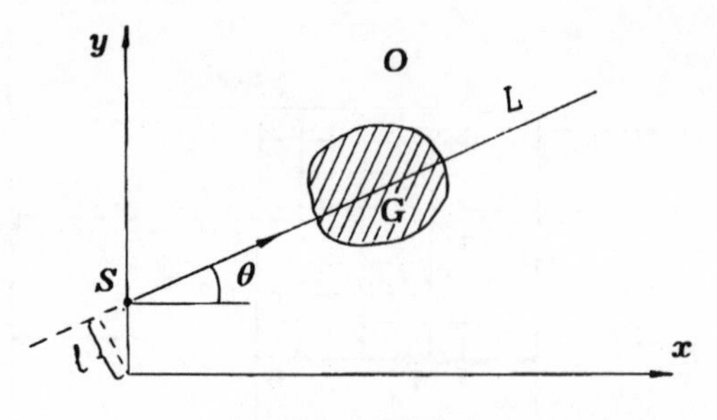

FIGURE 7.7.

To simplify the subsequent calculations, it is more convenient to represent the equalities like

$$\widetilde{q}_i = -\ln \mathcal{D}(i)/k_0 = \sum_{j=1}^{n} \widetilde{\mu}(i,j)\Delta x, \quad i \in \overline{1,r}. \tag{14}$$

The values $\widetilde{q}_i$ in the right-hand side of these equalities are calculated on the basis of information measured by the detector ($\mathcal{D}(i)$). The problem of recovering the object image is in determining the function $\widetilde{\mu}(i,j)$ (the coefficient of attenuation of the object) by measuring r its projections.

Designs of modern tomographic devices are very diverse (G. Herman, 1980). Thus in tomographic devices used for medical diagnostics, an object is irradiated by the source moving around it. Therefore, in general case, the information measured by a detector is a two-dimensional array

$$\mathcal{D}(\theta, l) = k_0 \exp\left(- \iint_G \mu(x,y)\delta(l - x\cos\theta - y\sin\theta)\, dx\, dy \right), \tag{15}$$

where the notation is shown in fig. 7.7,

$$-\infty < l < \infty, \qquad -\frac{\pi}{2} \leq \theta \leq \frac{\pi}{2}.$$

7.2.4 Radioastronomy

One of the main methods in investigating the celestial sphere is radio-interferometry.

If a section of the celestial sphere is considered as a plane, the brightness of its radiation can be characterized by a two-dimensional function $f(x,y)$.

In investigating the celestial sphere radio-interferometrically, detectors record the spectral composition (within the radiofrequency range) of this radiation. The brightness of radiation is related to its spectrum $F(\omega, \nu)$ by Fourrier's transformation (Kraus, 1966):

$$F(i\omega, i\nu) = \iint_{x,y \in Q} f(x, y) e^{-i(\omega x + \nu y)} \, dy \, dx, \tag{16}$$

where Q is the definition domain of $f(x, y)$,

$$i = \sqrt{-1}.$$

The function $F(i\omega, i\nu)$ is complex-valued, i.e.

$$F(i\omega, i\nu) = V(\omega, \nu) + iU(\omega, \nu), \tag{17}$$

where

$$\begin{aligned} V(\omega, \nu) &= \operatorname{Re} F(i\omega, i\nu), \\ U(\omega, \nu) &= \operatorname{Im} F(i\omega, i\nu) \end{aligned} \tag{17'}$$

The radio-interferometer measures both components V and U of $F(i\omega, i\nu)$ (16).

Usually, the section Q of the celestial sphere is represented by the pixels, and the brightness is represented as a function of numbers of the pixels. The frequencies ω and ν are also set in a discrete scale having the steps $\Delta\omega$ and $\Delta\nu$, respectively. A practical detector measures a real and an imaginary components of Fourrier's image $F(i\omega, i\nu)$ in some discrete points:

$$V(l\Delta\omega, k\Delta\nu) = \sum_{s,j} f(s\Delta x, j\Delta y) \cos(l\Delta\omega \, s\Delta x + k\Delta\nu \, j\Delta y), \tag{18}$$

$$U(l\Delta\omega, k\Delta\nu) = \sum_{s,j} f(s\Delta x, j\Delta y) \sin(l\Delta\omega \, s\Delta x + k\Delta\nu \, j\Delta y). \tag{19}$$

The problem arising here is that of recovering the brightness function $f(s, j)$ by using the information about the spectrum components $V(l, k)$ and $U(l, k)$.

7.2.5 Engineering geophysics

It is necessary to study physical and mechanical properties and structure of rocks for carrying out large-scale construction works and searching for minerals (Bakushinsky, Goncharsky, 1989).

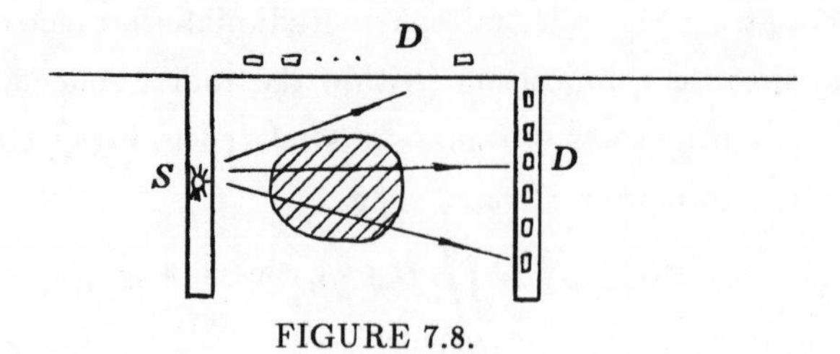

FIGURE 7.8.

In investigating various geological media, seismic methods are widely used, since the speed of seismic wave propagation and dampening is related to the chemical composition, density, porosity and other parameters of a medium.

Explosions are used as seismic wave sources. A speedy structure of a medium is recovered by the ray description of seismic waves, wherein the speed of seismic waves in a medium is assumed to vary little.

A system of observation consists of emitter holes where an explosive substance is put and detector holes where seismic wave receivers are set. Sometimes emitters and detectors are set on the surface (fig. 7.8).

The time τ of a wave travel from the emitter to the receiver is determined by the attenuation function $f(x, g) = 1/v(x, g)$, where v is the speed of wave propagation in a medium.

$$\tau_i(l, \theta) = \int_{L_i} f(l\cos\theta - t\sin\theta,\ l\sin\theta + t\cos\theta)\, dt, \tag{20}$$

where L_i is a trajectory of wave front propagation and

i is the number of the detector.

If L_i is supposed to be a straight line then the equation (20) is linear. The problem is to recover the function $f(x, y)$ of seismic wave dampening in a medium using the observations of τ.

7.3 Methods for estimating the image density function

The examples of image recovering in different branches of science show that the measured values and the desired density functions are related by the linear operators (2) and (3).

Taking into account a discrete character of data and noises of the equipment, the model of measurements can be represented in the following form

$$\begin{aligned} Tu &= q, \\ v &= q + \xi, \end{aligned} \tag{1}$$

where $u = \{u_1, \ldots, u_m\} \geq 0$ is a vector having the components formed out of the ordinates of the density function $\varphi(k, l)$ $(k \in \overline{1,p};\ l \in \overline{1,n})$;

$m = pn$ is the number of the function $\varphi(k, l)$ ordinates to be recovered;

$q = \{q_1, \ldots, q_r\} \geq 0$ is a vector having the components formed from the ordinates of the function $D(i, j)$ $(i \in \overline{1,I};\ j \in \overline{1,J})$;

$r = IJ$ is the number of measurements;

$T = [t_{ks}] \geq 0$ is a matrix having the coefficients formed out of the kernel $K(i, j, k, l)$ (4);

$\xi = \{\xi_1, \ldots, \xi_r\}$ is a vector characterizing the noise of the measurement which is supposed to be such that

$$\begin{aligned} \mathcal{M}\xi &= 0; \\ \operatorname{cov}\xi &= \sigma^2 E, \end{aligned} \tag{2}$$

where $\mathcal{M}$ is the mathematical expectation operator;

$\operatorname{cov}\xi$ is a matrix of co-variations; and

E is the unit matrix.

The matrix T in (1) is a characteristic of the measuring system. Though measuring systems are different in every particular applied investigation (that affects the coefficients of this matrix), there are two properties common to all measuring systems. In terms of the matrix T notation, these common properties are the following:

$$r < m, \tag{3}$$

$$\operatorname{rank} T = r. \tag{4}$$

The first property means that the number of measurements is less than the dimension of the vector characterizing the image density function. This property is very important, since most traditional methods of mathematical statistics turn out to be ineffective because of this. The second property means that the measuring system does not take "extra" measurements (if an "extra" measurement is understood as related to the others by a linear combination).

The problem is to make a good (in a sense) estimate of the vector u using the observed vector v, taking into account the properties of noise ξ (2).

This problem turns out to be a non-standard problem of estimating because of conditions (3) and (4).

In fact, let $\xi = 0$, i.e. the measurements are ideal. One would think that, in this case, a single vector u^0 corresponding to q should exist. However, it is not the case. It follows from (1) and (3) that there is a set $\mathfrak{U}$ of vectors u which result in ideal measurements q. Therefore for $\xi = 0$ the estimate of u is reduced to choice of an appropriate element from the set $\mathfrak{U}$.

If $\xi \neq 0$ then the set $\mathfrak{U}_\xi$ of vectors u generating by the measurements v will differ from the set $\mathfrak{U}$. Therefore for $\xi \neq 0$, it is desirable to choose an appropriate element u from a particular realization of the set $\mathfrak{U}_\xi$ which varies little when this set is deformed because of random errors in measurements.

A traditional approach to choosing an appropriate element from the set $\mathfrak{U}$ in case $\xi = 0$ or from the set $\mathfrak{U}_\xi$ in case $\xi \neq 0$ is to choose an auxiliary strictly-convex (concave) function defined on the set $R_+^m = \{u : u_s \geq 0,\ s \in \overline{1,m}\}$ and to minimize (maximize) it on the set

$$\mathcal{D} = \{u : Tu = v\}. \tag{5}$$

Since the construction of appropriate auxiliary function is a non-formalizable problem, it is necessary to make some assumptions concerning the desired properties of the solution to be found. The review of Censor, 1981 offers to distinguish three classes of such properties.

(a) The solution u should be in a neighborhood of hyperplanes (1).

(b) The solution u minimizes (maximizes) a natural criterion on the set (5).

(c) The solution u is sought as regularized.

The last trend is developed very actively by A. N. Tikhonov and his school (Bakushinsky, Goncharsky, 1989).

Consider the second trend in detail. One of its possible realization is based on ideas of the Bayes estimate. The object whose image we are going to construct is unknown. If a stochastic model is taken as that of "uncertainty", the vector u (1), which characterizes the object image, is stochastic too. Assume that there is a distribution density function $p(u)$ of the vector u and a distribution density function $p(\xi)$ of the measurement noise vector ξ. Given that the vectors u and ξ are independent, the common distribution density of the pair (u, ξ) is equal to

$$p(u, \xi) = p(\xi)p(u). \tag{6}$$

The Bayes estimate of u is

$$\widehat{u} = \arg\max p(\xi)p(u). \tag{7}$$

To realize this procedure, it is necessary to construct the functions $p(u)$ and $p(\xi)$. This is usually considered as a complex problem and therefore is replaced by some hypotheses about a kind of these functions, mainly about the function $p(u)$.

For example, $p(u)$ and $p(\xi)$ are assumed to be normal distributions with the covariance matrices C_u and C_ξ. Then, instead of maximization of $p(u, \xi)$ (6), the following quadratic form can be minimized:

$$(v - Tu)'C_\xi^{-1}(v - Tu) + (u - \overline{u})'C_u^{-1}(u - \overline{u}), \tag{8}$$

where $\overline{u} = Mu$.

A simpler variant of this approach is to use the least squares method:

$$\widehat{u} = \arg\min(v - Tu)'(v - Tu). \tag{9}$$

However, in passing from (8) to (9), the solution becomes nonunique because of the conditions (3) and (4), and solutions with negative components can arise which is unacceptable because of the model (1).

Criteria different from those generated by the Bayes approach are often used in papers on the recovery of images obtained by means of computer-aided tomography. In particular, the norm of the vector u, namely

$$\widehat{u} = \arg\min\{\|u\|^2,\ Tu = q,\ v = q + \xi,\ u \geq 0\}, \tag{10}$$

and its entropy, namely

$$\widehat{u} = \arg\max\left\{-\sum_{s=1}^{m} u_s \ln u_s,\ Tu = q,\ v = q + \xi\right\}, \tag{11}$$

are used.

Note that there is no condition of nonnegativity in (11), because the entropy function is defined on the non-negative ortant.

Assume that a rule for choosing the element u from the set $\mathfrak{U}_\xi$ is formulated, i.e. the estimate $\widehat{u}$ can be constructed using the observations v. This means that in carrying out R experiments on the same object, we can obtain a set of estimates

$$\widehat{u}^1[v^1], \ldots, \widehat{u}^R[v^R]. \tag{12}$$

The ρ-th experiment is regarded as the relations (1) in which the observation vector v^ρ is realized.

To estimate the quality of the rule $\widehat{u}[\nu]$, a pattern u^0 is required.

The distance between the pattern u^0 and estimates (5) can be characterized as follows:

$$V = \frac{1}{R} \sum_{\rho=1}^{R} F(u^0, \widehat{u}^\rho), \tag{13}$$

where F is a non-negative function.

Herman (1980) provides examples of functions used for estimating the quality of images obtained by means of computer-aided tomography in medical research:

$$F_1 = \left[\sum_{s=1}^{m} (u_s^0 - \widehat{u}_s^\rho)^2 \Big/ \sum_{s=1}^{m} (u_s^0 - \overline{u}^0)^2 \right]^{\frac{1}{2}}, \tag{14}$$

$$F_2 = \sum_{s=1}^{m} |u_s^0 - \widehat{u}_s^\rho|^2 \Big/ \sum_{s=1}^{m} |u_s^0|. \tag{15}$$

7.4 Physical justification of entropy estimation

Now there exists rather comprehensive experimental material confirming that the entropy estimation (like (3.11)) results in a significantly better quality of images to be recovered (see, for example, the review Gull, Scilling, 1984). However, there is no explanation to this fact so far, except for a philosophic- methodological one. Therefore it is useful to try to construct at least a physical explanation.

Consider a scheme of tomographic research shown in fig. 7.6. It has been mentioned above that the process of absorbing photons in the object is random, and the probability ρ of passing through the object and the probability

p of being absorbed in the object exist for every photon and beam direction. If there is no photon reflection and dispersion from the object boundaries then $\nu = 1 - \rho$; otherwise $\nu < 1 - \rho$.

Let us divide the object under investigation into pixels. Every photon achieving the pixel (i, j) can be characterized by the probability ν_{ij} of being absorbed in the pixel (i, j). The probabilities ν_{ij} satisfy the condition

$$\sum_{j=1}^{n} \nu_{ij} \leq 1. \tag{1}$$

The index i corresponds to the i-th direction of the beam in fig. 7.6.

Let the source generate N photons in the beam. Denote by N_{ij} the number of photons absorbed in the pixel (i, j). Then the absorption probability is

$$\nu_{ij} = \lim_{N \to \infty} \frac{N_{ij}}{N}. \tag{2}$$

Assume that the object under investigation has analogs, and one of these was studied, i.e. we found the function $\mu_0(i, j)$ characterizing a change of its attenuation coefficient. According to (2.10) and (2.11),

$$\mu_0(i, j) = -\ln(1 - \nu_{ij}),$$

where ν_{ij} is the probability of a sample object to absorb photons.

Hence the apriori absorption probabilities can be determined for the investigated object:

$$\nu_{ij} = 1 - e^{-\mu_0(i,j)}. \tag{3}$$

Assume that the apriori absorption probabilities ν_{ij} (3) are known. Consider the i-th layer of the investigated object irradiated by N photons (fig. 7.9). We represent the layer as a set $\mathcal{R}_{n+1}^{(i)}$ involving n pixels of the object and one additional pixel imitating a detector. Recall that every photon can either be absorbed in a pixel or pass through it and can only be absorbed in the additional pixel.

Let us represent the set $\mathcal{R}_{n+1}^{(i)}$ as (fig. 7.9)

$$\mathcal{R}_{n+1}^{(i)} = \mathfrak{S}_1^{(i)} \cup \mathcal{R}_n^{(i)}, \tag{4}$$

where $\mathfrak{S}_1^{(i)}$ is a subset containing the first pixel and

$\mathcal{R}_n^{(i)}$ is a subset containing the others n pixels.

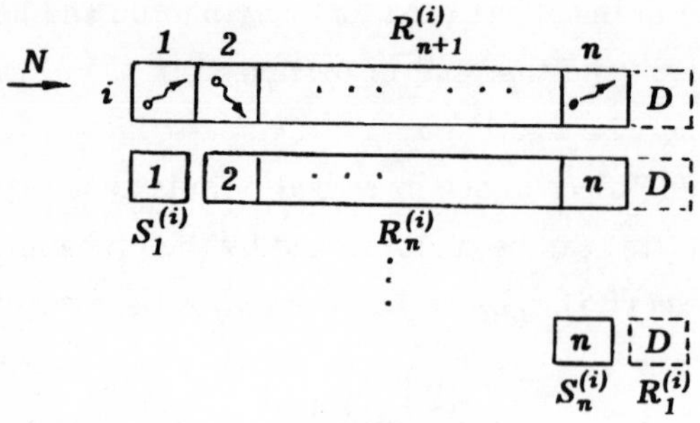

FIGURE 7.9.

The beam passing through the layer results in localizing N_{i1} photons in the set $\mathfrak{S}_1^{(i)}$ and $N - N_{i1}$ photons in the set $\mathcal{R}_n^{(i)}$. The probability of a photon getting to $\mathfrak{S}_1^{(i)}$ is equal to the absorption probability ν_{i1} and the probability of its getting to $\mathcal{R}_n^{(i)}$ is equal to the probability of passing through the first pixel $\omega_{in} = 1 - \nu_{i1}$.

The following stochastic event is regarded as a realization of the set $\mathcal{R}_{n+1}^{(i)}$ (4): the subset $\mathfrak{S}_1^{(i)}$ contains N_{i1} photons, the subset $\mathcal{R}_n^{(i)}$ contains $N_n = N - N_{i1}$ photons, i.e.

$$\text{Prob}\left(\mathcal{R}_{n+1}^{(i)}\right) = P_i(N_{i1}, N_n). \tag{5}$$

The distribution of photons among the subsets $\mathfrak{S}_1^{(i)}$ and $\mathcal{R}_n^{(i)}$ coincides with the classic scheme of Bernoulli trials with two outputs. Therefore

$$P_i(N_{i1}, N_n) = \frac{N!}{N_{i1}!\, N_n!} \widetilde{\nu}_{i1}^{N_{i1}} \omega_{in}^{N_n}, \tag{6}$$

where $\widetilde{\nu}_{i1} = \nu_{i1}$.

Consider now the set $\mathcal{R}_n^{(i)}$ in (4) (fig. 7.8) and represent it in the form

$$\mathcal{R}_n^{(i)} = \mathfrak{S}_2^{(i)} \cup \mathcal{R}_{n-1}^{(i)}, \tag{7}$$

where $\mathfrak{S}_2^{(i)}$ is a subset containing the second pixel; and

$\mathcal{R}_{n-1}^{(i)}$ is a subset containing the others $(n-1)$ pixels.

Note that

$$\omega_{in}^{N_n} = \text{Prob}\left(\mathcal{R}_n^{(i)}\right)$$

in (6).

We define the probability of realizing the set $\mathcal{R}_n^{(i)}$, similarly to (5), in the form

$$\operatorname{Prob}\left(\mathcal{R}_n^{(i)}\right) = P_i(N_{i2}, N_{n-1}). \tag{8}$$

Then

$$\omega_{in}^{N_n} = P_i(N_{i2}, N_{n-1}), \quad N_{n-1} = N - N_{i1} - N_{i2}. \tag{9}$$

The set $\mathcal{R}_n^{(i)}$ is arranged similarly to the set $\mathcal{R}_{n+1}^{(i)}$, and the process of distributing $N_n = N - N_{i1}$ photons passed through the first pixel is also described by the scheme of Bernoulli trials:

$$P_i(N_{i2}, N_{n-1}) = \frac{N_n!}{N_{i2}!\, N_{n-1}!} \widetilde{\nu}_{i2}^{N_{i2}} \omega_{i,n-1}^{N_{n-1}}, \tag{10}$$

where $\widetilde{\nu}_{i2} = (1 - \nu_{i1})\nu_{i2}$ is the probability of absorbing a photon in the second pixel;

$\omega_{i,n-1} = 1 - \widetilde{\nu}_{i2}$ is the probability of a photon passing through the second pixel.

By substituting (9) and (10) into (6), we obtain

$$P_i(N_{i1}, N_{i2}, N_{n-1}) = \frac{N_n!}{N_{i1}!\, N_{i2}!\, N_{n-1}!} \widetilde{\nu}_{i1}^{N_{i1}} \widetilde{\nu}_{i2}^{N_{i2}} \omega_{i,n-1}^{N_{n-1}}, \tag{11}$$

where

$$N_{n-1} = N_n - N_{i2}. \tag{11'}$$

Continuing the dichotomy of the sets $\mathcal{R}_{n-1}^{(i)}, \mathcal{R}_{n-2}^{(i)}, \ldots$ we will attain a situation when

$$\mathcal{R}_2^{(i)} = \mathfrak{S}_n^{(i)} \cup \mathcal{R}_1^{(i)}, \tag{12}$$

where $\mathcal{R}_2^{(i)}$ is a set containing the n-th pixel of the object and the additional pixel (detector);

$\mathfrak{S}_n^{(i)}$ is a subset containing the n-th pixel of the object; and

$\mathcal{R}_1^{(i)}$ is a subset containing the additional pixel.

The probability of N_{in} photons being localized in $\mathfrak{S}_n^{(i)}$ and $N_1 = N_2 - N_{in} = N - N_{i1} - \ldots - N_{in}$ photons being localized in $\mathcal{R}_1^{(i)}$ is equal to:

$$P_i(N_{in}, N_1) = \frac{N_2!}{N_{in}!\, N_1!} \widetilde{\nu}_{in}^{N_{in}} \omega_{i1}^{N_1}, \tag{13}$$

where

$$\widetilde{\nu}_{in} = [1 - \nu_{i,n-1}(1 - \nu_{i,n-2} \circ (\ldots) \ldots)] \nu_{in} \tag{14}$$

is the probability of a photon being absorbed in the n-th pixel;

$$\omega_{i1} = 1 - \widetilde{\nu}_{in} = \rho_i \tag{15}$$

is the probability of a photon passing through the i-th layer of the object.

Thus if we consider the event that N_{i1} of N photons irradiating the i-th layer of the object are absorbed in the first pixel, N_{i2} are absorbed in the second pixel, ..., N_{in} photons are absorbed in the n-th pixel and $N_{i,n+1} = N - N_{i1} - \ldots - N_{in}$ reach the detector, then the probability of this event is equal to

$$P_i(N_{i1}, \ldots, N_{in}, N_{i,n+1}) = \frac{N!}{N_{i1}! \ldots N_{in}!\, N_{i,n+1}!} \widetilde{\nu}_{i1}^{N_{i1}} \ldots \widetilde{\nu}_{in}^{N_{in}} \rho_i^{N_{i,n+1}}. \tag{16}$$

This function is a polynomial distribution of N particles among the $n+1$ pixel (Feller, 1957).

Since the photon absorption in one layer does not depend on this process in another layer, the distribution function of photons absorbed in the pixels of the object and those reached the detector can be represented as

$$P(\mathcal{N}) = \prod_{i=1}^{r} \frac{N!}{N_{i1}!\, N_{i2}! \ldots N_{in}!\, N_{i,n+1}!} \widetilde{\nu}_{i1}^{N_{i1}} \ldots \widetilde{\nu}_{in}^{N_{in}} \rho_i^{N_{i,n+1}}, \tag{17}$$

where $\mathcal{N}$ is a matrix with the elements N_{ij}, $i \in \overline{1,r}$; $j \in \overline{1,n+1}$.

The distribution function $P(\mathcal{N})$ (17) is known to have a "sharp" maximum for some realization $\mathcal{N} = \mathcal{N}^0$ of the matrix. This means that the probability of realizing the distribution $\mathcal{N}^0$ is maximum, and all the others distributions have a significantly less probability. Therefore it is logical to expect the distribution of photons among the pixels $\mathcal{N}^0 = [N_{ij}^0;\ i \in \overline{1,r};\ j \in \overline{1,n+1}]$ realized in given object to correspond to the maximum of probability (17), similarly to that for the distribution in the object characterized by the apriori probabilities of absorption $\nu = [\nu_{ij};\ i \in \overline{1,r};\ j \in \overline{1,n+1}]$.

Finding the logarithm of (17) and using the Stirling approximation of factorials in (17), we obtain

$$\ln P(\mathcal{N}) = C + \widetilde{H}(\mathcal{N}), \tag{18}$$

where

$$\widetilde{H}(\mathcal{N}) = -\sum_{i=1}^{r}\left(\sum_{j=1}^{n} N_{ij}\ln\frac{N_{ij}}{\widetilde{\nu}_{ij}e} + N_{i,n+1}\ln\frac{N_{i,n+1}}{\rho_i e}\right) \tag{19}$$

is the entropy of distribution $\mathcal{N}$.

Note that the last term in (19) is related to the number of photons reached the detector. Value $\mathcal{D}(i)$ (2.13) characterizing the darkening of the photographic plate in the i-th pixel is proportional to $N_{i,n+1}$ in the first approximation. Therefore

$$N_{i,n+1} = \frac{\mathcal{D}(i)}{h}, \tag{20}$$

where h is a proportionality coefficient depending on the properties of the photographic plate material.

Since $\mathcal{D}(i)$ is a measurable value, $N_{i,n+1}$ can be determined from (20), and the distribution entropy of photons absorbed in the object can be considered, instead of (19), namely

$$H_{\text{abs}}(\mathcal{N}) = -\sum_{i=1}^{r}\sum_{j=1}^{n} N_{ij}\ln\frac{N_{ij}}{\widetilde{\nu}_{ij}e}. \tag{21}$$

This is the generalized information entropy of Boltzmann (see Chapter 2). It is convenient to use the relative units below:

$$p_{ij} = \frac{N_{ij}}{N}, \tag{22}$$

which have the meaning (for $N \to \infty$, see (2)) of probabilities of absorbing photons in the pixel (i, j). Substituting this expression into (21) and taking into account that

$$\sum_{i=1}^{r}\sum_{j=1}^{n} p_{ij} = 1 - \sum_{i=1}^{r}\rho_i = 1 - \sum_{i=1}^{r}\frac{N_{i,n+1}}{N} = 1 - \frac{1}{hN}\sum_{i=1}^{r}\mathcal{D}(i), \tag{23}$$

we obtain

$$H_{\text{abs}}(\mathcal{P}) = C_0 + H(\mathcal{P}), \tag{24}$$

where

$$C_0 = N(\ln N + 1)\left(1 - \frac{1}{hN}\sum_{i=1}^{r}\mathcal{D}(i)\right); \tag{25}$$

$$H(\mathcal{P}) = -\sum_{i=1}^{r}\sum_{j=1}^{n} p_{ij}\ln\frac{p_{ij}}{\widetilde{\nu}_{ij}}. \tag{26}$$

Taking into account the connection between the attenuation coefficient μ_{ij} and absorption probability (2.11), the equalities (2.14) can be represented as

$$k\Delta x \sum_{j=1}^{n} p_{ij} = \tilde{q}_i.$$

Thus the most probable distribution of photons absorbed in the object is

$$\mathcal{P}^0 = \arg\max\left(H(\mathcal{P}) \mid \mathcal{P} \in \tilde{Q}\right), \tag{27}$$

where $H(\mathcal{P})$ is the entropy determined by the equality (26);

$$\tilde{Q} = Q \cap P^+;$$

$$Q = \left\{p_{ij} : \sum_{j=1}^{n} p_{ij} = q_i, \quad i \in \overline{1,r}\right\}; \tag{28}$$

$$P^+ = \left\{p_{ij} : 0 \le p_{ij} \le 1, \quad i \in \overline{1,r}; \quad j \in \overline{1,n}\right\};$$

$$\mathcal{P} = \left[p_{ij}; \quad i \in \overline{1,r}; \quad j \in \overline{1,n}\right], \qquad q_i = \tilde{q}_i / k\Delta x.$$

It is convenient to represent the variables in problem (27)–(28) in a single-index form for investigation and solution. Let us introduce the following notation:

$$x = \{x_1, \ldots, x_m\};$$

$$a = \{a_1, \ldots, a_m\};$$

$$x_s \to p_{ij}, \quad s \to (i,j), \quad s \in \overline{1,m}; \quad m = rn; \tag{29}$$

$$a_s \to \tilde{\nu}_{ij};$$

$$T = \left[t_{is}; \quad i \in \overline{1,r}; \quad s \in \overline{1,m}\right] =$$

$$= \left[\begin{array}{c|c|c|c} \overbrace{1\ldots1}^{n} & \overbrace{\;0\;}^{n} & \ldots & \overbrace{\;0\;}^{n} \\ & 1\ldots1 & \ldots & \\ 0 & 0 & \ldots & 1\ldots1 \end{array}\right].$$

The problem of recovering an image by its projections obtained in the "emitter-detector" system shown in fig. 7.6 can be then represented as

$$H(x) = -\sum_{s=1}^{m} x_s \ln \frac{x_s}{a_s} \to \max, \tag{30}$$

$$0 \leq x_s \leq 1, \quad s \in \overline{1,m}$$

$$Tx = q, \tag{31}$$

$$v = q + \xi. \tag{32}$$

According to (29), the elements t_{is} of matrix T are 1 or 0.

However, if the "emitter-detector" system has a structure different from those shown in fig. 7.6, the matrix T have arbitrary elements. We shall suppose them to be normed, i.e.

$$0 \leq t_{is} \leq 1, \quad i \in \overline{1,r}; \quad s \in \overline{1,m}. \tag{33}$$

In problem (30)–(32), the vector $\xi = \{\xi_1, \ldots, x_r\}$ simulates the measurement noises with the characteristics (3.2), while the vector $\nu = \{\nu_1, \ldots, \nu_r\}$ simulates the mean measured intensities.

Sometimes, instead of (31) and (32), it is more convenient to consider a system of the following form

$$\sum_{s=1}^{m} t_{is} x_s = v_i - \xi_i, \quad i \in \overline{1,r}. \tag{34}$$

Since we cannot measure the noise realizations in every experiment, it is difficult to take into account their influence on the estimation of vector x for a model of the "emitter-detector" system in the form (34). On the other hand, if the mean square error σ (3.2) of measuring system is known, and the measurement noises are supposed to be normal stochastic values, then (34) can be replaced by the following system of inequalities:

$$v_i - 3\sigma \leq \sum_{s=1}^{m} t_{is} x_s \leq v_i + 3\sigma, \quad i \in \overline{1,r}. \tag{35}$$

Thus, along with the problem (30)–(32), which is used, as a rule, for $\xi = 0$, the problem of recovering an image by its projections can be represented in the form

$$H(x) = -\sum_{s=1}^{m} x_s \ln \frac{x_s}{a_s} \to \max, \tag{36}$$

$$0 \leq x_s \leq 1, \quad s \in \overline{1,m}$$

$$Wx \leq d, \tag{37}$$

where

$$W = \begin{pmatrix} T \\ -T \end{pmatrix}, \qquad d = \begin{pmatrix} v + 3\sigma \\ -v + 3\sigma \end{pmatrix}. \tag{38}$$

We suppose the sets

$$\mathcal{D} = \{x : Tx = q\} \tag{39}$$

and

$$\widetilde{\mathcal{D}} = \{x : Wx \leq d\} \tag{40}$$

to be such that

$$\mathcal{D} \cap X \neq \varnothing \quad \text{and} \quad \widetilde{\mathcal{D}} \cap X \neq \varnothing, \tag{41}$$

where

$$X = \{x : 0 \leq x_s \leq 1, \quad s \in \overline{1, m}\}. \tag{42}$$

This assumption is based on the fact that the parameters of $\mathcal{D}$ and $\widetilde{\mathcal{D}}$ (the matrices T and W and the vectors q and d) result from the normalization of real parameters of "emitter-detector" system and measurements.

The properties of problems (30)–(32) and (36)–(38) are the following.

(1) *High dimension.* In most practical problems,

$$m \simeq 10^5 \div 10^6; \qquad t \simeq 10^3 \div 10^4.$$

(2) *Sparsity of T.* Most elements of T are zero. It is seen, for example, for T (29) describing the "emitter-detector" system with a parallel projection (fig. 7.6).

(3) *Structure features of T.* Though the matrix T (29) is structured, this cannot be used, for example, to decompose the problem.

(4) *Time constraint.* Calculation time is limited not only for financial reasons, but also because the images should be obtained in almost real time. Thus in medical research performed by means of X-ray tomographs, the irradiation time is strictly limited by the admitted radiation dose.

(5) *Computational resource constraint.* Since the problem has a large dimension, the computational resources are of significant importance, namely the memory, speed, the number of operations and their types, and so on.

A family of RA-methods (row-action methods) are used to solve the problems of image recovering (Herman, Lent, Rowland, 1973; Minerbo, 1979; Censor, 1981; Herman, 1982; Censor, 1983; Censor, 1986).

These methods use various procedures of successive projection on hyperplanes given by the equalities in (39) or subsets given the equalities in (40). The most general approach to realizing this idea appears to be formulated by L. M. Bergmann in 1967.

7.5 Multiplicanive algorithm for image recovering

Consider a problem of entropy estimation (4.30)–(4.32), supposing that $\xi = 0$:

$$H(x) = -\sum_{s=1}^{m} x_s \ln \frac{x_s}{a_s} \to \max; \tag{1}$$

$$\sum_{s=1}^{m} t_{is} x_s = q_i, \qquad i \in \overline{1,r}. \tag{2}$$

The parameters of constraint system (2) are such that (4.41) is satisfied.

Using the Lagrange method (see Chapter 2), we can obtain the solution of problem (1)–(2) in the following form:

$$x_s = \tilde{a}_s \prod_{j=1}^{r} z_j^{t_{js}}, \qquad s \in \overline{1,m}, \tag{3}$$

where $z_j = e^{-\lambda_j} \geq 0$, $j \in \overline{1,r}$; $\tilde{a}_s = a_s e^{-1}$, $s \in \overline{1,m}$; and $\lambda_1, \ldots, \lambda_r$ are the Lagrange multipliers.

The variables $z_1, \ldots, z_r$ are determined by the system of equalities (2) in which x_s is substituted from (3):

$$\vartheta_i(z) = \frac{q_i}{\sum_{s=1}^{m} t_{is} \tilde{a}_s \prod_{j=1}^{r} z_j^{t_{js}}} = 1, \qquad i \in \overline{1,r}. \tag{4}$$

To determine a non-negative solution of this system, let us use a multiplicative algorithm (see Chapter 4) with a coordinate-wise scheme of calculations. Before this, let us introduce the following notation

$$\varepsilon_i(z^n) = |1 - \vartheta_i(z^n)|; \tag{5}$$

$$i_+ = \max_i \varepsilon_i(z^n),$$

where z^n is a fixed point.

Consider an algorithm where a multiplicative step is made along one coordinate having the number i_+:

$$z_{i_+}^{n+1} = z_{i_+}^n \vartheta_{i_+}^\gamma (z^n); \tag{6}$$

$$z_i^{n+1} = z_i^n; \quad i \in \overline{1,r}, \quad i \neq i_+ .$$

Note that the whole matrix T is used in algorithm (6) in the constraint system (2) (see (4)). Let us proceed to the equalities (3) and represent them in the following form:

$$x_s^n = \widetilde{a}_s \prod_{j=1}^{r} (z_j^n)^{t_{js}}; \tag{7}$$

$$x_s^{n+1} = \widetilde{a}_s \prod_{j=1}^{r} (z_j^{n+1})^{t_{js}}. \tag{8}$$

Let us substitute z_j^{n+1} in (8) by their expressions from (6):

$$x_s^{n+1} = \widetilde{a}_s \prod_{\substack{j=1 \\ j \neq i_+}}^{r} (z_j^n)^{t_{js}} \left[z_{i_+}^n \vartheta_{i_+}^\gamma (z^n) \right]^{t_{i_+ s}} =$$

$$= \widetilde{a}_s \prod_{j=1}^{r} (z_j^n)^{t_{js}} \left[\vartheta_{i_+}^\gamma (z^n) \right]^{t_{i_+ s}}. \tag{9}$$

Now using (7) and (4), we obtain

$$x_s^{n+1} = x_s^n \left[\frac{q_{i_+}}{\sum_{s=1}^{m} t_{i_+ s} x_s^n} \right]^{\gamma t_{i_+ s}}, \quad s \in \overline{1,m}. \tag{10}$$

The algorithm obtained coincides with RA-method modification, namely with MART algorithm (see, for example, Herman, 1982). Only one row of matrix T is used in algorithm (10) for the next iteration with respect to prime variables in problem (1)–(2), with its number chosen according to (5) at each step. It should be noted that in the MART algorithm this rule is different and not connected with the estimate of the iterative process state ($i_+ = n \pmod{r} + 1$). However, the rule (5) can also be used in MART (Censor, 1981).

Hence it follows that MART is a modification of multiplicative algorithms introduced in Chapter 4 with a specific coordinate-wise scheme of calculations. On the other hand, every iteration in MART is the construction of entropy projection onto the hyperplane with the number i_+.

Understanding this connection, we obtain a regular method for synthesizing multiplicative algorithms which use iterations with respect to the prime variables in (1)–(2).

An entropy projection onto the hyperplane characterized by the maximum deviation (5) on the previous iteration is built on every iteration in algorithm (10).

Let us introduce

$$i_- = \min_i \varepsilon_i(z^n) \tag{11}$$

and consider the algorithm

$$\begin{gathered} z_{i_+}^{n+1} = z_{i_+}^n \vartheta_{i_+}^\gamma(z^n), \\ z_{i_-}^{n+1} = z_{i_-}^n \vartheta_{i_-}^\gamma(z^n), \\ z_i^{n+1} = z_i^n; \quad i \in \overline{1,r}, \quad i \neq i_+, \quad i \neq i_-. \end{gathered} \tag{12}$$

Similar to (7)–(9), we obtain

$$x_s^{n+1} = x_s^n \left[\frac{q_{i_+}}{\sum_{s=1}^m t_{i_+ s} x_s^n} \right]^{\gamma t_{i_+ s}} \left[\frac{q_{i_-}}{\sum_{s=1}^m t_{i_- s} x_s^n} \right]^{\gamma t_{i_- s}}, \quad s \in \overline{1,m}, \tag{13}$$

where i_+ and i_- are determined by the equalities (5) and (11), respectively.

Two rows of T are already used in the algorithm (12); however, the entropy projection onto the intersection of two hyperplanes characterized by the maximum and minimum deviations is potentially the algorithm accelerator.

Consider the problem (4.36)–(4.38), supposing that $\xi \neq 0$

$$H(x) = -\sum_{s=1}^m x_s \ln \frac{x_s}{a_s} \to \max. \tag{14}$$

$$\sum_{s=1}^m w_{is} x_s \leq u_i + \Delta_i = d_i, \qquad i \in \overline{1,2r}, \tag{15}$$

where

$$u_i = \begin{cases} v_i & (i \in \overline{1,r}); \\ -v_i & (i \in \overline{r+1,2r}); \end{cases} \tag{16}$$

$$\Delta_i = 3\sigma.$$

We assume the parameters of system of inequalities (15) to satisfy the condition (4.41).

Let us use the optimality conditions in terms of a saddle point of the Lagrange function:

$$L(x,\lambda) = H(x) + \sum_{i=1}^{2r} \lambda_i \left(d_i - \sum_{s=1}^{m} w_{is} x_{is} \right).$$

In problem (14)–(16), the following conditions are satisfied at the maximum point x^*:

$$\nabla_{x_s} L(x^*, \lambda^*) = 0, \qquad s \in \overline{1,m}; \tag{17}$$

$$\nabla_{\lambda_i} L(x^*, \lambda^*) \geq 0, \qquad i \in \overline{1,2r}; \tag{18}$$

$$\lambda_i^* \nabla_{\lambda_i} L(x^*, \lambda^*) = 0, \quad \lambda_i^* \geq 0, \quad i \in \overline{1,2r}. \tag{19}$$

From these conditions we obtain the following system of equations

$$x_s = \widetilde{a}_s \exp\left(-\sum_{j=1}^{2r} \lambda_j w_{js} \right), \qquad s \in \overline{1,m}; \tag{20}$$

$$\lambda_i \left(d_i - \sum_{s=1}^{m} w_{is} \widetilde{a}_s \exp\left(-\sum_{j=1}^{2r} \lambda_j w_{js} \right) \right) = 0, \quad \lambda_i \geq 0, \quad i \in \overline{1,2r}. \tag{21}$$

To solve this system, let us use a multiplicative procedure with the coordinate-wise scheme of calculations.

Let us introduce the notation

$$\delta_i(x^n, \lambda^n) = \left| \lambda_i^n \left(d_i - \sum_{s=1}^{m} w_{is} x_s^n \right) \right|. \tag{22}$$

It is an absolute value of deviation of the i-th equation in (21) at the point (x^n, λ^n).

Let

$$i_+ = \max_i \delta_i(x^n, \lambda^n). \tag{23}$$

Consider an algorithm where a multiplicative step is realized with respect to the coordinate having the number i_+:

$$\lambda_{i_+}^{n+1} = \lambda_{i_+}^{n}\left(1 + \gamma P_{i_+}(\lambda^n)\right), \tag{24}$$

$$\lambda_i^{n+1} = \lambda_i^n; \quad i \in \overline{1,2r}; \quad i \neq i_+; \quad \lambda_i^0 > 0; \quad i \in \overline{1,r},$$

where $P_i(\lambda^n) = -\sum_{s=1}^{m} w_{i_+s}\tilde{a}_s \exp\left(-\sum_{j=1}^{2r} \lambda_j^n w_{js}\right) + d_i;$ (25)

γ is a step coefficient;

i_+ is determined by the rule (23).

The whole matrix W is used in this algorithm and the iterations are realized with respect to the dual variables.

Now let us proceed to the system of equations (20) and represent it in the form

$$x_s^n = \tilde{a}_s \exp\left(-\sum_{j=1}^{2r} \lambda_j^n w_{js}\right), \quad s \in \overline{1,m}; \tag{26}$$

$$x_s^{n+1} = \tilde{a}_s \exp\left(-\sum_{\substack{j=1\\ j\neq i_+}}^{2r} \lambda_j^n w_{js} - \lambda_{i_+}^{n+1} w_{i_+s}\right), \quad s \in \overline{1,m}. \tag{27}$$

Let us substitute an expression for $\lambda_{i_+}^{n+1}$ from (24) into (27):

$$x_s^{n+1} = \tilde{a}_s \exp\left[-\sum_{j=1}^{2r} \lambda_j^n w_{js} - \gamma\lambda_{i_+}^n w_{i_+s} \times \right.$$

$$\left. \times\left(d_{i_+} - \sum_{s=1}^{m} w_{i_+s}\tilde{a}_s \exp\left(-\sum_{j=1}^{2r} \lambda_j^n w_{js}\right)\right)\right].$$

Taking (26) into account, we obtain:

$$\begin{aligned} x_s^{n+1} &= x_s^n \exp\left(-\varphi(\lambda_{i_+}^n, x^n, \gamma) w_{i_+s}\right), \quad s \in \overline{1,m}; \\ \lambda_{i_+}^{n+1} &= \lambda_{i_+}^n + \varphi(\lambda_{i_+}^n, x^n, \gamma); \end{aligned} \tag{28}$$

$$\varphi(\lambda_i, x, \gamma) = \gamma\lambda_i\left(d_i - \sum_{p=1}^{m} w_{ip}x_p\right); \tag{29}$$

$$x_s^0 > 0, \quad s \in \overline{1,m}; \quad \lambda_i^0 > 0, \quad i \in \overline{1,2r},$$

where i_+ is determined by the rule (23).

Thus the iterations are performed simultaneously with respect to the prime (28) and dual (29) variables in algorithm (28)–(29); with only one row of the matrix W being used in each of these blocks.

The algorithm (28)–(29), neglecting the notation, coincides with the well-known algorithm of entropy maximization under linear inequality constraints based on the Bergmann successive projection method (Censor et al., 1986).

Recall that the same was earlier the case for the algorithm (10).

This indicate a connection between the multiplicative scheme of iterative processes (more precisely, its coordinate-wise realization) and the Bergmann successive projection method when using them for solving entropy maximization problems on convex sets (polyhedrons).

The multiplicative scheme, however, seems to be methodically richer, since it provides for regular methods of improving the computational properties of algorithms.

Consider one of such modifications of the algorithm (28)–(29).

In choosing the appropriate equation from the system (19), the cyclic recurrence is realized in (28)–(29) by the rule (23). The computational properties of algorithm can also be improved by changing the rule of type (11)

$$i_- = \min_i \delta_i(x^n, \lambda^n). \tag{30}$$

To solve the system (19), consider the following algorithm

$$\begin{aligned} \lambda_{i_+}^{n+1} &= \lambda_{i_+}^n \left(1 + \gamma P_{i_+}(\lambda^n)\right); \\ \lambda_{i_-}^{n+1} &= \lambda_{i_-}^n \left(1 + \gamma P_{i_-}(\lambda^n)\right); \\ \lambda_i^{n+1} &= \lambda_i^n; \quad i \in \overline{1,2r}; \quad i \neq i_+; \quad i \neq i_-, \end{aligned} \tag{31}$$

where $P_i(\lambda)$ is determined by the equality (25).

From the system (20) we have

$$x_s^n = \tilde{a} \exp\left(-\sum_{j=1}^{2r} w_{js} \lambda_j^n\right); \tag{32}$$

$$x_s^{n+1} = \widetilde{a}_s \exp\left(- \sum_{\substack{j=1 \\ j \neq i_+; j \neq i_-}}^{2r} w_{js}\lambda_j^n - w_{i_+ s}\lambda_{i_+}^{n+1} - w_{i_- s}\lambda_{i_-}^{n+1} \right); \quad s \in \overline{1,m}. \quad (33)$$

Substituting λ_i^{n+1} from (31) into these expressions, we obtain:

$$x_s^{n+1} = x_s^n \exp\left(-\varphi(\lambda_{i_+}^n, x^n, \gamma)w_{i_+ s} - \varphi(\lambda_{i_-}^n, x^n, \gamma)w_{i_- s} \right), \quad s \in \overline{1,m}; \quad (34)$$

$$\begin{aligned} \lambda_{i_+}^{n+1} &= \lambda_{i_+}^n + \varphi(\lambda_{i_+}^n, x^n, \gamma); \\ \lambda_{i_-}^{n+1} &= \lambda_{i_-}^n + \varphi(\lambda_{i_-}^n, x^n, \gamma); \end{aligned} \quad (35)$$

$$\varphi(\lambda_i, x, \gamma) = \gamma\lambda_i \left(d_i - \sum_{\rho=1}^{m} w_{i\rho}x_\rho \right); \quad (36)$$

$$\lambda_i^0 > 0; \quad i \in \overline{1,2r}; \quad x_s^0 > 0; \quad s \in \overline{1,m}.$$

Two rows of the matrix W with the numbers i_+ (23) and i_- (30) are used at every step of this algorithm thus requiring a double capacity of RAM. But, on the other hand, if the maximum and minimum deviations $\delta_i(x^n, \lambda^n)$ (22) differ significantly ("ravineous effect"), the algorithm (34)–(35) is potentially superior to the algorithm (28)–(38) with respect to convergence rate.

Appendix

```
PROGRAM MSSN;
uses Crt;
const NumPrint=5;
      NameDir = 'c:\pop\dat\';
var z,ef,m1,ce,aber                   :array[1..30] of real;
    a,a1,a2,z2,n1,z1,nn,absnn         :array[1..60] of real;
    t,b1,b2,pen                       :array[1..30,1..60] of real;
    j,k,m,r,n,s,s1,sr,InitCond,
    NumGraf,Jcond,TypeModen,
    ClastAlgn,RegWritn,
    ClassModen,k1                     :integer;
    CountPrint                        :byte;
    d,m2,norm,gamma,delt,g,v,p,
    iniz,abermax,na                   :real;
    BoundPrint,StepPrint              :array[1..NumPrint] of word;
    Ent                               :array[1..3] of string[50];
    Constr                            :array[1..2] of string[50];
    NameFile2,NameFile3,
    NameMode1,NameFile4,
    NameMode3,NameMode                :string;
    fil,fot,rst,out                   :text;
    Ch,ClassMode,TypeMode,
    ClastAlg,RegWrite                 :char;
```

```
    procedure TEST;        { formulation of test-problem }
        var k,n:integer;
           begin
             for k:=1 to r do
              begin
                ce[k]:=0.;na:=0.;
                for n:=1 to m do
                 begin
                   a[n]:=abs(sin(2*3.14/sqrt(m)*n));
                   na:=na+a[n];
                   a2[n]:=1+a[n];
                   if n < r+1 then
                     begin
                       if n=k then b1[n,n]:=a2[n] else
                         b1[k,n]:=a2[n]*exp(-v*abs(k-n));
                         ce[k]:=ce[k]+b1[k,n];
                     end;
                   if n > r then
                     begin
                       b2[k,n]:=sqrt(k)*ln(a2[n]);
                       ce[k]:=ce[k]+b2[k,n];
                     end;
                 end;
              end;
            for n:=1 to m do
              begin
                a[n]:=a[n]/na;
                a1[n]:=g*a[n];
              end;
  write('          m=',m:3,' r=',r:3,' d=',d:2:1);
  write('   capacity of subsets=',g:4:1);writeln;
  writeln;
  if (InitCond=0) or ((InitCond=1) and (Jcond=1)) then
             begin
```

```
write(fot,'       m=',m:3,'  r=',r:3,'  d=',d:2:1);
write(fot,'   capacity of subsets=',g:4:1);
writeln(fot);writeln(fot);
write(rst,'       m='m:3,'   r=',r:3,'  d=',d:2:1);
write(rst,'   capacity of subsets=',g:4:1);
writeln(rst);writeln(rst);
          end;
    if RegWritn=1 then
       begin
writeln('                              matrix T');
if (InitCond=0) or ((InitCond=1) and (Jcond=1)) then
          begin
writeln(fot,'                             matrix T');
writeln(rst,'                             matrix T');
writeln(fot);writeln(rst);
          end;
writeln;
       end;
          for k:=1 to r do
           begin
             t[k,1]:=(d/g)*b1[k,1]/ce[k];
             for n:=2 to m do
               if n<=r then t[k,n]:=b1[k,n]/ce[k] else
               t[k,n]:=b2[k,n]/ce[k];
    if RegWritn=1 then
       begin
         for n:=1 to m do
          begin
write('     ',t[k,n]:4:3);
if (InitCond=0) or ((InitCond=1) and (Jcond=1)) then
             begin
write(fot,'   ',t[k,n]:4:3);
write(rst,'   ',t[k,n]:4:3);
             end;
```

```
          end;
    if (InitCond=0) or ((InitCond=1) and (Jcond=1)) then
               begin
    writeln(fot);writeln(rst);
               end;
    writeln;
          end;
              end;
    writeln;
    if (InitCond=0) or ((InitCond=1) and (Jcond=1)) then
               begin
    writeln(fot);writeln(rst);
               end;
        if RegWritn=1 then
          begin
    writeln('          prior probabilities A');
    if (InitCond=0) or ((InitCond=1) and (Jcond=1)) then
               begin
    writeln(fot,'      prior probabilities A');
    writeln(rst,'      prior probabilities A');
    writeln(fot);writeln(rst);
               end;
    writeln;
            for n:=1 to m do
             begin
    write('      ',a[n]:4:3);
    if (InitCond=0) or ((InitCond=1) and (Jcond=1)) then
               begin
    write(fot,     ',a[n]:4:3);
    write(rst,'    ',a[n]:4:3);
               end;
             end;
    if (InitCond=0) or ((InitCond=1) and (Jcond=1)) then
               begin
```

```
writeln(fot);writeln(rst);writeln(fot);writeln(rst);
              end;
       end;
writeln;writeln;
         end;

  procedure INPiniCOND;  {input of initial points }
      var k:integer;
         begin
           s:=0;s1:=0; CountPrint:=1;
    if InitCond=0 then  { hand input init.points}
           begin
          for k:=1 to r do
             begin
                readln(fil,z[k]); z1[k]:=z[k];
             end;
           end;
    if InitCond=1 then  {automatic input init.points}
           begin
         if Jcond<NumGraf then
                for k:=1 to r do z[k]:=iniz*pen[Jcond,k];
         if Jcond=NumGraf then
                for k:=1 to r do z[k]:=iniz;
         for k:=1 to r do z1[k]:=z[k];
           end;
         end;
  procedure MODEL;      {macrostate for MSS }
         begin
     if ClassModen=1 then nn[n]:=a1[n]/z2[n]
             else if ClassModen=3 then
             nn[n]:=a1[n]/(a[n]+(1-a[n])*z2[2])
             else nn[n]:=a1[n]/(z2[n]-a[n]);
     if ClassModen=2 then absnn[n]:=(a[n]/(1-a[n]))*g
             else absnn[n]:=a1[n];
```

```
        end;
    { MAIN PROGRAM }
begin
      assign ( fil,NameDir+'dat.dat');
      reset ( fil );
 readln(fil);
 readln( fil ,r,m,d,gamma,delt,g,Ch,
 ClassMode,TypeMode,ClassAlg,RegWrite);
 readln(fil);
 readln(fil,InitCond,iniz,p,v);
   for j:=1 to NumPrint do
 readln(fil,BoundPrint[j],StepPrint[j]);
   for k:=1 to 3 do
 readln(fil,Ent[k]);
   for k:=1 to 2 do
 readln(fil,Constr[k]);
   NumGraf:=(1-InitCond)+InitCond*(r+1);
   for j:=1 to r do
     begin
       for k:=1 to r do if j<=k then pen[j,k]:=1/p
                                else pen[j,k]:=1;
     end;

     case ClassMode of
      'B':ClassModen:=1;
      'E':ClassModen:=2;
      'F':ClassModen:=3;
     end;
     case TypeMode of
       'C':TypeModen:=1;
       'U':TypeModen:=2;
     end;
     case ClastAlg of
       'P':ClastAlgn:=1;
```

```
        'K':ClastAlgn:=2;
      end;
      case RegWrite of
        'S':RegWritn:=1;
        'L':RegWritn:=2;
      end;
      NameMode:='M'+ClassMode+TypeMode+ClastAlg+'  ';
      NameMode:=NameMode+Ent[ClassModen]+'  ';
      NameMode3:='          '+Constr[TypeModen];
      NameMode1:='M'+ClassMode+TypeMode+ClastAlg;
      NameFile2:=NameDir+'fot'+NameMode1+'.dat';
      NameFile3:=NameDir+'rst'+NameMode1+'.dat';
      NameFile4:=NameDir+'out'+NameMode1+'.dat';
      assign(fot,NameFile2);rewrite(fot);
      assign(rst,NameFile3);rewrite(rst);
      assign(out,NameFile4);rewrite(out);
 writeln(fot,'      ',NameMode);
 writeln(fot,'      ',NameMode3);
 writeln(fot);writeln(fot);
 writeln(rst,'      ',NameMode);
 writeln(rst,'      ',NameMode3);
 writeln(rst);writeln(rst);
 writeln('      'NameMode);
 writeln('      'NameMode3);
 writeln;writeln;
{MAIN CYCLE of INITIAL CONDITIONS CHOICE (Jcond)}
   for Jcond:=1 to NumGraf do
     begin
       INPiniCOND;

       TEST;
       if RegWritn=1 then
         begin
 writeln('            iteration');
```

```
writeln(fot,'            Iteration');
writeln;writeln(fot);
writeln('     ','gamma=',gamma:4:2,'   ','delt='
      ,delt:10:9);
writeln(fot,'     ','gamma=',gamma:4:2,'   ',
      'delt=',delt:10:9);
writeln;writeln(fot);
write(' s   ');
write(fot,'  s  ');
         for k:=1 to r do
           begin
             if TypeModen=1 then
                begin
write('      z[',k']    ');
write(fot,'      z[',k,']   ');
                end
                   else
                begin
write('      L[',k,']   ');
write(fot,'      L[',k,']   ');
                end;
           end;
write('    norm   ');write(fot,'      norm    ');
writeln;writeln(fot);
writeln;writeln(fot);
write('  0');write(fot,'  0');
         for k:=1 to r do
           begin
write(z1[k]:13:3);write(fot,z1[k]:13:3);
           end;
writeln;writeln(fot);
        end;
writeln;
writeln('       gamma=',gamma:4:2,'    delt='
```

```
          delt:10:9);
  writeln(rst);
  writeln(rst,'      gamma=',gamma:4:2,'   delt=',
          delt:10:9);
  writeln(rst);writeln;

      { MAIN CYCLE of ITERATION (begin)}

     repeat
      begin
        for n:=1 to m do
          begin
            n1[n]:=0
            for j:=1 to r do

              if TypeModen=1 then
                 n1[n]:=n1[n]+t[j,n]*ln(z[j])
                         else
                 n1[n]:=n1[n]+t[j,n]*z[j];
            z2[2]:=exp(n1[n]);

            MODEL;
          end;
        for k:=1 to r do
          begin
            ef[k]:=0;
            for n:=1 to m do
               ef[k]:=ef[k]+t[k,n]*nn[n];
          end;
        m2:=0;
        for k:=1 to r do   {parametr of STOP }
          begin
            if TypeModen=1 then
               m1[k]:=sqrt(ef[k]-1)
```

```
                  else
             m1[k]:=sqrt(z[k]*(ef[k]-1));
         m2:=m2+m1[k];
       end;
     norm:=sqrt(m2);

               { ALGORITHM }

     if (TypeModen=1) and (ClastAlgn=1) then
       for k:=1 to r do
           z[k]:=z[k]*exp(gamma*ln(ef[k]));
     if (TypeModen=1) and (ClastAlgn=2) then
       begin
         abermax:=0;
         for k:=1 to r do
           begin
             aber[k]:=abs(1-ef[k]);
             if aber[k]>abermax then
                begin
                   abermax:=aber[k];k1:=k;
                end
                        else abermax:=abermax;
           end;
         z[k1]:=z[k1]*exp(gamma*ln(ef[k1]));
         for n:=1 to m do
           begin
             n1[n]:=0;
             for j:=1 to r do
                 n1[n]:=n1[n]+t[j,n]*ln(z[j]);
             z2[2]:=exp(n1[n]);
             MODEL;
           end;
         for k:=1 to r do
           begin
```

```
          ef[k]:=0;
          for n:=1 to m do
               ef[k]:=ef[k]+t[k,n]*nn[n];
        end;
      for k:=1 to r do
        if k=k1 then z[k1]:=z[k1]
        else z[k]:=z[k]*exp(gamma*ln(ef[k]));
    end;
  if (TypeModen=2) and (ClastAlgn=1) then
    for k:=1 to r do
        z[k]:=z[k]*(1-gamma*(1-ef[k]));
  if (TypeModen=2) and (ClastAlgn=2) then
    begin
      abermax:=0;
      for k:=1 to r do
        begin
          abermax[k]:=abs(z[k]*(1-ef[k]));
          if aber[k]>abermax then
             begin
                abermax:=abermax[k];k1:=k;
             end
                    else abermax:=abermax;
        end;
      z[k1]:=z[k1]*(1-gamma*(1-ef[k1]));
      for n:=1 to m do
        begin
          n1[n]:=0;
          for j:=1 to r do
               n1[n]:=n1[n]+t[j,n]*z[j];
          z2[n]:=exp(n1[n]);
          MODEL;
        end;
      for k:=1 to r do
        begin
```

```
                  ef[k]:=0;
                  for n:=1 to m do
                        ef[k]:=ef[k]+t[k,n]*nn[n];
                end;
              for k:=1 to r do
                if k=k1 then z[k1]:=z[k1]
                 else z[k]:=z[k]*(1-gamma*(1-ef[k]));
            end;

                    {algorithm END }


          s:=s+1;
          if s > BoundPrint[CountPrint] then
             begin
               CountPrint:=CountPrint+1;
               if CountPrint > NumPrint then Halt(1);
             end;
          if ((s mod StepPrint[CountPrint]) = 0) and
             (RegWritn=1) then
                begin
  write( s:3);write(fot,s:3);
                  for k:=1 to r do
                    begin
  write(z[k]:13:3);write(fot,z[k]:13:3);
                    end;
  write(norm:13:3);write(fot,norm:13:3);
  writeln;writeln(fot);
                end;
        end;
      until (norm<delt);
      s1:=s;
{ main cycle of iteration (END)}

  writeln;writeln(fot);writeln(fot);
```

```
writeln('         result          ');
writeln;
write(rst,'          INITIAL POINTS          ');
write(rst           RESULT          ');
writeln(fot,'              RESULT           ');
writeln(rst);writeln(rst);
writeln(fot);writeln(fot);
 for k:=1 to r do
       begin
          if TypeModen=1 then
             begin
write(rst,'           z[',k,']=',z1[k]:5:1);
write('        z[',k,'] =',z[k]:18:8);
write(rst,'                 z[',z,'] =',z[k]:18:8);
write(fot,'         z[',k,'] =',z[k]:18:8);
             end
               else
             begin
write(rst,'           L[',k,']=', z1[k]:4:1);
write('        L[',k,'] =',z[k]:14:9);
write(rst,'                     L[',k,'] =',z[k]:14:9);
write(fot,'         L[',k,'] =',z[k]:14:9);
             end;
writeln;writeln(rst);writeln(fot);
       end;
writeln(rst);
   for k:=1 to r do
     begin
       if TypeModen=2 then
          begin
writeln('       z[',k,'] =',exp(z[k]):14:9);
write(rst,'                                  ');
writeln(rst,'       z[',k,'] =',exp(z[k]):14:9);
write(fot'                                   ');
```

```
writeln(fot,'        z[',k,'] =',exp(z[k]):14:9);
         end;
    end;
writeln;writeln;
writeln(rst);writeln(rst);
writeln(fot);writeln(fot);
write('        norm=',norm:9:8);
write('     s=',s:3);
write('           abs max');writeln;
write(rst,'        norm=',norm:9:8);
write(rst,'     s=',s:6);
write(rst,'                 absmax');writeln(rst);
write(fot,'        norm=',norm:9:8);
write(fot,'     s=',s:6);
write(fot,'                 absmax');writeln(fot);
writeln(rst);writeln(rst);
writeln;writeln(fot);writeln(fot);
    for n:=1 to m do
      begin
write('                             ');
write('nn[',n:2,'] =',nn[n]:10:4);
write('     ',absnn[n]:10:4);writeln;
write(rst,'                             ');
write(rst,'nn[',n:2,'] =',nn[n]:10:4);
write(rst,'     ',absnn[n]:10:4);writeln(rst);
write(fot,'                             ');
write(fot,'nn[',n:2,'] =',nn[n]:10:4);
write(fot,'     ',absnn[n]:10:4);writeln(fot);
writeln(out,nn[n]:10:4);
      end;
writeln(fot);writeln(fot);
writeln(rst);writeln(rst);
    if TypeModen=2 then
      begin
```

```
         for k:=1 to r do
           begin
 writeln('       f[',k,'] =',ef[k]:10:9);
 writeln(rst,'        f[',k,'] =',ef[k]:10:9);
 writeln(fot,'        f[',k,'] =',ef[k]:10:9);
           end;
       end;
     if InitCond=0 then
       begin
 writeln;write('Press any key');
         Ch:=ReadKey;
       end;
   end;  {MAIN CYCLE of INITIAL CONDITIONS CHOICE}
     close(fil);close(fot);close(rst);close(out);
end.
```

Bibliography

Alekseev V. M., Tikhomirov V. M., Fomin S. V., 1979. *Optimal Control* (in Russian), Nauka: Moscow.

Aliev A. S., Dubov Yu. A., Izmailov R. N., Popkov Yu. S., 1985. "Convergence multiplicative algorithm for solving of convex programming problem", *Dynamic of nonhomogeneous systems* (in Russian), Proc. VNIISI: Moscow, pp. 59–67.

Andersson A. E., Batten D. F., 1988. "Creative nodes, logistical networks and the future of the metropolis", *Transportation*, 14, pp. 281–293.

Andersson A. E., Batten D. F., Johanssen B., Nijkamp P., 1989. *Advances in Spatial Theory and Dynamics*, Elsevier: Amsterdam, New York, Oxford Tokyo.

Annual – 1971, 1971. *USSR Cement Industry* (in Russian), NII Cement: Moscow.

Annual –1976, 1976. *USSR Cement Industry* (in Russian), NII Cement: Moscow.

Annual – 1981, 1981. *USSR Cement Industry* (in Russian), NII Cement: Moscow.

Antipin A. S., 1989. "Continious and iterative processes with project-operator", *Proc. Problems of cibernetics* (in Russian), Moscow, pp. 5–43.

Arrow K. J., 1964. "Control in large organizations", *Management Science*, V. 10, No. 3.

Bakushinsky A. B., Goncharsky A. V., 1989. *Incorrect Problems (Numerical Methods and Applications)* (in Russian), Publishing house of MGU (Moscow State University).

Batten D. F., Roy J. R., 1982. "Entropy and economic modelling", *Environment and Planning A*, 14, pp. 1047–1061.

Batten D. F., 1983. *A Unifying Framework for Location-Production-Interaction Modelling*, Paper of Intern. Symp. on New Directions in Urban Modelling. – Waterloo, Canada, pp. 1–35.

Batty M., March L., 1976. "Dynamic Urban Models Based on Information-Minimizing", *Geographical Papers*, University of Reading, England, No. 48.

Bernusson J., Tittli A., 1982. *Interconnected Dynamical Systems: Stability, Decomposition and Decentralization.* Amsterdam, North-Holland Pbl. Comp.

Birkin M., Wilson A. G., 1986. "Industrial location models 1: a review and an integrating framework", *Environment and Planning A*, 18, pp. 175–205.

Boltzmann L., 1984. "On the link between the second beginning of mechanical calory theory and probability theory in theorems of thermal equilibrium" In: *L. Boltzmann. Selections". Series "Classics of Science"* (in Russian), Nauka: Moscow, pp. 190–236.

Brandeau M. L., Chiu S. S., 1989. *Location of Public and Private Facilities in a User-Optimizing Environment*, Stanford University.

Bregman L. M., 1967. "Proof of the convergence of Sheleikhovsky's method for a problem with transpotation constraints", *USSR Computational Mathematics and Mathematical Physics* 1, pp. 147–156.

Bregman L. M., 1967. "Relaxation method for locating a common point of convex sets and its application for solving problems of convex programming", *USSR Computational Mathematics and Mathematical Phisics* (in Russian) 7, pp. 200–217.

Censor Y., 1981. "Row-Action Methods for huge and sparse systems and their applications", *SIAM Review*, 23, No. 4, pp. 444–466.

Censor Y., 1983. "Finite Series-Expansion Reconstruction Methods", *Proceedings of the IEEE*, 71, No. 3, pp. 409–419.

Censor Y., De Pierro A. R., Elfving T., Herman G. T., Insem A. N., 1986. "On Iterative Methods for linearly Constrained Entropy Maximization", *Technical Report N MIPG 112 Medical Image Processing Group*, Dept. of Radiology, University of Pennsylvania.

Chernikov S. N., 1968. *Linear Inequalities* (in Russian), Nauka: Moscow.

Clarke M., 1981. "A note on the stability of equilibrium solutions of production – constrained spatial interaction models", *Environment and Planning A*, 13, pp. 601–604.

Clarke M., Wilson A. G., 1983. "The dynamics of urban spatial structure: progress and problems", *Journal of Regional Science*, 23, pp. 1–18.

Clarke M., 1985. "The role of attractiveness functions in the determination of equilibrium solutions to production – constrained spatial interaction models", *Environment and Planning A*, 17, pp. 175–183.

Collatz L., 1964. *Funktionalanalysis und Numerische Mathematik*, Berlin: Springer-Verlag.

Crouchley R., 1987. "An examination of the equivalence of three alternative mechanisms for estabilishing the equilibrium solutions of the production-constrained spatial interaction model", *Environment and Planning A*, 19, pp. 861–874.

De Groot S. R., Masur P., 1964. *Nonequilibrium thermodynamics* (in Russian), "Mir": Moscow.

Dubov Yu. A., Ikoeva N. V., Imelbayev Sh. S., Popkov Yu. S., Shmulyan B. L. et al., 1975. "Mathematical modelling of urban system development (survey)", *Avtomatika i telemekhanika* (in Russian), No. 11, pp. 93–128.

Dubov Yu. A., Ikoeva N. V., Imelbayev Sh. S., Popkov Yu. S., Shmulyan B. L. et al., 1976. "Optimal planning and problems of controlling the development of urban systems (survey and problems of investigation)", *Avtomatika i telemekhanika* (in Russian), No. 6, pp. 78–116.

Dubov Yu. A., Imelbayev Sh. S., Popkov Yu. S., 1983. "Multiplicative schemes for iterative optimisation algorithms", *Soviet Mathem. Dokl.*, 28, No. 2, pp. 524–526.

Dubov Yu. A., Imelbayev Sh. S., Popkov Yu. S., 1984. "Multiplicative algorithms in extremal problems", *Izvestiya of the USSR Academy of Sciences. Technical cybernetics* (in Russian), 6, pp. 129–137.

Eggermont P. P. B., 1988. "Multiplicative iterative algorithm for convex programming", *Cent. Math. and Computer Science*, No. AM-R88-08.

Erlander S., 1980. *Optimal Spatial Interaction and the Gravity Model*, Berlin: Springer-Verlag.

Feller W., 1964. *An Introduction to Probability Theory and Its Application* (in Russian), "Mir": Moscow.

Fotheringham A. S., 1985. "Spatial competition and agglomeration in urban modelling", *Environment and Planning A*, 17, pp. 213–230.

Gantmacher F. R., 1966. *Matrix theory* (in Russian), Nauka: Moscow, -576p.

Golts G. A., 1981. *Transport and settling* (in Russian), Nauka: Moscow.

Greenwood M. J., 1985. "Human migration: theory, models and empirical studies", *Journal of Regional Science*, 25, pp. 521–544.

Gull S. F., Skilling J., 1984. "Maximum entropy method in image processing", *IEE Proc.*, V. 131, Pt. F., No. 6, pp. 646–659.

Haken H., 1974. *Synergetics*, Springer-Verlag.

Haken H., ed., 1974. *Cooperative Effects, Progress in Synergetics*, North Holland, Amsterdam.

Handjan G. O., 1984. *On One Method for Determining Apriori Probabilities in Models of Urban Transportation Systems.* In: "Dynamics of non-homogeneous systems" (in Russian), VNIISI: Moscow, pp. 235–241.

Hansen S., 1974. *Entropy and Utility in Traffic Modelling*, Transportation and Traffic Theory, Sydney, ed. Buckely, DJ, AH and A. Reed, pp. 352–435.

Harris B., Wilson A. G., 1978. "Equilibrium values and dynamics of attractiveness terms in production-constrained spatial-interaction models", *Environment and Planning A*, 10, pp. 371–388.

Hartwick J. M., Spencer M., 1988. "An Oil-Exporting versus an Industrialized Region". In: "A. E. Andersson, D. F. Batten, B. Johansson, P. Nijkamp eds. *Advances in Spatial Theory and Dynamics*, Amsterdam, North-Holland.

Hatvany J. and others, 1981. "Results of a word survey of computeraided manufacturing", Proc. of the 8th Triennical World Congress IFAC, Kyoto, Japan.

Herman G. T., Lent A., Rowland S. W., 1973. "ART: Mathematics and Applications", *J. of Theoretical Biology*, 42, No. 1, pp. 1–32.

Herman G. T., 1980. *Image Reconstruction from Projectious: the Fundamentals of Computerised Tomography*, Academic Press, New York.

Herman G. T., 1982. "Mathematical optimization versus practical performance: a case study based on the maximum entropy criterion in image reconstruction", *Mathem. Progr. Study*, North-Holland, No. 20, pp. 96–112.

Horowits A. R., Horowits I., 1976. "The real and illusory virtues of entropy – based measure for business and economic analysis", *Decision Sciences*, 7, pp. 121–136.

Imelbayev SH. S., Shmulyan B. L., 1978. "Modelling of stochastic communication system". In: Wilson A.G. *Entropy methods for modelling complex systems* (in Russian), Nauka: Moscow, pp. 170–234.

Kittel Ch., 1977. *Statistical thermodynamics* (in Russian), "Mir": Moscow, -336p.

Kotelnikov A. P., 1984. "Analysis of methods for modelling products-exchange in construction material industry". In: *Dynamics of non-homogeneous systems* (in Russian), VNIISI: Moscow, pp. 210–219.

Kotelnikov A. P., 1985. "Invariance analysis of apriory information in modelling functional-spatial organisation of products-exchange system". In: *Dynamics of non-homogeneous systems* (in Russian), VNIISI: Moscow, pp. 111–117.

Krasnoselsky M. A., et al., 1969. *Approximate Solution of Nonlinear Operator Equations* (in Russian), Nauka: Moscow.

Krasnoselsky M. A., Zabreyko P. P., 1975. *Geometrical Methods of Non-Linear Analysis* (in Russian), Nauka: Moscow.

Kraus J.D., 1966. *Radio Astronomy*, Mc Graw-Hill, New York.

Landau L. D., Lifshitz Ye. M., 1964. *Statistical Phisics* (in Russian), Nauka: Moscow.

Leonardi G., 1978. "Optimal facility location by accessibility maximizing", *Enwironment and Planning A*, 10, pp. 1287–1305.

Leontovich A. M., 1975. "One problem concerning self-assembly of segments", *Information transmission problems*, 11, No. 4, pp. 97–105.

Leontovich M. A., 1935. "Main equation of kinetic theory of gases from the point of view of stochastic process theory", *Journal of experimental and theoretical physics* (in Russian), 5, issue 3, 4, pp. 211–231.

Lombardo S. T., Rabino G. A., 1986. "Calibration procedures and problems of stability in nonlinear dynamic spatial interaction modelling", *Environment and Planning A*, 18, pp. 341-350.

Lyusternic L. A., Sobolev V. I., 1965. *Elements of Functional Analysis* (in Russian), Nauka: Moscow.

Mesarovich M., Mako D., Takahara I., 1973. *Theory of Hierarchical Systems* (Russian Edition), "Mir": Moscow.

Milner B. Z., 1975. *Management of Production* (in Russian), Economica: Moscow.

Minerbo G., 1979. "MENT: A Maximum Entropy Algorithm for Reconstructing a Source from Projection Data", *Computer Graphics and Image Processing*, 10, pp. 48–68.

O'Kelly M. E., 1981. "Generalized information measure", *Environment and Planning A*, 13, pp. 681–689.

Ovsievich B. L., 1979. *Model of Forming of Management Structure* (in Russian), Nauka: Leningrad.

Paulov Jan., 1985. "Gravitacny model:analyticky nastroj structurneho vyskumu v geografii", *Geographica*, No. 25, pp. 79–99.

Persson H., 1986. "Algorithm 12: Solving the entropy maximization problem with equality and unequality constraints", *Environment and Planning A*, 18, pp. 1555–1700.

Pervozvansky A. A., Gaitsgori V. G., 1979. *Decomposition, Aggregation and Approximate Optimization* (in Russian), Nauka: Moscow.

Pittel B. G., 1967. "One simplest model of collective behaviour", *Problems of information* (in Russian), issue 3, III, pp. 38–52.

Polak E., 1974. *Computational Methods in Optimization (A Unified Approach)* (in Russian), "Mir": Moscow.

Poljak B. T., 1983. *Introduction to Optimization* (in Russian), Nauka: Moscow.

Popkov Yu. S., 1976. "Principles of Constructing Models of Urban Systems". In: *Proceedings of the ist Winter School for modelling cities* (in Russian), Moscow.

Popkov Yu. S., Kopeykin A. B., Shmulyan B. L., 1977. "Spatial model of an urban system". In: *Using applied system analysis for simulating and controlling urban development* (in Russian), Stroyizdat: Moscow.

Popkov Yu. S., 1977. "Informational analysis of structures with probabilistic hierarchy (of an urban system)". In: *Proceedings of papers on the VII-th All-Union Conference devoted to control problems* (in Russian), Minsk, pp. 115–121.

Popkov Yu. S., 1978. "Macrosystem theory and the book by A. G. Wilson". Preface to the translation of the book by A. G. Wilson, *Entropy methods for modelling complex systems* (in Russian), Nauka: Moscow, pp. 7–14.

Popkov Yu. S., 1979. *Macrosystem Approach in Modelling Spatial Organization of Cities and Urban Agglomerations* (in Russian), Preprint VNIISI: Moscow, 48pp.

Popkov Yu. S., Ryazantsev A. N., 1980. *Spatio-Functional Models of the Demographic Processes*, Moscow, USSR. N.Y.: United Nations Fund for population activity. -15p.

Popkov Yu. S., Shmulyan B. L., Gutnov A. E., Imelbayev Sh. S., Dubov Yu. A. et al., 1980. "A spatio-functional model of an urban system and some results of its experimental employment". In: *Modelling of citi control processes*, issue 2. Moscow: Executive Commitee of the Moscow town Soviet of People's Deputies. Major scientific Centre of Calculations (in Russian), pp. 3–15.

Popkov Yu. S., 1980. "Simulation and structural-property analysis of settlement systems", *Ekonomika i matematicheskie metody* (in Russian), XVI, issue 6, pp. 1138–1152.

Popkov Yu. S., 1980. "Simulation and analysis of structural properties of human settlement systems by means of entropy maximizing models", *Environment and Planning A*, 12, pp. 1165–1190.

Popkov Yu. S., Shmulyan B. L., 1981. "Methods of modelling and analysing spatio-functional models of urban systems" In: *"Dostizhenia i perspektivi". Moscow: Intern. Centre of scient. and techn. inf., System Analysis Com. of the Presidium of the USSR AC. of SC.* (in Russian), issue 18 (Towns and systems of settling), No. 3.

Popkov Yu. S., Shmulyian B. L., 1982. "Principles of formulation and analysis of urban models", *Sistemi Urbani*, Guida editori, 3, pp. 253–266.

Popkov Yu. S., 1983. "Mathematical models of macrosystem equilibrium states: principles of constructing, analysis of properties, algorithms of functioning". In: *Dynamika neodnorodnykh sistem.* Seminar proceedings (in Russian), VNIISI: Moscow, pp. 203–218.

Popkov Yu. S., Posokhin M. V., Gutnov A. E., Smulyian B. L., 1983. *System Analysis and Problems of Urban Development* (in Russian), Nauka: Moscow, -512pp.

Popkov Yu. S., 1984. *Probabilistic Hierarchical Structures (Principles of Modelling, Models for Ensembles of Structures)* (in Russian), Preprint VNIISI: Moscow, -54pp.

Popkov Yu. S., 1984. "A multiplicative algorithm for solving the problem of maximizing the generalised Fermi-entropy with linear constraints-equalities". In: *Dinamika neodnorodnykh sistem* (in Russian), VNIISI: Moscow, pp. 52–60.

Popkov Yu. S., 1984. *Probabilistic Hierarchical Structures (Models for Forming Structures)* (in Russian), Preprint of VNIISI: Moscow, -59pp.

Popkov Yu. S., 1984. "Mathematical models for stationary states of macrosystems", *Avtomatika i telemekhanika* (in Russian), No. 6, pp. 129–137.

Popkov Yu. S., 1984. "Macromanagement of spatial models of urban systems", *Environment and Planning A*, V. 16(12), pp. 1544–1546.

Popkov Yu. S., 1985. *Elements of Macrosystem Theory and Its Applications* (in Russian), Preprint of VNIISI: Moscow, 79pp.

Popkov Yu. S., 1986. "Mathematical models of macrosystem stationary states: principles of formulation, analysis of properties and algorithms", *Large scale systems*, 10, pp. 1–20.

Popkov Yu. S., 1986. "Simulation and analysis of human settlement patterns". In: *Systems research II. Methodological Problems*, Pergamon Press, pp. 142–158.

Popkov Yu. S., 1986. "Macrosystems theory and regional planning", *Environment and Planning A*, 18, pp. 291–292.

Popkov Yu. S., 1987. "Parametrical properties of stationary state models for macrosystems". Proceedings: *Optimum control in dynamic macrosystems* (in Russian), VNIISI: Moscow, issue 19, pp. 27–36.

Popkov Yu. S., 1988. "Multiplicative algorithm for Fermi-models of macrosystems", *Avtomatika i telemekhanika* (in Russian), No. 3, pp. 70–80.

Popkov Yu. S., 1988. "Global parametrical properties of stationary state models for a class of macrosystems". Proceedings of the II All-Union school *Applied problems of macrosystem control* (in Russian), VNIISI: Moscow, pp. 20–29.

Popkov Yu. S., 1988. "On nonstochastic macrosystems models", *Environment and Planning A*, 20, No. 8, pp. 997–998.

Posokhin M. V., Gutnov A. E., Popkov Yu. S., Shmulyan B. L., 1979. "Urban system model and its application to spatio-functional analysis of the city of Moscow". In: *Dostizhenia i perspektivy.* Moscow: Int. Centre of sc. and techn. inf., System Analysis Com. of the Pres. of USSR Ac. of Sc. (in Russian), issue 11 (Towns and systems of settling), pp. 22–34.

Posokhin M. V., Gutnov A. E., Popkov Yu. S., Shmulyian B. L., 1980. "A systems model of the citi and experiments in using it for a functional and spatial analysis of Moscow", *Environment and Planning B*, 7, pp. 107–119.

Prastacos P., 1986. "An integrated laud-use-transportation model for the San Francisco Region: 1. Design and mathematical structure", *Environment and Planning A*, 18, pp. 307–322.

Prokhorov Yu. V., 1967. *Probability Theory* (in Russian), Nauka: Moscow.

Reif F., 1977. *Statistical Physics (Berkeley Physics Course)*, V. 5, Mc Graw Hill Book Company.

Roy J. R., 1980. "External zones in the gravity model – some further developments", *Environment and Planning A*, 12, pp. 1203–1206.

Roy J. R., Lesse P. F., 1981. "On appropriate microstate descriptions in entropy modelling", *Transportation Research B*, 15, pp. 85–96.

Roy J. R., Brotchie J. F., 1984. "Some supply and demand considerations in urban spatial interaction models", *Environment and Planning A*, pp. 1137–1147.

Roy J. R., Lesse P. F., 1985. "Modelling commodity flows under uncertainty", *Environment and Planning A*, 17, pp. 1271–1274.

Roy J. R., 1987. "An alternative infirmation theory approach for modelling spatial interaction", *Environment and Planning A*, 19, pp. 385–394.

Segal A., 1979. "Optimal distributed routing for virtual lines-witched data networks", *IEEE Trans. on Commun.*, V. COM-27, No. 1, January, pp. 201–209.

Singh M. G., Tittli A., 1978. *Systems: Decomposition, Optimization and Control*, Pergamon Press, London.

Sheleykhovsky A. G., 1946. *Composition of an Urban Map as a Transport Problem* (in Russian), Moscow.

Shmulyan B. L., Imelbayev Sh. S., 1978. "Models of passenger travel in a city". In: *Transport systems* (in Russian), VNIISI: Moscow, pp. 7–22.

Snickars F., Weibull J. W., 1977. "A minimum information principle – theory and practice", *Regional Science and Urban Economics*, 7, pp. 137–168.

Theil H., 1967. *Economics and Information Theory*, Amsterdam: North Holland.

Tikhonov A. N., 1952. *Systems of Differential Equations with an Infinitesimal Parameter at Derivations*, Mathematical collection (in Russian), 31(73), issue 3.

Tikhonov A. N., Goncharsky A. V., Stepanov V. V., 1987. "Incorrect problems of image processing", *Moscow.-Proceedings of the USSR Academy of Sciences* (in Russian), 294, No. 4, pp. 832–837.

Uhlenbeck G. E., Ford G. W., 1963. *Lectures in Statistical Mechanics*, AMS, Providence.

Volterra V., 1931. *Theorie Mathematique de la Lutte pour la Vie*, Gauthier-Villars, Paris.

Volterra V., 1976. *Mathematical Theory of Struggle for Existence* (in Russian), Nauka: Moscow.

Walsh P. K., Gibbert R. W., 1980. "Developments of an entropy model for residential location with maximum zonal population constraints", *Environment and Planning A*, 12, pp. 1253–1268.

Weidlich W., Haag G., eds., 1988. *Interregional Migration: Dynamic Theory and Comparative Analysis*, Springer-Verlag: Berlin.

Wilson A. G., 1967. "Statistical theory of spatial distribution models", *Transportation Research*, 1, pp. 253–269.

Wilson A. G., 1970. *Entropy in Urban and Regional Modelling*, Pion Ltd.: London, -256p.

Wilson A. G., 1970. "Interregional Commodity Flows: Entropy Maximizing Approacher:, *Geographical Analysis*, 2, pp. 255–282.

Wilson A. G., 1974. *Urban and Regional Models in Geography and Planning*, New-York, John Wiley. -280p.

Wilson A. G., 1978. *Entropy Methods in Complex System Modelling* (in Russian), Nauka: Moscow.

Lecture Notes in Control and Information Sciences

Edited by M. Thoma

1992–1995 Published Titles:

Vol. 167: Rao, Ming
Integrated System for Intelligent Control.
133 pp. 1992 [3-540-54913-7]

Vol. 168: Dorato, Peter; Fortuna, Luigi; Muscato, Giovanni
Robust Control for Unstructured Perturbations: An Introduction.
118 pp. 1992 [3-540-54920-X]

Vol. 169: Kuntzevich, Vsevolod M.; Lychak, Michael
Guaranteed Estimates, Adaptation and Robustness in Control Systems.
209 pp. 1992 [3-540-54925-0]

Vol. 170: Skowronski, Janislaw M.; Flashner, Henryk; Guttalu, Ramesh S. (Eds)
Mechanics and Control. Proceedings of the 4th Workshop on Control Mechanics, January 21-23, 1991, University of Southern California, USA.
302 pp. 1992 [3-540-54954-4]

Vol. 171: Stefanidis, P.; Paplinski, A.P.; Gibbard, M.J.
Numerical Operations with Polynomial Matrices: Application to Multi-Variable Dynamic Compensator Design.
206 pp. 1992 [3-540-54992-7]

Vol. 172: Tolle, H.; Ersü, E.
Neurocontrol: Learning Control Systems Inspired by Neuronal Architectures and Human Problem Solving Strategies.
220 pp. 1992 [3-540-55057-7]

Vol. 173: Krabs, W.
On Moment Theory and Controllability of Non-Dimensional Vibrating Systems and Heating Processes.
174 pp. 1992 [3-540-55102-6]

Vol. 174: Beulens, A.J. (Ed.)
Optimization-Based Computer-Aided Modelling and Design. Proceedings of the First Working Conference of the New IFIP TC 7.6 Working Group, The Hague, The Netherlands, 1991.
268 pp. 1992 [3-540-55135-2]

Vol. 175: Rogers, E.T.A.; Owens, D.H.
Stability Analysis for Linear Repetitive Processes.
197 pp. 1992 [3-540-55264-2]

Vol. 176: Rozovskii, B.L.; Sowers, R.B. (Eds)
Stochastic Partial Differential Equations and their Applications. Proceedings of IFIP WG 7.1 International Conference, June 6-8, 1991, University of North Carolina at Charlotte, USA.
251 pp. 1992 [3-540-55292-8]

Vol. 177: Karatzas, I.; Ocone, D. (Eds)
Applied Stochastic Analysis. Proceedings of a US-French Workshop, Rutgers University, New Brunswick, N.J., April 29-May 2, 1991.
317 pp. 1992 [3-540-55296-0]

Vol. 178: Zolésio, J.P. (Ed.)
Boundary Control and Boundary Variation. Proceedings of IFIP WG 7.2 Conference, Sophia-Antipolis, France, October 15-17, 1990.
392 pp. 1992 [3-540-55351-7]

Vol. 179: Jiang, Z.H.; Schaufelberger, W.
Block Pulse Functions and Their Applications in Control Systems.
237 pp. 1992 [3-540-55369-X]

Vol. 180: Kall, P. (Ed.)
System Modelling and Optimization. Proceedings of the 15th IFIP Conference, Zurich, Switzerland, September 2-6, 1991.
969 pp. 1992 [3-540-55577-3]

Vol. 181: Drane, C.R.
Positioning Systems - A Unified Approach.
168 pp. 1992 [3-540-55850-0]

Vol. 182: Hagenauer, J. (Ed.)
Advanced Methods for Satellite and Deep Space Communications. Proceedings of an International Seminar Organized by Deutsche Forschungsanstalt für Luft-und Raumfahrt (DLR), Bonn, Germany, September 1992.
196 pp. 1992 [3-540-55851-9]

Vol. 183: Hosoe, S. (Ed.)
Robust Control. Proceesings of a Workshop held in Tokyo, Japan, June 23-24, 1991.
225 pp. 1992 [3-540-55961-2]

Vol. 184: Duncan, T.E.; Pasik-Duncan, B. (Eds)
Stochastic Theory and Adaptive Control. Proceedings of a Workshop held in Lawrence, Kansas, September 26-28, 1991.
500 pages. 1992 [3-540-55962-0]

Vol. 185: Curtain, R.F. (Ed.); Bensoussan, A.; Lions, J.L.(Honorary Eds)
Analysis and Optimization of Systems: State and Frequency Domain Approaches for Infinite-Dimensional Systems. Proceedings of the 10th International Conference, Sophia-Antipolis, France, June 9-12, 1992.
648 pp. 1993 [3-540-56155-2]

Vol. 186: Sreenath, N.
Systems Representation of Global Climate Change Models. Foundation for a Systems Science Approach.
288 pp. 1993 [3-540-19824-5]

Vol. 187: Morecki, A.; Bianchi, G.; Jaworeck, K. (Eds)
RoManSy 9: Proceedings of the Ninth CISM-IFToMM Symposium on Theory and Practice of Robots and Manipulators.
476 pp. 1993 [3-540-19834-2]

Vol. 188: Naidu, D. Subbaram
Aeroassisted Orbital Transfer: Guidance and Control Strategies.
192 pp. 1993 [3-540-19819-9]

Vol. 189: Ilchmann, A.
Non-Identifier-Based High-Gain Adaptive Control.
220 pp. 1993 [3-540-19845-8]

Vol. 190: Chatila, R.; Hirzinger, G. (Eds)
Experimental Robotics II: The 2nd International Symposium, Toulouse, France, June 25-27 1991.
580 pp. 1993 [3-540-19851-2]

Vol. 191: Blondel, V.
Simultaneous Stabilization of Linear Systems.
212 pp. 1993 [3-540-19862-8]

Vol. 192: Smith, R.S.; Dahleh, M. (Eds)
The Modeling of Uncertainty in Control Systems.
412 pp. 1993 [3-540-19870-9]

Vol. 193: Zinober, A.S.I. (Ed.)
Variable Structure and Lyapunov Control
428 pp. 1993 [3-540-19869-5]

Vol. 194: Cao, Xi-Ren
Realization Probabilities: The Dynamics of Queuing Systems
336 pp. 1993 [3-540-19872-5]

Vol. 195: Liu, D.; Michel, A.N.
Dynamical Systems with Saturation Nonlinearities: Analysis and Design
212 pp. 1994 [3-540-19888-1]

Vol. 196: Battilotti, S.
Noninteracting Control with Stability for Nonlinear Systems
196 pp. 1994 [3-540-19891-1]

Vol. 197: Henry, J.; Yvon, J.P. (Eds)
System Modelling and Optimization
975 pp approx. 1994 [3-540-19893-8]

Vol. 198: Winter, H.; Nüßer, H.-G. (Eds)
Advanced Technologies for Air Traffic Flow Management
225 pp approx. 1994 [3-540-19895-4]

Vol. 199: Cohen, G.; Quadrat, J.-P. (Eds)
11th International Conference on Analysis and Optimization of Systems – Discrete Event Systems: Sophia-Antipolis, June 15–16–17, 1994
648 pp. 1994 [3-540-19896-2]

Vol. 200: Yoshikawa, T.; Miyazaki, F. (Eds)
Experimental Robotics III: The 3rd International Symposium, Kyoto, Japan, October 28-30, 1993
624 pp. 1994 [3-540-19905-5]

Vol. 201: Kogan, J.
Robust Stability and Convexity
192 pp. 1994 [3-540-19919-5]

Vol. 202: Francis, B.A.; Tannenbaum, A.R. (Eds)
Feedback Control, Nonlinear Systems, and Complexity
288 pp. 1995 [3-540-19943-8]